TASER ELECTRONIC CONTROL DEVICES
AND SUDDEN IN-CUSTODY DEATH

ABOUT THE AUTHOR

Howard E. Williams served twenty-five years with the Austin (Texas) Police Department, advancing through the ranks from patrol officer to Commander before becoming Chief of the San Marcos (Texas) Police Department in 2003. During his career, Chief Williams commanded the Training Division of the Austin Police Department, where he was responsible for developing cadet and officer training programs, including programs in the use of force and the use of less-lethal weapons.

Chief Williams holds a Master Peace Officer's license from the Texas Commission on Law Enforcement Officers Standards and Education (TCLEOSE), and has served on TCLEOSE's Education Committee. Additionally, he also holds a Master of Science degree in Criminal Justice from Texas State University. He is also a graduate of the Leadership and Command College at the Law Enforcement Management Institute of Texas at Sam Houston State University. Chief Williams is an adjunct instructor in the Criminal Justice Department at Texas State University, where he instructs classes in management and forensic evidence, and he has been a guest lecturer for criminology classes at the University of Texas. He is also the author of two other books in criminal investigations, *INVESTIGATING WHITE-COLLAR CRIME: Embezzlement and Financial Fraud*, Second Edition and *ASSET FORFEITURE: A Law Enforcement Perspective*.

TASER ELECTRONIC CONTROL DEVICES AND SUDDEN IN-CUSTODY DEATH

Separating Evidence from Conjecture

By

HOWARD E. WILLIAMS

CHARLES C THOMAS • PUBLISHER, LTD.
Springfield • Illinois • U.S.A.

Published and Distributed Throughout the World by

CHARLES C THOMAS • PUBLISHER, LTD.
2600 South First Street
Springfield, Illinois 62704

This book is protected by copyright. No part of
it may be reproduced in any manner without
written permission from the publisher.
All rights reserved.

©2008 by CHARLES C THOMAS • PUBLISHER, LTD.

ISBN 978-0-398-07775-4 (hard)
ISBN 978-0-398-07776-1 (paper)

Library of Congress Catalog Card Number: 2007021989

With THOMAS BOOKS *careful attention is given to all details of manufacturing and design. It is the Publisher's desire to present books that are satisfactory as to their physical qualities and artistic possibilities and appropriate for their particular use.* THOMAS BOOKS *will be true to those laws of quality that assure a good name and good will.*

Printed in the United States of America
CR-R-3

Library of Congress Cataloging-in-Publication Data

Williams, Howard E.
  Taser electronic control devices and sudden in-custody death : separating evidence from conjecture / by Howard E. Williams.
    p. cm.
  On t.p. the registered trademark symbol "R" is superscript following Taser.
  Includes bibliographical references and index.
  ISBN 978-0-398-07775-4 (hard) -- ISBN 978-0-398-07776-1 (pbk.)
  1. Restraint of prisoners--United States--Case studies. 2. Prisoners--United States--Death--Case Studies. 3. Stun guns--United States--Evaluation. 4. Nonlethal weapons--United States--Evaluation. I. Title.

HV8778.W55 2007
363.2'3--dc22

2007021989

PREFACE

Basic research is what I'm doing when I don't know what I'm doing.
(Wernher Von Braun, Attributed)

In 2001, I was the Commander of the Northeast Area Command of the Austin (Texas) Police Department (APD), and I was the first Commander at the APD to purchase TASER® electronic control devices for the patrol officers.[1] Special Weapons and Tactics (SWAT) officers had them, but line officers did not. It seemed to me that more line officers would unexpectedly encounter violently resisting subjects than would SWAT officers. Shortly after we trained the officers and deployed the TASER M26 devices, I began to hear success stories relating how well they worked. I was so sure of the effectiveness of the weapons that, when I became Chief of the San Marcos (Texas) Police Department in 2003, I purchased TASER X26 devices for the line officers there. Once again, I soon heard the success stories. I also noticed that lost work time due to on-the-job injuries decreased more than 70 percent in the first year following purchase of the TASER electronic control devices. The next year, lost time fell to zero. Officers were also taking fewer prisoners to the hospital for treatment following arrests. I was satisfied that conducted energy weapons were a beneficial addition to the arsenal.

Soon, however, I began to hear concerns regarding the deaths that had followed the use of a TASER device. I read the Amnesty International reports that questioned the safety of TASER electronic control devices and the many news media articles that insinuated, or in some cases alleged, that TASER products were the cause of many deaths. I also read articles defending TASER devices, including some that employees at TASER International, Inc. had written. However, I noticed that many of the criticisms I read were rife with conjecture. I could understand questioning the safety and efficacy of the weapons that the police use, but, being an old cop, I did not want Amnesty International or TASER International telling me what to think. I

---

1. TASER® is a trademark of TASER International, Inc. registered in the United States. All rights reserved.

wanted to see the evidence. As I have often told my students at Texas State University, where I am adjunct faculty in the Criminal Justice Department, "You do not have to trust what anyone tells you. Research it yourself."

So, I determined to take my own advice. Not being sure exactly what I needed to learn, with apologies to the memory of Doctor Von Braun, I began with some basic research. I decided that one cannot draw conclusions regarding the use of TASER devices without first knowing how the devices work, knowing what other researchers and scientists have learned about the workings of electronic control devices, knowing the facts of what happened in each case, and knowing what the medical examiner or coroner said about each death. The amount of information necessary to make an informed decision was not available in the reports that I had read.

Finding journal articles and published reports of scientists and other researchers was not too difficult. The library resources at Texas State University made that part of the research simple enough. Obtaining the relevant police and autopsy reports was another matter. The public information laws of the various states made that task much more difficult, and, in some cases, made it impossible. Nevertheless, I gathered as much information as I could from the primary sources, and when primary sources of information were not available, I relied on news reports. I then analyzed the information by comparing what happened in each incident to what we know about the phenomenon of sudden death and the workings of TASER electronic control devices. I tried to set aside my preconceived ideas and simply examine each case on the merits of its own evidence. Was there sufficient evidence to confirm the TASER devices as the cause of death? Was there sufficient evidence to exclude the TASER devices as the cause of death?

I believe that the evidence presented in *TASER ELECTRONIC CONTROL DEVICES AND SUDDEN IN-CUSTODY DEATH: Separating Evidence from Conjecture* will show why, after completing the research, I am still a proponent of the use of TASER devices. Of course, TASER electronic control devices are weapons, and I will never refer to them as anything but weapons. The police should never take lightly using weapons on the citizenry, but the judicious use of force is necessary to enforce the law and to maintain order. As with any new tool, police officers will have to adapt and learn to use the tool with the proper discretion. As with any new technology, police managers will have to define policies and procedures to govern the use of TASER electronic control devices and other conducted energy weapons, but policy discussions are for another book.

TASER International does not need me to defend their products. They are quite capable of defending themselves. My purpose is only to try to separate the evidence from conjecture. The evidence makes the case that TASER devices are not instruments of death, and I believe that the only conclusion the evidence supports is that TASER electronic control devices are safe weapons. At least, they are as safe as weapons can be.

In the spirit of full disclosure, I have no financial interest in TASER International, Inc. I do not own stock, I am not on their payroll, and I have received no financial support from anyone in conducting the research for/or the writing of this book.

Additionally, while researching this book, I may not have discovered cases that occurred between 1983 and 2006. If anyone knows of a case that I should examine for future research, please contact me at my e-mail address: howardewilliams@ msn.com.

CONTENTS

Page

Preface v

Chapter
1. SUDDEN IN-CUSTODY DEATH 3
    Excited Delirium 6
        Restraint Asphyxia 8
        Metabolic Acidosis 9
        Rhabdomyolysis 9
    Handling Excited Delirium Patients 9
    Catecholamine 11
    Sympathomimetics 11
    Neuroleptics 12
    Correlation versus Causation 13
    Summary 15
    Key Terms 16

2. CONDUCTED ENERGY WEAPONS 17
    A Short History of TASER International, Inc. 17
    Electronic Characteristics 19
    Physiological Effects 19
    Literature Reviews 22
    Technical Studies 24
    Medical Studies 26
    Adverse Studies 31
    TASER Devices as Nonlethal Force 31
    Summary 33
    Key Terms 33

3. CASE STUDIES 34
    First and Second Generation TASER Weapon Cases 35
        Vincent Alvarez 36
        Raul Guevara, Jr. 36

Larry Donnell Gardner 37
Cornelius Garland Smith 37
Lannie Stanley McCoy 38
Joseph Rodriguez 38
Anthony Manwell Williams, III 39
Robert Zapata 39
Robert (Charles) Herbert Bobier 40
Yale Lary Wilson 40
Miguel Contreras 41
Mario Antonio "Tony" Gastelum 41
Stewart Alan Vigil 42
William McCall 42
Edward Breen 43
Charles Eugene Miles 43
Jose Torres 44
Anthony Puma 44
Jeffrey Michael Leonti 45
Duane Jay Johnson 45
Douglas L. Danville 46
Douglas Charles 47
Max Leyza Garcia 47
Donnie Ray Ward 48
James Ricard 48
Clarice A. Younger 49
David Martinez 49
Michael James Bryant 50
Vital Montilla 50
Daniel Scott Gizowski 51
Richard Wayne Harris 51
LeeGrand Griffin 52
Bruce Klobuchar 52
Byron Williams 53
Scott Jaron Norberg 53
James Quentin Parkinson 54
Kimberly LaShon Watkins-Oliver 54
Andrew Hunt, Jr. 55
Garner Roosevelt Hicks, Jr. 55
Mark Andrew Brown 56
Michael Labmeier 56
David Torres Flores 57
Third and Fourth Generation TASER Weapon Cases 58
Enrique Juarez Ochoa 58
Mark Lorenzo Burkett 58
Steven Vasquez 59
Marvin Hendrix 59
Vincent Delostia 60

Anthony Spencer	60
Henry William Canady	61
Richard Joseph Baralla	62
Eddie Rene Alvarado	62
Nicholas Aguilar	63
Jason Nichols	64
Fermin Rincon	64
Clever Craig, Jr.	65
Gordon Randall Jones	65
Johnny Lozoya	66
Frederick Steven Webber	66
Stephen L. Edwards	67
Ronald Edward Wright	68
Christopher Smith	68
Corey Calvin Clark	68
Joshua Alva Hollander	69
Timothy Roy Sleet	69
David Sean Lewandowski	70
Troy Dale Nowell	70
John Lee Thompson	71
Walter Curtis Burks, Jr.	72
Gordon Benjamin Rauch	72
Ray Charles Austin	73
Glenn Richard Leyba	74
Roman Gallius Pierson	74
Dennis D. Hammond	75
Louis N. Morris, Jr.	75
James Lee Borden	76
Michael Sharp Johnson	77
Kerry Kevin O'Brien	77
Lewis Sanks King	78
Curtis Lamar Lawson	78
David Glowczenski	79
Raymond Siegler	80
William D. Lomax, Jr.	80
Curt Lee Rosentangle	81
Terry L. Williams	82
Phillip LeBlanc	82
Melvin Samuel	83
Robert Harold Allen	83
Anthony Alfredo Diaz	84
Eric Wolle	85
Henry John Lattarulo	85
Frederick Jerome Williams	86
Daryl Lavon Smith	86
Anthony Carl Oliver	87

Jerry W. Pickens 88
James Arthur Cobb, Jr. 88
Jacob John Lair 89
Abel Ortega Perez 89
Kris J. Lieberman 90
Eric Bernard Christmas 90
Demetrius Tillman Nelson 91
Willie Smith, III 91
Milton Francisco Salazar 92
Keith Tucker 92
David Riley 93
Ernest J. Blackwell 94
Anthony Lee McDonald 94
William Malcolm Teasley 95
Richard "Kevin" Karlo 96
Michael Lewis Sanders 96
Lawrence Samual Davis 97
Jason David Yeagley 97
Michael Robert Rosa 98
Samuel Ramon Wakefield 99
Andrew Lamar Washington 99
John Alex Merkle 100
Dwayne Anthony Dunn 100
Christi Michele Ball 101
Greshmond Gray 101
Robert Guerrero 102
Keith Raymond Drum 103
Ricardo Reyes Zaragoza 103
Jessie Robert Tapia 104
Charles Christopher Keiser 105
Byron W. Black 106
Patrick Fleming 106
Kevin Downing 107
Lyle Lee Nelson 107
Douglas G. Meldrum 108
David J. Cooper 108
Jeanne Marie Hamilton 109
Ronnie James Pino 109
Timothy Bolander 110
Christopher Dearlo Hernandez 111
Gregory Saulsbury 111
Dennis S. Hyde 112
Carl Nathaniel Trotter 113
Jerry John Moreno 114
James Edward Hudson 114
Jeffrey A. Turner 115

Ronald Alan Hasse 115
Robert Fidalgo Camba 116
Joel Don Casey 117
Robert Clark Heston, Jr. 118
Shirley Andrews 118
David Levi Evans, Jr. 119
Willie Michael Towns 119
Mark Young 120
Milton Woolfolk, Jr. 120
Eric J. Hammock 121
James Floyd Wathan, Jr. 121
Ricky Paul Barber 122
John Cox 123
Jesse Cleon Colter, III 123
Keith Edward Graff 124
Lawrence Berry 124
Stanley Wilson 125
Vernon Anthony Young 125
Leroy Pierson 126
Randy Martinez 126
Lee Marvin Kimmel 127
Richard James Alvarado 128
Walter Lamont Seats 128
Richard Thomas Holcomb 128
Nazario Javier Solorio 129
Ravan Jermont Conston 130
Russell Walker 131
Horace Owens 131
Michael Anthony Edwards 132
Shawn Christopher Pirolozzi 132
Robert Earl Williams 133
Melinda Kaye (Fairbanks) Neal 134
Carolyn J. Daniels 134
Phoarah Kareem Knight 135
Tommy Valentine Gutierrez 135
Rockey Bryson 136
Kevin Ray Omas 136
Otis Gene Thrasher 137
Ernesto Valdez, Jr. 137
Carlos Casillas-Fernandez 138
Michael Leon Crutchfield 138
Maurice Cunningham 139
Eliseo Maldanado 140
Terrence Thomas 140
Eric Michael Mahoney 141
Brian Patrick O'Neal 141

Dwayne Zachary 142
Olson Agoodie 143
Frank Gilman Edgerly 143
Robert E. Boggon 144
Brian Lichtenstein 144
Shawn A. Norman 145
David Anthony Cross 146
Timothy Michael Torres, Jr. 146
Patrick Aaron Lee 147
Michael Lesean Clark 147
Mary Eleanor Malone Jeffries 148
Timothy Glen Mathis 148
David Michael Croud 149
Steven Michael Cunningham 150
Jose Maravilla Perez, Jr. 150
Miguel Serrano 151
Cedric Stemberg-Barton 151
Joshua Brown 152
Jose Angel Rios 152
Hansel Cunningham, III 153
Barney Lee Green 153
Tyler Marshall Shaw 154
Tracy Rene Shippy 155
Kevin Dewayne Wright 155
Jeffrey Dean Earnhardt 156
Michael Stanley Tolosko 156
Howard Starr 157
David L. Moss, Jr. 157

4. ANALYSIS OF TASER CASE STUDY DATA 159
   Group One Analysis 159
   Group Two Analysis 162
   Summary 164

Appendices
   Appendix A: Unexpected Deaths Following Application
      of TASER–2006 165
   Appendix B: Unexpected Deaths by Jurisdiction–1983
      through 2006 169
   Appendix C: List of Fatal Shootings 1987 to 2006 173
Glossary 177
Bibliography 185
Index 205

TASER ELECTRONIC CONTROL DEVICES AND SUDDEN IN-CUSTODY DEATH

Chapter 1

SUDDEN IN-CUSTODY DEATH

Death is terrifying because it is so ordinary. It happens all the time.
Susan Cheever, *Home Before Dark*

No single incident can spark as much controversy and create as many public relations difficulties for a law enforcement executive than an in-custody death, especially when someone who appears strong and healthy suddenly dies. Now, more than ever, the public, the news media, and the courts scrutinize the actions of law enforcement officers. Nowhere is that scrutiny more apparent than in use of force encounters, and the scrutiny intensifies dramatically when a death is involved. Unexpected deaths generally create great public controversy regarding police tactics and the legitimate need for the use of force.

The most significant controversy of the past decade is the use of electromuscular disruption technology to subdue aggressive or resisting subjects. Electromuscular disruption technology is a less-lethal technology that uses pulses of electricity to incapacitate suspects. Conducted energy weapons are defensive systems that deploy an electromuscular disruption technology to affect the sensory and motor functions of the somatic nervous system, the part of the peripheral nervous system associated with the voluntary control of body movements through the action of skeletal muscles and with the reception of external stimuli. Although many different types and brand names of conducted energy weapons exist, because of their popularity with law enforcement agencies across the United States, the controversy has centered on TASER devices, which are manufactured by TASER International, Inc. of Scottsdale, Arizona.

Unlike most discussions about police tactics, which attract little public attention, the debate over the use of TASER devices is a very public battle. Newspapers consistently publish headlines announcing deaths supposedly related to the use of TASER electronic control devices. Amnesty International and the American Civil Liberties Union are waging campaigns, if not to ban law enforcement from using electromuscular disruption technology, at least to impose a moratorium on using conducted energy weapons. Critics of the technology argue that research on conducted energy weapons is lacking and that the numbers of deaths following the use of such weapons are evidence that they are not safe. Supporters counter that conducted energy weapons are safer than traditional less-lethal weapons, such as batons and chemical sprays, and that they reduce the incidence of injury to officers and to the public. The crux of the debate is whether the TASER devices are responsible for the sudden unexpected deaths that have followed their use.

We know little about the frequency or causes of sudden in-custody deaths; however, we do know that people die unexpectedly while in police custody, with and without the use of a TASER device. To understand whether TASER devices significantly contribute to sudden in-custody deaths, one must first understand something of the phenomenon of sudden death. The World Health Organization, a specialized agency of the United Nations that acts as a coordinating authority on international public health, compiles the internationally recognized International Classification of Diseases. The International Classification of Diseases (ICD) defines sudden death as cardiorespiratory collapse occurring within twenty-four hours of the beginning of symptoms (ICD 798.2). Roberts (1986) defines sudden death as one that is nonviolent or nontraumatic, that is unexpected, that is witnessed, and that is instantaneous or occurs within a few minutes of an abrupt change in a previous clinical state. In contrast to sudden death, instantaneous death is death with immediate collapse without preceding symptoms (DiMarco, 2003). Most in-custody deaths fall within Roberts' definition of sudden death, which excludes police-related shootings, since a shooting is violent and traumatic, and death is an entirely expected consequence of the use of deadly force.

Although sudden death has recently become a hot topic for law enforcement executives, it is not a new or uncommon phenomenon. Each year in the United States, more than 300,000 individuals die suddenly from some form of cardiovascular disease, alone (DiMarco, 2003). Sudden death is a complicated and complex process, and it is not reasonable to expect law enforcement officers to diagnose medical conditions that may lead to sudden death. In fact, most predisposing factors of sudden death are not visible to officers. Only a review of a person's medical history and a physical examination will reveal those factors. Research has discovered several predisposing factors that increase the risk for sudden death:

- Obesity;
- Cardiomegaly (an enlarged heart);
- Coronary atherosclerosis (coronary artery disease);
- A previous heart attack;
- Myocarditis (an inflammation of the muscles of the heart);
- A fibrotic heart;
- Under the influence of illicit drugs, although some instances of sudden death can occur months after the last ingestion of illicit drugs;
- Too much or too little neuroleptic medication (drugs used to treat psychoses), or the medications are not working;
- Failure to take prescribed drugs;
- Diabetes and hypoglycemia (a lower than normal amount of glucose in the blood);
- Hyperthyroidism;
- A head injury, or a previous head injury;
- Dehydration;
- Underlying psychiatric disease; and
- Alcohol intoxication or withdrawal (Peters, 2006c).

Several clinical and autopsy-based studies have reported triggering of sudden cardiac death with exercise. Data supporting the concept that vigorous physical activity can trigger ventricular fibrillation have come from emergency medical records showing that 11 to 17 percent of adults collapsed during or immediately after exertion, but the amount of exertion is rarely quantified (Pinto & Josephson, 2004). In one review in Seattle, 11 percent of 316 consecutive victims of sudden death died during or immediately after exercise. In Miami, 17 percent of 150 patients had exertion-related sudden death. In Seattle, researchers estimated the incidence of exercise-related sudden death at 5.4 per 100,000. During vigorous activity, the incidence is 5 times higher for men who exercise frequently and 56 times higher for men who exercise infrequently (Fletcher, Flipse, & Oken, 2004). Clearly, the risk of sudden death following violent exertion, such as a struggle with police or straining against restraints, increases manifold for people with predisposing factors for sudden death.

The case studies in Chapter 3 show that, in the 213 custody deaths following the application of a TASER electronic control device from 1983 through 2005, the medical examiner or coroner observed one or more predisposing factors of sudden death in at least 187, or 87.8 percent, of those cases. In at least 75 cases, or 35.2 percent, the medical examiner or coroner observed more than one predisposing factor. Because not all autopsy reports were available, the exact number of cases with predisposing factors for sudden death is unknown.

The major difficulty with studying in-custody deaths is that, unlike crime records reporting, no central database exists to collect data on in-custody deaths. Consequently, it is difficult to ensure that data are complete and accurate. Emergency physicians at the University of Minnesota faced that daunting task when they conducted a twelve-month surveillance study of in-custody deaths in 2004 and 2005 (Ho, Miner, Heegaard & Reardon, 2005). The study relied on news reports to collect data on subject gender, age, behavior, force and weapons used to affect the arrest, time of collapse proximal to the arrest, and the presence of illicit substance abuse. They conducted follow-up interviews with law enforcement agencies when necessary to clarify information. The physicians reported on 162 in-custody deaths: 21 were instantaneous deaths, 85 deaths occurred in less than one hour, and 56 deaths occurred between one and forty-eight hours after the initial contact with law enforcement officers. Males accounted for 156 deaths, or 96.3 percent, and females accounted for six. The mean age of the deceased was 35.7 years, with an age range of 15 to 75 years. Just prior to arrest, 102, or 63 percent exhibited bizarre behavior and 101, or 62.3 percent, had confirmed illicit drug use (Ho et al., 2005).

Perhaps the most interesting finding of the study was the type of force that law enforcement officers used. In 22 cases, the law enforcement officers used no force beyond the application of handcuffs. In 111 cases, the suspect and the officers engaged in significant hands-on force. In 84 cases, the officers used intermediate or less-lethal weapons, including 20 uses of chemical spray, 14 uses of impact weapons, and 50 applications of a TASER device. Because officers sometimes used more than one level of force in a given incident, the numbers total more than the 162 cases. In every case, the police used handcuffs to secure the subject (Ho et al., 2005).

The authors of the study noted that in none of the 21 instantaneous deaths was a TASER electronic control device involved, which is an important observation. If the electrical discharge from the device disrupted the rhythm of the heart, as many critics claim, then collapse and death should have been almost instantaneous, within 5 to 15 seconds. Since death did not occur instantaneously following any of the TASER device applications in the University of Minnesota study, the cause of death must have been something other than the electrical impulses of a conducted energy weapon.

Unlike in the United States, the government in the United Kingdom, for years, has required the reporting of custody deaths. Before the introduction of TASER devices in the United Kingdom, deaths in custody occurred at the rate of approximately one a week. The subjects who died following the application of TASER devices in the United States and the subjects who died in custody in the United Kingdom shared many of the same risk factors for sudden death (Bleetman & Steyn, 2003).

The case studies in Chapter 3 show that, in the 213 in-custody deaths following the application of a TASER electronic control device from 1983 through 2005, the demographics closely match those of the University of Minnesota study. Males accounted for 204 deaths, or 95.8 percent, and females accounted for nine. The mean age of the deceased was 35.5 years of age, and the age range was 17 to 65. Just prior to arrest, 156, or 73.2 percent, exhibited bizarre behavior, and 145, or 68.1 percent, had confirmed illicit drug use. Table 1, Sudden Death Study Comparisons, demonstrates how similar the findings of the University of Minnesota sudden death study are to the findings of the TASER device case studies in Chapter 3.

Table 1
SUDDEN DEATH STUDY COMPARISONS

	University of Minnesota Study	*Case Studies from Chapter 3*
Gender 　Male 　Female	 96.3 % 3.7 %	 95.8 % 4.2 %
Mean Age	35.7 years	35.5 years
Age Range	15 to 75 years	17 to 65 years
Bizarre Behavior	63.0 %	73.2 %
Illicit Drug Use	62.3 %	63.8 %

EXCITED DELIRIUM

Writing in the *American Journal of Insanity* in 1849, Doctor Luther Bell, who ran the McLean Asylum for the insane in Massachusetts, described the symptoms of some of his patients as being confused, having no tolerance to light, making low muttering noises, being suspicious of food being filthy or poisoned, and having a dull impression of impending danger. He recognized the patients' propensities for violence, noting that they would attack "anyone who approaches . . . with a blind fury," "struggle with the utmost desperation, irrespective of the numbers or strength of those who may be endeavoring to restrain [them]," and have "no disposition to yield to an overpowering force, noticeable in some degree in the blindest fury of the most intense forms (Peters, 2006a)." Bell reported that the onset of symptoms was rapid, usually about one week, and that three out of four patients died within two or three weeks. The behavioral issues that Bell described are similar to behaviors that officers note today in sudden in-custody death cases.

The University of Minnesota physicians noted that in 102 of the 162 cases they studied, the subject engaged in what they described as bizarre behavior just prior to the arrest (Ho et al., 2005). One of the most notable conditions of sudden in-custody death cases is excited delirium. Excited delirium, alternatively known as Bell's mania, lethal catatonia, and agitated delirium, is characterized by an acute onset of bizarre and violent behavior. Combativeness, hyperactivity, unexpected strength, paranoid delusions, incoherent shouting, hallucinations, and hyperthermia often accompany the condition (Wetli, Mash & Karch, 1996; Farnham & Kennedy, 1997). Underlying causes of excited delirium include:

- Bipolar disorder, a psychiatric diagnostic category describing a class of mood disorders in which the person experiences clinical depression and/or mania, hypomania, and/or mixed states;
- Chronic schizophrenia, a psychiatric diagnosis that describes a mental disorder characterized by impairments in the perception or expression of reality and/or by significant social or occupational dysfunction;
- Intoxication with sympathomimetics, a class of drugs whose effects mimic those of a stimulated sympathetic nervous system, such as increased cardiac output, dilated bronchioles, and constricted blood vessels including cocaine, methamphetamine, and intoxication with anticholinergics, members of a class of pharmaceutical compounds which serve to reduce the effects mediated by acetylcholine in the central nervous system and peripheral nervous system;
- Cocaine intoxication;
- Alcohol withdrawal; and
- Head trauma (Park, Korn & Henderson, 2001).

Little information exists on the frequency of incidents because prospective studies are rare and retrospective studies show a bias toward fatal cases, but Doctor Vincent DiMaio, chief medical examiner in Bexar County, Texas, estimates that excited delirium kills as many as 800 people nationwide each year (Glick, 2006). Case reports seem to cluster in summer months and in areas noted for high temperature and high humidity (Ruttenber, McAnally & Wetli, 1999), and a Quetelet Index, or body mass index, in the upper three quartiles seems to increase the risk for fatal excited delirium (Ruttenber, Lawler-Heavner, Yin, Wetli, Hearn & Mash, 1997).

Fatal excited delirium appears to consist of four distinct and sequential phases: elevated body temperature, agitated delirium, respiratory arrest, and death (Wetli, Mash & Karch, 1996). Patients initially appear agitated to grossly psychotic, and they exhibit feats of superhuman strength, especially during attempts to restrain them. In one study, an average of four officers was required to subdue each suspect, with a range of three to six officers. Shortly after the patients were restrained, their violent struggling ceased, and witnesses noted that they had developed a labored or shallow breathing pattern. Some patients died moments later, but death typically occurred within one hour of first contact with police. More than 75 percent of patients died either at the scene or during initial transportation (Ross, 1998).

In another study, emergency medical services personnel witnessed eighteen cases of excited delirium death. The medical personnel described initial cardiac rhythms in thirteen of those cases. In contrast to patients with acute cocaine toxicity, ventricular dysrhythmia, a group of conditions in which the muscle contraction of the heart is irregular, or is faster or slower than normal, occurred in only one patient. Asystole, a state of no cardiac electrical activity, hence no contractions of the myocardium and no cardiac output or blood flow, was the most common presenting rhythm. Interestingly, toxicology investigations of these patients demonstrated cocaine levels similar to those found in recreational cocaine users and lower than those found in individuals who died of acute cocaine intoxication. Factors that were frequently associated with sudden death of restrained excited delirium victims included evidence of forceful struggle (100%), stimulant drug use (78%), established natural disease states (56%), and obesity (56%). Medical personnel described cessation of struggle against restraints and the onset of shallow or labored breathing in each of the sudden deaths (Stratton, Rogers, Brickett, & Gruzinski, 2001; Sztajnkrycer & Baez, 2005).

Critics claim that the American Medical Association does not recognize excited delirium as a medical or psychiatric condition. However, the National Association of Medical Examiners has recognized it for more than a decade, and medical examiners in most major cities currently apply the diagnosis (Paquette, 2003). Consequently, considerable debate exists regarding the use of excited delirium to explain sudden death during restraint. Some critics claim that excited delirium, first applied to the deaths of suspects in police custody in the 1980s, is nothing but a cover-up for police brutality (Floyd, 2006). However, the police do not diagnose excited delirium, the medical community does. Doctor Bell was the first to describe the symptoms of patients under his medical care, and medical examiners apply the same diagnoses in cases other than in-custody deaths.

Darrell Porter, 50, a former All-Star catcher for four Major League Baseball teams and the Most Valuable Player in the 1982 World Series, died in 2002 in Kansas City, Missouri. A passerby found Porter dead next to his car parked near the Missouri River. There were no signs of a struggle or indications of foul play, and there was no police or emergency medical service intervention before he died. According to the medical examiner, Porter had a level of cocaine in his system typical of someone who used cocaine recreationally. The medical examiner concluded that Porter did not die of a cocaine overdose. Instead, he listed the official cause of death as excited delirium (*Houston Chronicle*, 2002).

In 2004, Natalee Drissell, 20, was found dead the morning after Thanksgiving in the woods outside a Christian conference center in the Colorado mountains. She was naked, and her arms and legs were bruised from running into trees and brush. A single set of footprints in the snow led to where she had fallen. After a long investigation, the coroner ruled that the cause of death was excited delirium. The autopsy revealed no drugs in Drissell's system, but the coroner listed six factors to support the diagnosis of excited delirium: found nude in the snow, no internal injuries, negative toxicology, abrasions and contusions consistent with running into trees and brush, recent arrival to elevation from near sea level, elevated urine glucose. There was no sign that anyone used any force against Drissell, and no one else was present when she died (Frankel, 2004).

In December 2005, Ramona Knapp, a 51-year-old-woman who was in a California psychiatric care facility, died one day after the medical staff restrained her. Knapp was admitted to the facility the previous day for an acute psychosis. When she became combative with the medical staff, they held her down and administered a sedative, then they noticed that she was not breathing. She died the following day at a local hospital (Lillis, 2006). The coroner concluded that Knapp died of excited delirium. No police were present, and Knapp died following restraint by trained medical professionals.

The case studies in Chapter 3 show that, in the 213 in-custody deaths following the application of a TASER device from 1983 through 2005, the medical examiner or the coroner recorded excited delirium as either the cause of death or as a significant contributing factor in at least 54, or 25.4 percent, of those cases. In most cases, excited delirium was listed in combination with other factors as the cause of death. Because not all autopsy reports were available, the exact number of excited delirium deaths is unknown. Although no one fully understands the mechanisms of sudden death as it relates to excited delirium, there are several theories: restraint asphyxia, metabolic acidosis, rhabdomyolysis, and catecholamine-induced sudden death.

Restraint Asphyxia

Restraint asphyxia, also known as postural asphyxia or positional asphyxia, arises from being unable to breathe correctly, thus creating a severely deficient supply of oxygen to the body. Each year, a small but significant number of people die suddenly and without apparent reason during restraint by police and medical staff. Although restraint asphyxia may be a factor in some deaths, the diagnosis is controversial. During the struggle to arrest a violently resisting subject or to restrain a patient for emergency medical care, officers usually put the subject on his stomach and hold him down so they can apply handcuffs or other restraints. However, a person lying on his stomach may have trouble breathing with pressure applied to his back. The remedy seems to be getting the pressure off the person's back, but, during a violent struggle between an officer and a suspect, the solution is not that simple. Often, a cycle of resistance and restraint compounds the situation. The officer restrains the person face down to limit that person's ability to harm anyone, including himself. As the person's breathing becomes labored, he struggles harder to get up. As the person struggles more, the officer applies more weight to the person's back to keep him from getting up. The more weight the officer applies, the more severe the compression of the chest and the more trouble the person has breathing. The more difficulty the person has breathing, the more violently he resists, so the officer applies more pressure. People experiencing excited delirium often demonstrate extraordinary strength, and it often takes several officers applying pressure on the person to hold him down. Such pressure may interfere with the movement of the diaphragm, making breathing difficult, if not impossible. Interfering with proper breathing can lead to hypoxia, a below normal level of oxygen in the tissues. Hypoxia can disturb the body's chemistry and may lead to a fatal cardiac dysrhythmia (Reay, 1996).

Three factors contribute to increasing oxygen demands and decreasing oxygen delivery in people with excited delirium: (1) the state of excited delirium, combined with a confrontation with law enforcement officers or emergency medical personnel, increases catecholamine stress on the heart; (2) hyperactivity with struggling against both officers and restraints increases oxygen delivery demands on the heart and lungs; and (3) breathing is hindered by restraint in certain positions, making it difficult to expand the chest wall and depress the diaphragm (O'Halloran & Lewman, 1993). A study conducted in Los Angeles County between 1992 and 1998 included 22 cases of sudden in-custody deaths. Researchers found that, in each case, officers had restrained the subject in a prone position with hobble restraints (Stephens, Jentzen, Karch, Wetli, & Mash, 2004a).

Generally, a medical examiner diagnoses restraint asphyxia based on the historical events preceding the episodic event of physical struggle. Most cases of fatal restraint asphyxia studied have been those when law enforcement personnel transported a person in the prone position. However, other positions, including a bent neck with flexion toward the chest, and external airway obstruction or neck compression when the victim was not able to release himself from the compromising position, have also been contributing factors (Robinson & Hunt, 2005).

Studies do exist that question whether restraint asphyxia can so affect a person's breathing as to cause hypoxia (Chan, Neuman, Clausen, Eisele & Vilke, 2004; Vilke, Michalewicz, Kolkhorst, Neuman & Chan, 2005). Current data neither support nor refute restraint asphyxia secondary to prone positioning and the use of hobble restraints as a definite causal factor in sudden death (Sztajnkrycer & Baez, 2005). However, the case studies in Chapter 3 show that in 213 in-custody deaths following the application of a TASER device from 1983 through 2005, the medical examiner or coroner recorded restraint asphyxia as either the cause of death or as a significant contributing factor in at least 10, or 4.7 percent, of those cases.

Metabolic Acidosis

Metabolic acidosis, a state in which the blood pH is low due to increased production of hydrogen ions by the body or the inability of the body to form bicarbonate in the kidney, can be associated with cardiovascular failure following exertion. Drug use, particularly when combined with physical exertion, can lead to profound metabolic acidosis (Hick, Smith, & Lynch, 1999). The increased levels of exercised-induced lactic acid and the alteration in pain sensation from psychosis and delirium can potentially result in severe acidosis with maximal sympathetic discharge (Hick, Smith & Lynch, 1999). The causes of acidosis are diverse, and the consequences can be severe, including coma and death.

Rhabdomyolysis

Excited delirium and cocaine-associated rhabdomyolysis may be part of a syndrome found in chronic cocaine users. Rhabdomyolysis is the breakdown of skeletal muscle due to mechanical, physical, or chemical injury, the principal result being acute renal failure due to accumulation of muscle breakdown products in the bloodstream, several of which are injurious to the kidney. Formerly, the more common causes of acute rhabdomyolysis were from crush injuries during wartime and natural disasters, but alcohol and many of the common drugs of abuse have become frequent causative agents in up to 81 percent of cases of rhabdomyolysis (Coco & Klasner, 2004). Rhabdomyolysis has been noted in a number of patients with cocaine-induced excited delirium, and it is often associated with hyperthermia, agitation, hyperactivity, and bizarre behavior. Patients surviving initial cardiac arrhythmias may develop cocaine-associated rhabdomyolysis, particularly if they have hyperthermia (Ruttenber et al., 1999).

HANDLING EXCITED DELIRIUM PATIENTS

In the past, law enforcement officers were encouraged to be patient and to wait out emotionally disturbed individuals who were not an immediate threat to themselves or to others. However, experience has shown that excited delirium can create hyperthermia, a body temperature as high as 110 degrees Fahrenheit, which is a life-threatening condition. Additionally, patients often grow more violent and less predictable as their body temperature rises. Now, officers train to assure delivery of urgent medical treatment (*New York Times*, 1989). Since the etiology of fatal excited delirium is not known, specific care management guidelines do not exist. However, there are general guidelines for handling such subjects.

First, emergency responders must learn to recognize the signs of excited delirium and provide proper emergency medical treatment as soon as possible. Several indicators should cause an officer to suspect excited delirium.

- Patients may exhibit an exaggerated version of the fight-or-flight response, run at the first opportunity, and seem to be attracted to running in traffic. Their aggression is unpredictable, and often unprovoked.
- A patient may possess extraordinary strength and may seem indifferent or unresponsive to pain stimuli.
- Destroying property, especially breaking glass, is a common symptom, although no one seems to know why. Such destructive incidents usually involve making a lot of noise.
- Patients often exhibit animal-style behavior, including grunting, biting, and scratching. Their speech seems unintelligible or nonsensical.
- Excited delirium patients may exhibit a condition commonly known as "8-ball eyes." Their eyes are so wide-open that the white of the eye is visible all around the iris.
- Undressing or getting naked may indicate a psychosis, or it may be a result of hyperthermia.

Often, profuse sweating and a feverish temperature accompany nudity.
- Patients experiencing excited delirium usually exhibit impaired thinking, including the acute onset of paranoia, and they are oblivious to obvious injuries.
- Most people can sustain maximum exertion for about two minutes before the body's internal mechanisms cause it to quiet down. A patient in excited delirium can exert maximum effort far longer because the normal feedback mechanisms are impaired.
- A patient may appear psychotic and may manifest rapid changes in emotions, such as going from a state of hyperactive delirium to a state of flat delirium, a condition known as mixed delirium.
- A patient may be disoriented about place, time, purpose, and even self.
- Excited delirium patients often have muscle rigidity. When an officer grabs an arm for handcuffing, the arm will not easily go behind the back, which may not indicate intentional resistance, but may be due to the muscle rigidity caused by the subject's medical condition.
- A patient may hallucinate, hear voices, talk to people who are not there, or talk to inanimate objects.
- Patients often violently resist being controlled and restrained.
- Patients may claim to be unable to breathe during or after being subdued.
- Patients are often easily distracted and have a lack of focus.
- Excited delirium patients often have delusions of grandeur.
- Witnesses often describe the patient as having "snapped" or being "flipped out" (Peters, 2006c).

It is important to note that law enforcement officers cannot make a clinical diagnosis of a person. Only a qualified medical or health care professional can do that. Rather, the officers' focus should be on the identification of behavioral clues that indicate an individual is a high-risk candidate for sudden in-custody death. These behavioral clues may also indicate that the individual is in a state of excited delirium.

Law enforcement personnel who intervene during an incident of excited delirium should follow certain precautionary measures during the restraining process. Officers should remain cognizant of the amount of time a restraint is applied, the method of restraint, and the body position of the restrained individual during transport. Continued struggling and prolonged restraint may potentially cause a violent individual to suffer severe exhaustion resulting in sudden death. To reduce the chances of sudden death, officers should summon an ambulance to the scene of all incidents involving individuals who display signs of excited delirium. When the situation permits, responding law enforcement officers should simply contain the individual until additional officers and emergency medical workers arrive. Officers should consider restraining methods only as a last resort, and they should avoid placing subjects in the prone position, if possible. The prone position may aggravate a person's anxiety level and prevent observation and the monitoring of vital signs, including the subject's level of consciousness. To prevent incidents of restraint asphyxia, police personnel should try to avoid pressing down on the person's chest, abdomen, or back (Parent, 2006).

Emergency medical personnel should institute continuous pulse oximetry, a noninvasive method that allows health care providers to monitor the oxygenation of a patient's blood, both to detect hypoxia and to document the absence of restraint asphyxia. Medical personnel should also immediately initiate cardiac monitoring, given the potential for rapid deterioration, and quickly determine blood glucose for all patients with altered mental status. Benzodiazepines are normally the first-line treatment of psychomotor agitation in cocaine and other stimulant toxicity. Neuroleptic medications, such as Haloperidol, are usually contraindicated because of their adverse effects on temperature regulation, lowered seizure threshold and potential for dysrhythmia (Sztajnkrycer & Baez, 2005).

Finally, although rarely documented, elevated core body temperatures have been noted in many excited delirium patients. Medical personnel should record the individual's temperature as soon as possible, and, if feasible, begin aggressive temperature control measures, analogous to those used in caring for heat-stroke patients. Emergency services personnel should regard the cessation of struggling by an excited delirium patient as an ominous, near-terminal event. Should the subject develop shallow or labored breathing, emergency medical personnel should immediately begin aggressive evaluation and reassessment. The initial decompensation in excited delirium patients appears to be respiratory arrest, rather than cardiac arrest (Sztajnkrycer & Baez, 2005).

CATECHOLAMINE

Researchers have studied the roles of catecholamines and stress in cases of sudden death, and they widely accept that stress may increase the mortality of preexisting heart conditions (Mirchandani, Rorke, Sekula-Perlman & Hood, 1994). Catecholamines are chemical compounds in the blood that cause general physiological changes that prepare the body for physical activity, such as the fight-or-flight response. Studies suggest that catecholamines pharmacokinetically enhance the toxicity of cocaine leading to seizures, respiratory arrest, and cardiac arrest in rats. Studies also suggest that patients with excited delirium may be sensitive to catecholamines due to their history of cocaine use or schizophrenia. Neuronal catecholamine release may cause arrhythmias, and, in a patient with excited delirium, the additional stress of confrontation with law enforcement officials or medical personnel may lead to cardiopulmonary arrest (Mirchandani et al., 1994).

SYMPATHOMIMETICS

Sympathomimetics are a class of drugs whose effects mimic those of a stimulated sympathetic nervous system, such as increased cardiac output, dilated bronchioles, and constricted blood vessels. Sympathomimetics include cocaine and methamphetamine. Cocaine exerts its effects on the body through several distinct pathways. As a central nervous system stimulant, it modulates the effects of brain neurotransmitters, especially dopamine, which accounts for the euphoria associated with cocaine use and the rapid development of tolerance and addiction (Randall, 1992). Cocaine causes the adrenal glands to release epinephrine and blocks the reuptake of norepinephrine, resulting in a state of increased physical stimulation (Karch, 1996). Finally, cocaine has effects on the electrical conduction systems of the heart similar to those of tricyclic antidepressants. The concurrent use of alcohol and cocaine produces a new compound, cocaethylene, which lasts longer in the body and has even more potent toxic effects (Randall, 1992).

Of all the deaths reviewed by medical examiners and coroners, a large number involve the use of cocaine. A survey of medical examiners' offices by The Drug Abuse Warning Network indicated that, in 2000, there were 4,043 reported deaths related to the use of cocaine (Substance Abuse and Mental Health Services Administration, 2002). Interestingly, drug levels in those cases did not necessarily relate to drug toxicity or poisoning. A diagnosis of cocaine-induced excited delirium is possible when investigative information supports a history of chronic cocaine use. Whenever the victim of sudden death has a clinical or investigative history of acute psychosis, such as paranoid behavior, undressing, violent behavior, and often with hyperthermia, and a complete investigation and forensic autopsy does not reveal a pathologic process as the proximate or underlying cause of death, then a central nervous system active drug should be suspected. Cocaine may be present in low levels, or only the metabolite may be detected in urine or spinal fluid. Chronic drug use is necessary to induce the changes in the neurochemistry that lead to excited delirium, not high levels of cocaine. The presence of hyperthermia (core body temperature higher than 103 degrees Fahrenheit) is supportive of a cocaine-induced event (Stephens, Jentzen, Karch, Wetli, & Mash, 2004b). Clinical symptoms such as seizure, hyperthermia or excited delirium all point to cocaine as the proximate cause of the death, but other obvious causes of death must be carefully ruled out through a careful scene investigation, meticulous forensic autopsy, and a review of the medical information (Stephens, Jentzen, Karch, Wetli, & Mash, 2004b).

Methamphetamine is a strong psychomotor stimulant that mimics the actions of certain neurotransmitters that affect mood and movement. Methamphetamine causes a release of dopamine and serotonin, which produce the euphoria that users feel. Even after the initial stimulation subsides, the brain remains in an alert state and keeps the user's body on edge. After the effects have worn off, the brain is depleted of its dopamine, and depression is a common result. In addition, methamphetamine may generate many of the same toxic effects seen with other stimulants such as cocaine. High doses or chronic use have been associated with increased nervousness, irritability, paranoia, and occasionally violent behavior, while withdrawal from high doses generally leads to severe depression. Chronic abuse produces a psychosis similar to schizophrenia and is characterized by paranoia, picking at the skin, self-absorption, auditory and visual hallucinations, and sometimes episodes of violence.

Clinical manifestations and complications of amphetamines are similar to those from cocaine use and may be indistinguishable except for the duration of effect of amphetamines, which tends to be longer. Compared to cocaine, amphetamines are less likely to cause seizures, dysrhythmia, and myocardial ischemia, most likely due to the sodium channel-blocking effects and to the thrombogenic effect of cocaine. Psychosis appears to be more likely with amphetamines than cocaine, most likely due to the more prominent dopaminergic effects of amphetamines. Tachycardia and hypertension are the most common manifestations of cardiovascular toxicity. Most patients present to the emergency departments of hospitals, however, because of the central nervous system manifestations. These patients are anxious, volatile, aggressive, and may have life-threatening agitation. Visual and tactile hallucinations and psychoses are common.

Death from amphetamine toxicity most commonly results from hyperthermia, dysrhythmia, and intracerebral hemorrhage. Direct central nervous system effects may result in seizures. Tachycardia, hypertension, and vasospasm may lead to cerebral infarction, intraparenchymal and subarachnoid hemorrhage, myocardial ischemia or infarction, aortic dissection, acute lung injury, obstetrical complications, fetal death, and ischemic colitis. Dysrhythmias vary from premature ventricular complexes to ventricular tachycardia and ventricular fibrillation. Agitation, increased muscular activity, and hyperthermia can result in metabolic acidosis and rhabdomyolysis. Unless these systemic signs and symptoms are reversed, multiorgan failure and death ensue (Chaing, 2006).

The case studies in Chapter 3 show that in 213 in-custody deaths following the application of a TASER device from 1983 through 2005, the medical examiner or coroner recorded the use of sympathomimetics as the cause of death or as a significant contributing factor in at least 134, or 62.9 percent, of those cases. Because autopsy and toxicology reports were not available in every case, the exact number of cases involving sympathomimetic drugs is not known.

NEUROLEPTICS

Neuroleptic drugs, or simply neuroleptics, are also known as antipsychotics. They are a group of drugs used to treat psychoses. Common conditions for which neuroleptics might be used include schizophrenia, mania, and delusional disorder, although neuroleptics might be used to counter psychosis associated with a wide range of other diagnoses. Neuroleptics also have some effects as mood stabilizers, leading to their frequent use in treating mood disorder, particularly bipolar disorder, even when no signs of psychosis are present. A serious side effect is neuroleptic malignant syndrome, a life-threatening, neurological disorder most often caused by an adverse reaction to neuroleptics. The first symptom to develop is usually muscular rigidity, followed by high fever and changes in cognitive functions. Other symptoms can vary, but may include unstable blood pressure, confusion, delirium, and muscle tremors, several of the same symptoms noted in excited delirium cases. Once symptoms appear, they rapidly progress and can reach peak intensity in no more than three days.

Postural hypotension may occur during tricyclic antidepressant therapy. Other cardiovascular effects of the drugs include various arrhythmias such as palpitation, tachycardia (including ventricular tachycardia), bradycardia, ventricular fibrillation, sudden death, hypertension, stroke, and congestive heart failure. Patients with preexisting cardiovascular disease may be especially sensitive to the cardiotoxicity of tricyclic antidepressants. In cardiac patients receiving therapeutic doses of tricyclic antidepressants, the drugs have been reported to increase the incidence of sudden death. Although myocardial infarction has been attributed to therapy with tricyclic antidepressants, a causal relationship has not been established (McEvoy, 2006).

The case studies in Chapter 3 show that in 213 in-custody deaths following the application of a TASER device from 1983 through 2005, the medical examiner or coroner recorded the use of neuroleptics as the cause of death or as a significant contributing factor in at least 11, or 5.2 percent, of those cases. Because autopsy and toxicology reports were not available in every case, the exact number of cases involving neuroleptic drugs is not known.

CORRELATION VERSUS CAUSATION

Undeniably, from 1983 through 2006, in the United States, at least 310 people died unexpectedly following the application of a TASER device. The following three tables demonstrate some demographics of those deaths. Table 2, Custody Deaths Following Application of TASER Devices, details the number of deaths each year. Table 3, Unexpected Custody Death Cases by State, details the numbers of deaths in each state that has a death documented. Table 4, Unexpected Custody Death Cases by Gender, details the gender, racial and ethnic information of each deceased through 2005. Although the evidence does not support allegations that TASER devices are the cause of those deaths, the conjecture of many critics of electronic control devices is that the rising number of deaths is proof that the TASER devices contribute to a rising death toll.

First, there is no evidence that the number of custody deaths is rising, but the clear trend is that deaths following the application of TASER devices are rising. As the number of TASER devices available to police departments rises, the number of deaths also rises. In 2001, only about 1,000 of the 18,000 police departments in the United States used TASER electronic control devices. By 2007, approximately 10,500 police departments either were field-testing TASER devices or had adopted them for use, and there were almost 270,000 TASER devices in the hands of law enforcement officials (Mills, 2007). Naturally, as more officers begin using electronic control devices instead of other types of restraints when subduing subjects, the number of deaths following the application of TASER devices rises, but to date, no causal relationship between the use of a TASER device and sudden death has been shown.

Correlation is not the same as cause and effect. Correlation shows only the strength of relationship between two or more variables—as the number of conducted energy weapons available to officers rises, the number of sudden deaths each year following the application of a conducted energy weapon also rises. Many critics of electromuscular disruption technology cite that temporal relationship of the use of a TASER device prior to an individual's death as proof that the device caused the death, but this is an unscientific linking of two events. Although there is a temporal dimension to causality, temporal relationship alone does not establish causality.

Cook and Campbell (1979) note that causal inference depend upon three factors: First, the cause must precede the effect. Second, the cause and effect must be related. Third, other explanations of the cause and effect relationship must be eliminated. Causality can be inferred by examining observed associations between two or more events or variables, if there is:

- Strength of association - B occurs much more often with A than with other possible causes;
- Consistency - A is found with B by many studies in different places;

Table 2
UNEXPECTED CUSTODY DEATHS FOLLOWING APPLICATION OF TASER DEVICES
1983 THROUGH 2006

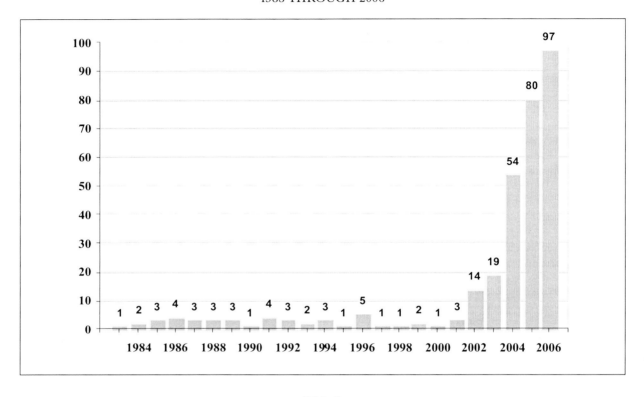

Table 3
UNEXPECTED CUSTODY DEATHS FOLLOWING APPLICATION OF TASER DEVICES BY STATE
1983 THROUGH 2006

State	Cases	State	Cases	State	Cases	State	Cases
California	84	Colorado	7	Minnesota	4	Kansas	2
Florida	45	Nevada	7	Missouri	4	Kentucky	2
Texas	25	New York	6	South Carolina	4	New Mexico	2
Ohio	17	Oklahoma	6	Tennessee	4	Utah	2
Arizona	11	Pennsylvania	6	Connecticut	3	Wisconsin	2
Louisiana	9	Alabama	5	Maryland	3	Montana	1
Washington	9	Indiana	5	Mississippi	3	Nebraska	1
Georgia	8	North Carolina	5	Oregon	3	Rhode Island	1
Illinois	8	Michigan	4	Arkansas	2		

Table 4
UNEXPECTED CUSTODY DEATHS FOLLOWING APPLICATION OF
A TASER DEVICE BY GENDER AND RACE – 1983 THROUGH 2006

Total Cases = 213	Female = 9	African/American	4
		American Indian	0
		Hispanic	0
		White	4
		Not Discernable from Available Records	1
	Male = 204	African/American	70
		American Indian	4
		Hispanic	37
		White	81
		Not Discernable from Available Records	12

- Specificity – a particular link exists between A and B;
- Temporality – A occurs before B, not the reverse;
- Biological gradient – if A occurs more often, then B occurs more often;
- Plausibility – a known mechanism exists to link A and B;
- Coherence – the A-B relationship fits with what else we know about A and B;
- Experiment – change A and observe what happens to B; and
- Analogy – A and B resemble the well-established pattern noted in C and D (Miles and Huberman, 1994).

The few studies available on sudden in-custody deaths show a much higher correlation between sudden death and heart disease, sudden death and the use of sympathomimetic drugs, and sudden death and bizarre behavior than between sudden death and the use of a TASER electronic control device. Additionally, literally tens of thousands of people who have been shocked with a TASER device survived without ill effects. Consequently, there is little strength of association between sudden death and the use of a TASER device, effectively precluding the argument of a causal relationship between the two events. Additionally, although a few studies indicate that the impulses from an electronic control device, under the right conditions, might interfere with the sinus rhythm of the heart, many other studies indicate that it does not. This lack of consistency and specificity also works to defeat an argument of causality.

SUMMARY

An unexpected in-custody death usually generates controversy regarding police tactics on use of force, and perhaps the greatest controversy of the past decade has been use of the TASER electronic control device. When someone dies following the application of a TASER device, before the cause of death is known, the reports of another death have special interest groups calling for a ban on the use of TASER devices. However, before coming to a conclusion on the role of the devices, one must first understand something of the complicated and complex phenomenon of sudden death.

Each year in the United States, more than 300,000 people die from sudden cardiac death. In more than three out of four sudden deaths following the use of a TASER device from 1983 through 2005, medical examiners and coroners noted in the deceased at least one of the predisposing factors that increase the risk of sudden death. Several clinical and autopsy-based studies have reported the triggering of sudden cardiac death with exercise, so it is simple to understand how the risk of sudden death following violent exertion, such as struggling with the police or straining against restraints, increases manifold for people with predisposing factors. Sudden death can be traced to many factors: excited delirium, restraint asphyxia, metabolic acidosis, rhabdomyolysis, catecholamine release, and the use of sympathomimetic and neuroleptic drugs, but no evidence exists to establish a cause and effect relationship between the use of the TASER and sudden custody deaths.

KEY TERMS

Catecholamine	Electronic Control Device	Rhabdomyolysis
Causation	Excited Delirium	Sudden Death
Conducted Energy Weapon	Instantaneous Death	Sympathomimetic
Correlation	Metabolic Acidosis	
Electromuscular Disruption Technology	Neuroleptic	
	Restraint Asphyxia	

Chapter 2

CONDUCTED ENERGY WEAPONS

Tom's electric rifle was well adapted for this work, as he could regulate the charge to merely stun, no matter at what part of the body it was directed.

Victor Appleton, *Tom Swift and His Electric Rifle*

Despite constant demands on the police service to avoid lethal force and to find less lethal means to subdue resisting or aggressive individuals, the most controversial development in law enforcement in the past decade has been the use of conducted energy weapons that were designed for that specific purpose. Conducted energy weapons deploy electromuscular disruption technology to affect the sensory and motor functions of the somatic nervous system (International Association of Chiefs of Police, 2006). These weapons temporarily incapacitate someone by using an electric current to disrupt the body's ability to control muscle movement. Forty years of technology have produced several products, but the best known and most popular are the TASER® electronic control devices: the TF-76, the TASERTRON, the AIR TASER 34000, the ADVANCED TASER M26, and the TASER X26.[1] Although other manufacturers may produce comparable devices, the TASER M26 and X26 models are the only ones currently in wide use by law enforcement agencies around the globe that have had extensive domestic and international scientific testing conducted on their safety, efficacy, and risk utility (Peters, 2006b).

---

1. TF-76, TASERTRON, AIR TASER, M26, and X26 are trademarks of TASER International, Inc. TASER® and ADVANCED TASER® are trademarks of TASER International, Inc. registered in the United States. All rights reserved.

A SHORT HISTORY OF TASER INTERNATIONAL, INC.

Jack Cover, a National Aeronautics and Space Administration (NASA) scientist who worked on the Apollo moon-landing program, invented the original TASER electronic control device, the TF-76, in 1974. Cover named his invention after the Thomas A. Swift Electric Rifle, a fixture of the Tom Swift books of the 1930s (Riordan, 2003). The TASER TF-76 model fired two darts up to a distance of 15 feet. These darts remained attached to the hand-held device by small, thin, insulated wires. Cover submitted a summary report on the weapon to the U.S. Consumer Product Safety Commission in 1975. In the report, he stated that years of studies and tests on the effects of electricity on the body had shown the TASER TF-76 device was not lethal to normally healthy people. However, he emphasized that neither the weapon's three-watt output nor the noninjury or no-harmful-after-effect aspect of the TASER TF-76 device could ever be guaranteed. Cover wrote, "There is no weapon, technique or procedure for subduing, constraining or dispersing that does not involve some risk of injury to healthy persons or of death, especially if the individual has a heart ailment (Kelley, 1990)."

Based partly on Cover's analysis, the Product Safety Commission's deputy medical director,

Doctor Leo T. Duffy, declared the TASER TF-76 to be a nonlethal device when "used as directed on the average, healthy adult." He described the gun as a "dangerous weapon," but he said that it had only a fraction of the power necessary to shock the heart into losing its rhythm and causing death (Kelley, 1990).

Because the original TASER device used a gunpowder charge to launch the darts, the United States Bureau of Alcohol, Tobacco, and Firearms (ATF) classified the TASER TF-76 model as a firearm. However, the TF-76 version looked like a flashlight instead of a gun. Because it did not fit the specifications for either a pistol or a long-gun, ATF classified the TF-76 device as a Title 2 weapon—the same as a short-barrel shotgun or rifle. The Title 2 classification meant that only people with special permits could purchase the TASER TF-76 device. Effectively, only law enforcement agencies could purchase the TF-76 model, and the civilian market was severely limited.

Shortly after the Title 2 classification, TASER Systems, the California-based company that made the TF-76 device, collapsed. The company went in and out of bankruptcy protection until it was sold to an investor who produced the devices as a sideline. The company reemerged as TASERTRON, but it struggled over the next decades and sold only a few devices in the law enforcement marketplace, mostly to the Los Angeles Police Department. The TASERTRON devices were originally offered in seven-watt versions and later in eleven-watt models.

In September 1993, brothers Tom and Rick Smith formed ICER Corporation, whose mission was to develop less-lethal electronic weapons. The Smith brothers contacted the original inventor, Jack Cover, who joined them to develop the next generation of TASER devices that would use a compressed air or compressed nitrogen propulsion system. Shortly thereafter, they changed the name of the company to AIR TASER, Inc. In December 1994, they launched the second generation of TASER devices, the AIR TASER 34000. The design intention of the AIR TASER 34000 device was to use the same electrical output as the original TASER TF-76 model, but with a compressed air propulsion system that would comply with federal firearm statutes and allow for private citizen sales. The AIR TASER 34000 model also implemented an innovative technology called Anti-Felon Identification (AFID). AFID used serialized confetti tags dispersed from every cartridge at the time of firing. The tags enabled law enforcement to trace persons who misused a TASER device. AIR TASER signed a noncompete agreement that recognized TASERTRON had exclusive rights to the underlying technology for use in the law enforcement and military markets in North America and precluded AIR TASER, Inc. from selling to law enforcement or military agencies in North America until the patent expired in 1998.

In 1995, following a disastrous demonstration of the AIR TASER 34000 device to the Czech police, wherein several police cadets fought through the pain compliance effects of the device, the company determined to develop a more effective device that would interfere with voluntary muscle control. The result of their research was the third generation of TASER devices, the ADVANCED TASER M26. Earlier devices caused some degree of stimulation of the somatic nervous system, but there was little or no interference with or impairment of muscular control. The ADVANCED TASER M26 was designed to cause significant, uncontrollable muscle contractions capable of incapacitating even the most focused and aggressive combatants. The company introduced the ADVANCED TASER M26 in late 1999. By then, the company had changed its name to TASER International, Inc. In addition to the AFID system, the ADVANCED TASER M26 implemented a new accountability control technology, the data port. The data port recorded the time and date of every trigger pull to allow law enforcement agencies to monitor use of the device.

In 2003, TASER International introduced the fourth generation device, the TASER X26. The X26 implemented a newer, more efficient electrical stimulation pulse that the company called Shaped Pulse™ technology. The new technology allowed for a more efficient power supply. The advances enabled TASER International to produce the X26, a model 60 percent smaller and 60 percent lighter than the M26. However, the X26 delivered an incapacitating effect that caused muscular contractions five percent stronger than the M26 (Smith, 2006).

ELECTRONIC CHARACTERISTICS

The TASER M26 device has 50,000 peak open circuit arcing voltage, with 5,000 peak loaded voltage, at 3.6 milliamps, with an energy pulse of 1.76 joules nominal at peak capacitor, and 0.5 joules delivered into the load (Southwell, 2003). The TASER X26 model has 50,000 peak open circuit arcing voltage, with 1,200 peak loaded voltage, at 2.1 milliamps, with an energy per pulse of 0.36 joules nominal at peak capacitor, and 0.07 joules delivered into the load (Southwell, 2004). For comparison, external cardiac defibrillators typically generate approximately 360 joules. Table 5, TASER Control Devices Electrical Characteristics, lists the electronic specifications of the M26 and X26 models.

Voltage, also called electromotive force, is the pressure behind the flow of electrons. One volt (V) is the amount of force required to send one ampere of current through a resistance of one ohm. High voltage, in and of itself, is not necessarily dangerous. A strong static electricity shock can exceed 30,000 volts, and modern Van de Graff generators can generate more than 1,000,000 volts with no more harmful effect than to make one's hair stand on end. The ampere (A) is a unit of electric current or the amount of electric charge. Amperes, or amps, measure the flow rate, or how many electrons flow each second. A milliampere (mA) is 1/1000 ampere. The ohm (Ω) is a unit of electrical impedance. A coulomb (C) is the amount of electric charge carried by a current of one ampere flowing for one second. A joule (J) is the work required to move an electric charge of one coulomb through an electrical potential difference of one volt.

The most pervasive misconception about TASER electronic control devices is the 50,000 volt myth. Practically every media report on TASER devices references the 50,000 volt shock that the device supposedly delivers. Both the M26 and X26 have 50,000 peak open circuit voltage at the main capacitor. The purpose of such high voltage is to permit a spark to cross a gap between clothing and a person's body should the probes not contact the skin. However, neither device delivers 50,000 volts into the load, or a person's body. In the drive stun mode (when the weapon is applied directly to the skin), during probe contact with the skin, and once the spark has crossed a gap, the TASER M26 model delivers a peak loaded voltage of 5,000 volts, and the TASER X26 model delivers a peak loaded voltage of 1,200 volts.

PHYSIOLOGICAL EFFECTS

Most previous stun devices created an electrical pulse to cause pain. The idea behind such weapons was to create enough pain to dissuade a person from resisting, but such technology was seldom effective on people who were intoxicated from alcohol or illicit drugs, who had mental illness, or who were determined enough to fight through the discomfort. The technology in modern TASER devices, however, does not rely on pain or injury for its incapacitating effects. The TASER electronic control device uses electrical stimuli to interfere with the signals sent by the command and control systems of the body, at the peripheral nervous system level, temporarily to impair a subject's ability to control his own body.

The human nervous system is the command and control system of the body. It has three primary elements: the central nervous system, the sensory nervous system, and the somatic nervous system. The central nervous system represents the largest part of the nervous system, including the brain and the spinal cord. Together with the peripheral nervous system, it has a fundamental role in the control of behavior. The sensory system is that part of the nervous system responsible for processing sensory information. The sensory nervous system consists of sensory receptors, neural pathways, and those parts of the brain involved in sensory perception. The

Table 5
TASER ELECTRONIC CONTROL DEVICES ELECTRICAL CHARACTERISTICS

Electrical Output Characteristic	ADVANCED TASER M-26	TASER X-26
Waveform	50 kHz damped sine wave pulse	A single cycle 100 kHz arcing phase followed by monophasic 100 microsecond (μs) stimulation phase shaped pulse
Waveform and Stimulation Capability		1 ampere (A) average current and 400 volt (V) average during the 100 microsecond (μs) pulse
Pulse Rate	20 pulses per second (pps) ± 25% with NiMH rechargeable cells; 15 pps ± 25% with Alkaline cells	19 pulses per second (pps)
Pulse Duration	40 microseconds (μs) full waveform; 10 microseconds (μs) primary phase	100 microseconds (μs)
Total Per Second Discharge Time	0.0008 seconds at 20 pulses per second (pps)	0.0019 seconds
Peak Open Circuit Arcing Voltage	50,000 volts (V)	50,000 volts (V)
Peak Loaded Voltage	5,000 volts (V)	1,200 volts (V)
Average Voltage - (One Second Baseline)	1.3 volts (V)	0.76 volts (V)
Average Current - (One Second Baseline)	3.6 milliamperes (mA) or 0.0036 amperes (A) average rectified current; 1.7 mA current from main phase	2.1 milliamperes (mA) or 0.0021 amperes (A) average rectified current; 1.9 mA current from main phase
Energy per Pulse - Nominal at Main Capacitor	1.76 joules (J)	0.36 joules (J)
Energy per Pulse - Delivered into Load	0.50 joules (J)	0.07 joules (J)
Main Phase Delivered Charge	85 microcoulombs (μC)	100 microcoulombs (μC)
Power Rating - Nominal at Main Capacitor	26 watts (W)	7 watts (W)
Power Rating - Delivered into Load	10 watts (W)	1.3 watts (W)
Power Source	8 - AA NiMH cells (1.2 V per cell); or, 8 - AA Alkaline cells (1.5 V per cell)	Digital Power Magazine - two three-volt (3 V) cells

commonly recognized sensory systems are those for vision, hearing, touch, taste and smell. Many news articles and reports discussing the effects of the TASER devices claim that, to incapacitate a person, the electric pulse affects the central nervous system (see, for example: Amnesty International, 2004; Davis, 2004; Ruggieri, 2005). However, TASER devices do not affect the central nervous system, they affect the peripheral nervous system, more specifically, the somatic nervous system.

The peripheral nervous system is the part of the nervous system consisting of the nerves and neurons that reside or extend outside the central nervous system to serve the limbs and organs. The peripheral nervous system is divided into the autonomic nervous system and the somatic nervous system. The autonomic nervous system controls involuntary bodily functions and homeostasis, such as the cardiovascular, digestive and respiratory functions, salivation, and perspiration. The somatic nervous system is the part of the peripheral nervous system that is associated with the voluntary control of body movements through the action of skeletal muscles and the reception of external stimuli. The somatic nervous system consists of afferent fibers that receive information from external sources and efferent fibers that are responsible for muscle contraction.

Third and fourth generation TASER devices use very short duration low energy electrical pulses that are similar to the pulses neurons use to communicate. The body's neurons conduct electrical stimuli to and from the brain. When a neuron is in its resting state, electrically charged ions cross the cell membrane so a net positive charge collects outside the membrane, and a net negative charge collects inside the membrane. In this resting state, the membrane serves as a charged capacitor. When the nerve cell is stimulated, channels in the membrane open up temporarily, allowing the positive ions temporarily to rush across the membrane. The voltage potential across the membrane briefly flips polarity as the charge balance reverses. As this process, known as action potential, occurs in one section of the cell membrane, the change in the electric fields causes the adjacent section of the membrane to depolarize. The result is a chain reaction of action potentials cascading down the length of the neuron, thereby carrying an electric impulse.

For each neuron, there is a threshold stimulation level. Upon attaining that threshold, an action potential occurs. Each neuron can only deliver one magnitude of impulse, so action potentials come in only one magnitude. There are not different intensities of action potential, and there is no such thing as a partial or weak action potential. Whether a muscle contraction will be strong or weak is not a function of the magnitude of the impulses of the connected neurons. The difference is in the pattern of impulses delivered.

Action potentials traveling down the motor neurons of the somatic branch of the nervous system cause the skeletal muscle fibers at which they terminate to contract. The junction between the terminal of a motor neuron and a muscle fiber is the neuromuscular junction, also known as the myoneural junction. The terminals of motor axons contain thousands of vesicles filled with acetylcholine, a chemical transmitter in the central and peripheral nervous systems. When an action potential reaches the axon terminal, hundreds of vesicles discharge acetylcholine onto a specialized area of a postsynaptic membrane on the fiber. This area contains a cluster of transmembrane channels that acetylcholine opens and lets sodium ions diffuse in.

The interior of a resting muscle fiber has a resting potential of about $^-95$ millivolts. The influx of sodium ions reduces the charge, creating an end-plate potential. When the end-plate potential reaches the threshold voltage of approximately $^-50$ millivolts, sodium ions flow in and create an action potential in the fiber. The action potential sweeps down the length of the fiber just as it does in an axon. No visible change occurs in the muscle fiber during, or immediately following, the action potential. This period, called the latent period, lasts from 3–10 milliseconds. Before the latent period is over, the enzyme acetylcholinesterase breaks down the acetylcholine in the neuromuscular junction, the sodium channels close, and the field clears for the arrival of another nerve impulse. The resting potential of the fiber is restored by an outflow of potassium ions. The brief period of approximately one to two milliseconds needed to restore the resting potential is the refractory period. The process of muscles contracting takes approximately 50 milliseconds. Relaxation of the fiber takes another 50 to 100 milliseconds. Because the refractory period is so much shorter than the time needed for contraction and relaxation, the fiber can be maintained in the contracted state when it is stimulated frequently enough, that is, more than 50 stimuli per second (Ganong, 2005).

When shocks are given at one per second, the muscle responds with a single twitch. At 5–10 shocks per second, the individual twitches begin to fuse together, a phenomenon called clonus. At 50 shocks per second, the muscle goes into the smooth, sustained contraction of tetanus. Clonus and tetanus are possible because the refractory period is much briefer than the time needed to complete a cycle of contraction and relaxation. The amount of contraction is greater in clonus and tetanus than in a single twitch.

TASER devices deliver very short duration electrical pulses at a rate of 15–20 pulses per second. As mentioned earlier, the first and second generation devices, the TASER TF-76 and the AIR TASER 34000, delivered a charge in each pulse only sufficient to stimulate the sensory nerves close to the skin. Very little motor nerve stimulation occurred, making those weapons relatively ineffective against focused, motivated, or pain-resistant subjects. The M26 and X26 models deliver a similar train of electrical pulses, also at 15–20 pulses per second. However, the M26 and X26 devices deliver more electrical charge in each pulse. This higher charge results in the stimulation of deeper nerves, such as motor nerves. As a result, the motor nerves between the two electrodes discharge at a rate of roughly 20 pulses per second. This stimulation rate is sufficient to cause clonus, when the individual twitches fuse together into a sustained contraction. However, 20 pulses per second are well below the 50–60 pulses per second required to cause complete tetanus.

Both nerve cells and muscle cells can be stimulated with electricity because both nerve and muscle cells use action potentials during stimulation.

The mechanism of stimulation from TASER devices is not direct electrical stimulation of muscle tissue, but stimulation of motor nerves, which then stimulate muscles in a nerve-mediated mechanism. This mechanism has been demonstrated in laboratory testing, wherein a test animal was administered a drug that blocked the neuromuscular junction. Before the drug administration, the application of the TASER device caused significant muscular contractions. After the drug administration, the application caused insignificant muscle reaction, demonstrating that the motor nerves mediate the mechanism of effect and that it is not a direct electrical stimulation of the muscle tissue (Smith, 2006).

In a study at the Air Force Research Laboratory, researchers found that the intensity of the muscle contractions caused by the ADVANCED TASER M26 model could be increased to more than 250 percent of the level of contraction from the field production M26. Accordingly, the M26 generates a muscle contraction 40 percent less than the maximal contraction attainable with waveforms that are more aggressive. The X26 model delivers a contraction roughly 5 percent greater than the M26, a level still well below 50 percent of the maximal contractions found in the Air Force Research Laboratory study (Jauchem, 2004). Although the contractions that TASER devices induce are sufficient to interfere with and to impair voluntary movement and to cause incapacitation in most applications, they are still within the normal operating range of voluntary muscle movements associated with strenuous activities such as weight lifting, wrestling, or other athletic or exertion activities.

LITERATURE REVIEWS

In 2001, the British medical journal, *The Lancet*, published an article reviewing the effects of TASER devices. That article reported that, because of the difference in excitability between nerves and cardiac muscle, and because the heart is distant from the skin, myocardial stimulation was extremely unlikely in normal use of such devices. The authors noted it was clear from the literature that, when properly used as a method of restraining violent people, TASER devices were less likely than guns to cause injury and death, and they were also generally more effective than other methods of restraint. The authors also concluded that, to that time, the deaths following the use of a TASER device had occurred in people who were out of control and who had taken potentially fatal drugs. In such cases, they opined, it was quite possible that the deaths would have occurred regardless of whether a TASER device was used (Fish & Geddes, 2001).

In 2003, TASER International commissioned two consultants in England to conduct an extensive medical literature review of electromuscular disruption technology and conducted energy weapons. The researchers reviewed 79 publications that included reports on the usage of electronic restraint devices and the medical hazards of electricity. Those publications included 35 relevant peer-reviewed articles on the occurrence and purported mechanism of injuries and deaths associated with conducted energy weapons. However, all the published medical literature on the effects of TASER devices were based on experience with the original TASER TF-76 and AIR TASER 34000 models, and not on the more recent devices. The authors published the following conclusions:

- The medical risks of conducted energy weapons compared favorably with those of conventional methods of controlling noncompliant and violent subjects. However, they admitted that it was impossible accurately to calculate how much electrical energy a TASER device delivered into the human body.
- Convincing evidence directly implicating TASER weaponry to sudden deaths did not exist.
- Risk factors for death in subjects subjected to the application of a TASER device appeared to be no different from known risk factors for other deaths in custody (e.g., drugs, exhaustion, and bizarre behavior leading to arrest).
- The risk of harm may have been higher when using TASER devices on patients with preexisting heart and neurological diseases, but those risks were largely theoretical and had not been demonstrated in field applications or laboratory testing.
- The risks to patients with implanted pacemakers and defibrillators were probably very small.
- The potential for significant injury existed for probes striking the eye, open mouth, neck, genitals, and large blood vessels in the groin.
- The TASER electronic control device delivered electricity to incapacitate the subject and end physical resistance, and likely psychological resistance, to arrest. Much useful data was gained from volunteers, but much more research was required to record the effects of TASER devices on physiological variables and electrocardiograph tracings.
- TASER devices were unlikely to cause permanent physical problems in healthy individuals.
- Litigation could be expected following custody deaths or from those who would claim later to have developed medical or psychological conditions. Based on the available medical literature, however, there was no solid basis for potential claimants to pursue successful litigation with the exception, perhaps, of posttraumatic stress disorder (Bleetman & Steyn, 2003).

In 2004, the Canadian Association of Chiefs of Police approached the Canadian Police Research Centre to conduct a comprehensive review of the existing scientific research and data and to provide a national perspective on the safety and use of conducted energy weapons. In 2005, the research team published its findings:

- Definitive research or evidence that implicated a causal relationship between the uses of conducted energy weapons and death did not exist.
- Existing studies indicated that the risk of cardiac harm to subjects from conducted energy weapons was very low.
- Excited delirium, although not a universally recognized medical condition, was gaining

increased acceptance as a main contributor to deaths proximal to conducted energy weapon use.
- The issue related to multiple conducted energy weapon applications and its impact on respiration, pH levels, and other associated physical effects, offered a plausible theory on the possible connection between deaths, conducted energy weapon use, and people who exhibited the symptoms of excited delirium.
- The increased use of conducted energy weapons was related to a decrease in the use of lethal force in some jurisdictions and was related to substantial decreases in police officer and subject arrest-related injuries.
- Although each use of force incident needed to be judged separately, the increased use of conducted energy weapons in nonlethal incidents was generally appropriate.
- Originally marketed and accepted as an alternative to lethal force, use of conducted energy weapons had grown to include incidents when intermediate weapons, but not lethal weapons, should have been used.
- It would be unwise and counterproductive for any police service or government body to develop policies and procedures that explicitly specify in what kinds of circumstances a conducted energy weapon may or may not be used (Manojlovic, Hall, Laur, Goodkey, Lawrence, Shaw, St-Amour, Nuefeld, & Palmer, 2005).

In 2005, The Potomac Institute for Policy Studies hosted a two-day conference to assemble experts from various fields, including medical and health effects, safety and regulatory issues, policy, and industry practices, to discuss what was known about electromuscular disruption technology and to offer insight and suggestions on filling the current gaps in knowledge. Based on the available evidence, and on accepted criteria for defining product risk versus efficacy, conference attendees reported that, when electromuscular disruption technology was appropriately applied, it was relatively safe and clearly effective. The only known field data suggested that the odds were, at worst, one in one thousand that a conducted energy weapon would contribute to (not to imply cause) a death. The one-in-one thousand figure was not different from the odds of death when conducted energy weapons were not used, but other multiple force measures were. According to the attendees, a more defensible figure was one in one hundred thousand (McBride & Tedder, 2005).

TECHNICAL STUDIES

In 2003, the Victoria (Australia) Police Department commissioned a study of the electrical safety aspects of the ADVANCED TASER M26 device. A biomedical engineer at the Alfred Hospital in Melbourne conducted the tests. The project was limited to a literature review, theoretical calculation of the TASER M26 model's output, verification that the output complied with applicable Australian medical device standards, practical testing of TASER M26 device output into known test resistances, and a comparison of actual measured TASER M26 device output to calculated outputs of common high electrical output medical devices and domestic equipment. The literature review concluded that:

- no proven connection had been reported between the use of the TASER M26 device and subsequent deaths of offenders;
- there were no comprehensive studies relating to long-term physiological effects on subjects who have been hit by a TASER probe;
- there was no comparative electrophysiology literature indicating that the TASER M26 device's electrical output was beyond currently published acceptable limits; and
- secondary effects such as fire, muscle spasm, and falls needed to be considered.

Following the tests, the engineer concluded that:

- the TASER M26 device's output was less than two percent of the normalized current likely to produce ventricular fibrillation;
- tissue damage from the TASER M26 device

was expected to be less than that from other routinely used patient treatment devices;
- test firing of the TASER M26 model confirmed the theoretical calculations; and
- from an electrical safety viewpoint, the TASER M26 model presented an acceptable risk when used by trained law enforcement officers in accordance with the manufacturers directions for use (Southwell, 2003).

The following year, 2004, the Victoria (Australia) Police Department commissioned a study of the electrical safety aspects of the TASER X 26 device, and the same biomedical engineer at the Alfred Hospital in Melbourne conducted the tests. Once again, the project was limited to a literature review, theoretical calculation of the TASER X26 electronic control device's electrical output, verification that the output complied with applicable Australian medical device standards, practical testing of TASER X26 device output into known test resistances, and a comparison of actual measured TASER X26 device's output to calculated outputs of common high electrical output medical devices and domestic equipment. The literature review concluded that:

- no proven connection had been reported between the use of the TASER X26 device and subsequent deaths of offenders;
- there were no comprehensive studies relating to long-term physiological effects on subjects who have been hit by a TASER probe;
- there was no comparative electrophysiology literature indicating that the TASER X26 device electrical output was beyond currently published acceptable limits; and
- secondary effects such as fire, muscle spasm, and falls needed to be considered.

Following the tests, the engineer concluded that:

- the TASER X26 device's output was less than one percent of the normalized current likely to produce ventricular fibrillation;
- tissue damage from the TASER X26 model was expected to be less than that from other routinely used patient treatment devices;
- test firing of the TASER X26 device confirmed the theoretical calculations; and
- from an electrical safety viewpoint, the TASER X26 model presented an acceptable risk when used by trained law enforcement officers in accordance with the manufacturers directions for use (Southwell, 2004).

In 2005, the Joint Non-Lethal Weapons Human Effects Center of Excellence (HECOE) published a report on a human effectiveness and risk characterization study for electromuscular incapacitation devices. The study reflected the results from a three-workshop process with sequential workshops held for data gathering and sharing, peer consultation, and independent external review of the report. The study included the ADVANCED TASER M26 and TASER X26 models. The HECOE study characterized probability estimates for intended and potential unintended effects of the devices. The study identified key potential unintended effects, such as ocular injury from probe strikes, seizures, ventricular fibrillation, or fall injuries. The researchers evaluated other potential effects during the study, but they did not assess those risks further because they were of limited severity (e.g., minor lacerations) or the available data did not support their occurrence (e.g., cancer or reproductive effects). The researchers integrated information developed in the dose-response and exposure assessment to provide quantitative and qualitative estimates of effectiveness and risk probabilities.

The HECOE study determined that the probability of inducing a complete electromuscular disruption ranged from 74 percent to 52 percent, depending on distance to the target. Severe unintended effects were likely to have a very low probability. Probability estimates were up to 0.04 percent for eye strikes and 0.15 percent for fall injuries, again depending on distance to the target. Ventricular fibrillation was not expected to occur in an otherwise healthy population, although experimental data were too limited to evaluate probabilities for susceptible populations or for alternative patterns of exposure. However, the report noted that no cases of ventricular fibrillation had been reported in training or field exposure conditions (Maier, Nance, Price, Sherry, Reilly, Klauenberg, & Drummond, 2005).

MEDICAL STUDIES

Part of the HECOE risk study entailed extrapolating adverse data from animal studies to human thresholds, an accepted medical study practice. The HECOE noted that, despite formulating risk factors, uncertainty in assessing the degree of human variability remained. The body weight adjustment they used did not account for all the variability in response to electrical effects. Furthermore, the relative importance of body weight, as opposed to some other correlate to body weight, was unclear from the available data. The study could not say what specific anatomical features, such as body fat, skin thickness, distance from the skin surface to the heart, or heart size, were responsible for changes in cardiac effects from the electromagnetic incapacitation charge. The degree of correlation among those parameters in human populations was not evaluated in the study because the critical predictors were unknown.

No specific data were identified for elderly populations, or for comparison of sensitivity between healthy adults and individuals with underlying heart conditions, or abnormalities in the physiologic environment within the body such as hypoxia, acidosis, electrolyte abnormalities, cardiac-sensitizing medications, or chemical exposure. In extremes of these conditions, the arrhythmogenic threshold could become so low that malignant arrhythmias including ventricular fibrillation could arise spontaneously. The study noted that animal data suggested that the normal operating TASER output did not induce ventricular fibrillation in pigs given drugs that sensitize the heart, although that was a limited subset of all of the possible known variables.

Based on this analysis, and assuming that the TASER probe hit the target and established a circuit that crossed the heart, the ventricular fibrillation threshold, relative to the X26 output, could be

Table 6
PREDICTED THRESHOLD FOR VENTRICULAR FIBRILLATION ABOVE NORMAL TASER X26 OUTPUT

Body Weight in Pounds	Predicted Threshold for Ventricular Fibrillation[1]	
	Typical Human	Sensitive Human
10	2.4	1.5
20	3.6	2.1
40	5.8	3.5
60	8.1	4.8
80	10	6.1
120	13	8.1
160	16	10
200	19	11
240	22	13
280	24	15

2. The value shown represents the fold increase in TASER X-26 output above normal operating output to exceed the ventricular fibrillation threshold for typical or sensitive humans of a given body weight.

calculated from the dose-response curves for normal persons without potentially sensitizing conditions. Predicted thresholds for typical, or 50th percentile, and sensitive, or fifth percentile, humans at a given body weight are provided in Table 6, Predicted Threshold for Ventricular Fibrillation above Normal X26 TASER Output.

Based on these threshold estimates one could conclude that, for large children and adults, even those who were sensitive responders, the risk of inducing ventricular fibrillation was very small. Consequently, a large safety margin existed. For example, the ventricular fibrillation threshold for a

40-pound child was 3.5 times greater than the normal TASER X26 operating output. The threshold for a 200-pound person was 19 times greater than the normal operating output for the typical person and 11 times for a person in the sensitive population. For very small children, however, where the margin was limited (e.g., approximately 1.5 times above normal output), the data were insufficient to conclude that there would be no ventricular fibrillation risk. Due to assumptions made in selecting uncertainty factor adjustments and the absence of specific threshold information in young children, the elderly, individuals with underlying heart conditions, or individuals with concurrent drug use, the HECOE reported that it was unknown whether there were highly sensitive individuals who could have a ventricular fibrillation response at the intended electromuscular incapacitation output, but results from the dose-response data were consistent with the existing field use data (Maier et al., 2005).

In a 2005 study published in the peer-reviewed journal *Pacing and Clinical Electrophysiology* (PACE), researchers, including two from TASER International, conducted a study on nine pigs to test the effects of the TASER pulse on cardiac safety. The researchers designed a prospective controlled study to compare the standard TASER X26 discharge with the discharge necessary to initiate ventricular fibrillation. They anesthetized the animals and continuously monitored their arterial blood pressure, blood saturation, respiration, and heart rate. They positioned the TASER probes to ensure that the charge was delivered across the thorax. They first applied the standard TASER charge to the animal. If the charge did not initiate ventricular fibrillation, they increased the charge until it did. Following the first induction, they decreased the charge until there was no induction of ventricular fibrillation by five successive charges of equal strength. The researchers defined the lowest discharge that produced ventricular fibrillation at least once as the minimum fibrillating discharge level. The maximum safe level was the highest discharge that could be applied five times without inducing ventricular fibrillation. They defined the ventricular fibrillation threshold as the average between the minimum fib-

Table 7
EXPERIMENTAL OUTCOMES FOR 19 PULSES PER SECOND DISCHARGES

Body Weight in Kilograms	Maximum Safe Level	Minimum Fibrillating Level	Safety Index
30	14	16	15
37	16	20	18
42	20	24	22
48	28	32	30
49	20	24	22
54	28	32	30
81	40	44	42
83	28	32	30
117	36	48	42

rillating discharge and the maximum safe level (McDaniel, Stratbuckler, Nerheim & Brewer, 2005). See Table 7, Experimental Outcomes for 19 Pulses per Second Discharges, for the results.

The PACE study found that a neuromuscular discharge that could produce ventricular fibrillation required 15–42 times the standard discharge of the TASER X26 device. The findings were very similar to the HECOE study in determining that the threshold rose with body weight, although the HECOE study predicted lower threshold levels.

In a study conducted in 2005 and published in the peer-reviewed journal *Academic Emergency Medicine*, researchers sought to examine the effects of TASER device application in resting adult volunteers to determine whether there was evidence of induced electrical dysrhythmia or direct cellular damage that would indicate a causal relationship between application of the device and in-custody death. Sixty-six human subjects underwent 24-hour monitoring after a standard application. The subjects had medical histories that included hypertension, hypercholesterolemia, mitral valve prolapse, hypothyroidism, congestive heart failure, previous myocardial infarction, cerebrovascular disease, asthma, diabetes, and gout. The researchers collected blood samples before exposure, immediately after exposure, and again at 16 and 24 hours after exposure. A subpopulation of 32 subjects had 12-lead electrocardiography performed at similar time intervals. Researchers analyzed blood samples for markers of skeletal and cardiac muscle injury and renal impairment, and a cardiologist, blinded to the study, read the electrocardiograms.

The study found that there was no significant change from baseline at any of the four times for serum electrolyte levels and the blood urea nitrogen/creatinine ratio. The researchers noted an increase in serum bicarbonate and creatine kinase levels at 16 and 24 hours, and they noted an increase in serum lactate level immediately after exposure that decreased at 16 and 24 hours. Serum myoglobin level increased from baseline at all three times. Myoglobin is released from damaged muscle tissue (a contributing factor in rhabdomyolysis), which has very high concentrations of myoglobin. The kidneys filter the released myoglobin, but myoglobin is toxic and may cause acute renal failure. All troponin levels measured were less than 0.3 nanograms per milliliter, except for a single value of 0.6 nanograms per milliliter in a single subject. The researchers evaluated this subject, and they identified no evidence of acute myocardial infarction or disability. Troponin is a complex of three proteins that is integral to muscle contraction in skeletal and cardiac muscle, but not to smooth muscle. Certain subtypes of troponin are very sensitive and specific indicators of damage to the heart muscle. At baseline, the cardiologist interpreted 30 of 32 electrocardiograms as normal. The two abnormal electrocardiograms were abnormal at baseline and remained the same at all four times.

The researchers concluded that, in a resting adult population, the TASER X26 pulse did not affect the recordable cardiac electrical activity within a 24-hour period following a standard five-second application. The authors were unable to detect any induced electrical dysrhythmias or significant direct cardiac cellular damage that related to sudden and unexpected death proximal to TASER X26 exposure. Additionally, they found no evidence of dangerous hyperkalemia or induced acidosis (Ho, Miner, Lakireddy, Bultman & Heegaard, 2006).

In 2005 and 2006, researchers at the University of California–San Diego conducted a series of three experiments on law enforcement officers undergoing training on the use of TASER devices. The first experiment, the results of which appeared in the peer-reviewed journal *Academic Emergency Medicine*, was a prospective pilot study to evaluate for cardiac changes using 3-lead electrocardiographic monitoring immediately before, during, and after firing of the TASER X26 device. The mean duration of the electrocardiograph tracing was 16.3 seconds. The researchers collected data on 20 volunteer subjects. The mean shock was 2.4 seconds, with a range of 1.2 to 5 seconds. Other than sinus tachycardia, a rapid heartbeat, the researchers observed no dysrhythmias. There was no change in morphology, QRS duration, or QT interval between the before and after data, although there was a 12 millisecond decrease in the PR interval. The mean heart rate increased 15 beats per minute. The researchers concluded in this pilot study that there were no significant cardiac dysrhythmias in healthy subjects immediately after receiving a TASER X26 shock. Additionally, there were no morphologic, rhythm, or interval changes other than a small decrease in PR interval and an increase in heart rate (Levine, Sloane, Chan, Vilke, & Dunford, 2005).

In the second study, the results of which were also published in *Academic Emergency Medicine*, the researchers enrolled 76 subjects, 9 of whom they later excluded because of equipment malfunction. Of the remaining 67 subjects, the mean shock duration was 2.2 seconds, with a range of 0.9 to 5.0 seconds. The researchers found that the mean heartbeat increased 19.4 beats per minute, but they observed no significant changes in the QRS morphology or any aberrantly conducted beats. They did find one subject who had a single premature ventricular contraction before and after the TASER shock, but no other subject had any other dysrhythmia except sinus tachycardia (Levine, Sloane, Chan, Vilke, & Dunford, 2006a).

In the third study, the researchers enrolled 58 subjects, 9 of whom they later excluded because of ECG lead displacement. In the remaining 49 subjects, the mean shock duration was 2.3 seconds, with a range of 1.2 to 5.0 seconds. Again, the subjects displayed an increase in mean heart rate of 20 beats per minute, but there were no significant changes in PR, QRS or QTc intervals. One subject had rare unifocal premature ventricular contractions pre and post discharge of the TASER device, but the researchers found no other dysrhythmia except sinus tachycardia (Levine, Sloane, Chan, Vilke, & Dunford, 2006b).

In 2005, researchers at the Air Force Research Laboratory conducted an experiment to determine whether repeated exposure to electromuscular incapacitating devices could result in rhabdomyolysis and other physiological responses, including acidosis, hyperkalaemia, and altered levels of muscle enzymes in the blood. Their results were published in the peer-reviewed journal *Forensic Science International*. The researchers performed experiments to investigate the effects of repeated exposures of the TASER X26 device on muscle contraction and resultant changes in blood factors in an anaesthetized swine model. They used 10 animals, six of which they exposed to the TASER charge for 5 seconds, followed by a 5-second period of no exposure, repeatedly for 3 minutes. In five of the animals, after a 1-hour delay, they added a second 3-minute exposure period. The remaining four animals they used for an additional pilot study.

The researchers reported that all four limbs of each animal exhibited contraction, even though the

Table 8
SINUS RHYTHM LABELS

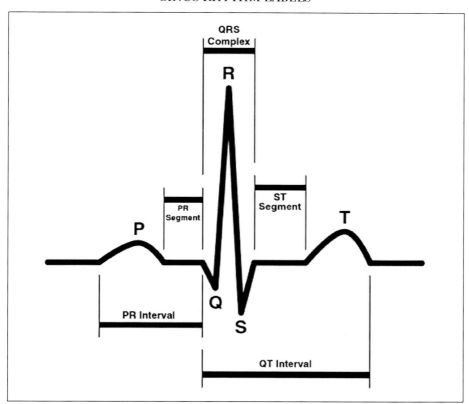

electrodes were positioned in areas at some distances from the limbs. The degree of muscle contraction generated during the second exposure period was significantly lower than that in the first exposure series. This finding was consistent with previous studies showing that prolonged activity in skeletal muscle will eventually result in a decline of force production. They discovered some similarities in blood sample changes in the experiments consistent with previous studies of muscular exercise. Thus, they concluded that problems concerning biological effects of repeated TASER exposures may be related, not directly to the electric output *per se*, but rather to the resulting contraction of muscles, related interruption of respiration, and subsequent sequelae. Transient increases in hematocrit, potassium, and sodium were consistent with previous reports in the literature dealing with studies of muscle stimulation or exercise, but it was doubtful that these short-term elevations would have any serious health consequences in a healthy individual. Blood pH significantly decreased for one hour following exposure, but subsequently returned toward a normal level. Leg muscle contractions and decreases in respiration each appeared to contribute to the acidosis. Lactate was highly elevated, with a slow return to baseline. The researchers concluded that, on the basis of the results of the current study, the repeated use of electromuscular incapacitating devices in a short period of time is feasible, with the caveat that some medical monitoring of subjects may be required (Jauchem, Sherry, Fines & Cook, 2006).

Additional studies have shown that the presence of cocaine in the system may work to increase cardiac safety in people subdued by the application of TASER devices. In 2006, researchers tested various TASER voltages with paired probes applied at five different sites on anesthetized pigs. They applied the current using a modified TASER device that was capable of producing currents up to 100 times greater than the devices used in law enforcement. Five of the animals were also infused intravenously with high doses of cocaine so that the researchers could study the interplay between the drug and the TASER current. They found that standard discharge from the TASER X26 model did not induce ventricular fibrillation at any of the tested body sites. In addition, cocaine attenuated the effect of the shocks by 50 percent to 150 percent above the baseline safety margin.

The researchers concluded that it takes about four to eight times the strength of a normal TASER discharge along the axis of the heart to induce ventricular fibrillation, and, as the probes move away from the heart, the amount of electrical current that is required to induce ventricular fibrillation becomes much higher. The research also suggested that in the absence of an appropriate substrate, such as preexisting myocardial ischemia, infarct, metabolic abnormalities, or cardiomyopathy, cocaine may not cause arrhythmias and may in fact protect against them. Cocaine appears to exert significant sodium channel blocking to increase ventricular fibrillation thresholds (Osterweil, 2006).

In an experiment to test the respiratory effects of the TASER device, researchers subjected 52 subjects to the electrical current from a TASER X26 model while those subjects were wearing a respiratory measurement device. Thirty-four subjects underwent a 15-second continuous exposure. Eighteen underwent three 5-second burst exposures. The researchers found that, in the continuous application group, the baseline mean tidal volume of 1.1 liters increased to 1.8 liters during application, the baseline end-tidal carbon dioxide level went from 40.5 millimeters of mercury to 37.3 millimeters of mercury after exposure, the baseline end-tidal oxygen level went from 118.7 millimeters of mercury to 121.3 millimeters of mercury after exposure, and the baseline respiratory rate went from 15.9 breaths per minute to 16.4 breaths per minute after exposure. Instead of their breathing being impaired, the test subjects breaths were faster and deeper than normal. The carbon dioxide level in their blood decreased, and the oxygen level increased.

In the 5-second burst group, the baseline mean tidal volume increased to from 1.1 liters to 1.85 liters during application, the baseline end-tidal carbon dioxide level went from 40.9 millimeters of mercury to 39.1 millimeters of mercury after exposure, the baseline end-tidal oxygen level went from

123.1 millimeters of mercury to 127.0 millimeters of mercury after exposure, and the baseline respiratory rate went from 13.8 breaths per minute to 14.6 breaths per minute after exposure. Again, instead of their breathing being impaired, the test subjects breaths were faster and deeper than normal, and again, the carbon dioxide level in their blood decreased, and the oxygen level increased. The researchers concluded that the prolonged TASER application did not impair respiratory parameters in that population of volunteers (Ho, Dawes, Bultman, Thacker, Skinner, Johnson, & Miner, 2007).

ADVERSE STUDIES

In 2006, researchers at the University of Toronto conducted experiments on the cardiac safety of the ADVANCED TASER M26 and the TASER X26 devices using six anesthetized pigs. The researchers had noticed that in previous studies using only surface electrocardiograph monitoring, the electromagnetic interference of the discharge had prevented examining the cardiac electrophysiological consequences of the pulse. The researchers elected to introduce intracardiac catheters and blood pressure transducers into the animals. They tested two models of the TASER devices, the M26 and the X26, by applying the charges in two vectors. The thoracic vector ran parallel to the long axis of the heart on the chest wall. The nonthoracic vector ran away from the heart across the abdomen. They also investigated two different lengths of discharge, 5 seconds and 15 seconds.

The researchers studied 150 discharges into the pigs; of those, 94 were thoracic discharges and 56 were nonthoracic. During the nonthoracic discharges, the researchers observed no stimulation of the myocardium. However, they did observe myocardial stimulation in 74 percent of the thoracic discharges. The researchers defined myocardial stimulation as change in electrogram morphology and rate during the discharge and perturbation of arterial blood pressure. The X26 model was more likely to stimulate the myocardium than the M26 model, and the 15-second duration was more likely to stimulate the myocardium than the 5-second discharge. None of the discharges resulted in ventricular fibrillation or tachycardia.

To simulate the conditions of excited delirium, the researchers infused the pigs with epinephrine and applied 16 discharges in the thoracic vector. Thirteen of those discharges resulted in stimulation of the myocardium. Of those 13 discharges, one resulted in ventricular fibrillation and one resulted in a nonsustained ventricular tachycardia, which spontaneously terminated. The researchers noted that they constructed the model to the worst-case scenario of probe placement and that it would be unlikely for the probe placement in real scenarios to duplicate those positions. They also noted that the results were useful in excluding as the cause of arrhythmia discharges that were not vectored across the heart or that initiated long after the discharge (Nanthakumar, Billingsley, Masse, Dorian, Cameron, Chauhan, Downar, & Sevaptsidis, 2006).

In 2006, researchers at the University of Wisconsin-Madison presented a poster presentation of their research into the cardiac safety of the TASER X26 device. In a test of 10 anesthetized pigs, they were able to induce ventricular fibrillation in all 10 animals by varying the distance of the probe from the heart. The research was just the first step in a line of research they intended to perform. The tests showed that the TASER pulse could trigger ventricular fibrillation at an average of about 17 centimeters, with the range being from zero to 20 centimeters. However, much more research was needed before the researchers could extrapolate their data to a human model (Will, Honyu, O'Rourke, & Webster, 2006).

TASER DEVICES AS NONLETHAL FORCE

Too often, law enforcement officers and administrators tout TASER technology as an alternative to the use of deadly force. As the available studies indicate, TASER devices are not deadly force weapons, and officers should never rely solely on the use of electronic control devices in lieu of deadly force when deadly force is appropriate. The resolution of lethal threats requires lethal force options (Conner, 2006). As with any weapon, TASER devices are not always effective. On occasion, the devices fail to subdue aggressive or violently resisting subjects. From 1987 through 2006, in at least 131 documented incidents, after efforts to subdue someone with a TASER electronic control device failed, officers shot and killed that person. See Table 9, Fatal Shootings Following Application of TASER Devices–1987 through 2006. For a list of those incidents, see Appendix C, List of Fatal Shootings 1987–2006. In many other cases, officers resorted to deadly force when a TASER device failed, but the subjects survived. In 131 of the 213 case studies cited in Chapter 3, or 61.5 percent, TASER devices were ineffective in subduing violent or aggressive persons. In 2004, in El Paso, Texas, an officer tried to subdue a subject with a TASER electronic control device, and, when the device was ineffective, the subject shot and killed the officer (*Associated Press*, 2004, September 26).

The United States Department of Defense defines nonlethal weapons as weapons that are explicitly designed and primarily employed to incapacitate personnel or material, while minimizing fatalities, permanent injury to personnel, and undesired damage to property and the environment (Maier et al., 2005). By that definition, TASER devices are nonlethal weapons. They are designed and primarily employed to incapacitate personnel while minimizing fatalities and permanent injury, but no weapon is perfectly safe. One person can kill another with water, given the right circumstances. It is unreasonable to expect an electronic control device to subdue violent and aggressive people without ever inflicting an injury, but the same is true

Table 9
FATAL SHOOTINGS FOLLOWING APPLICATION OF TASER DEVICES - 1987 THROUGH 2006

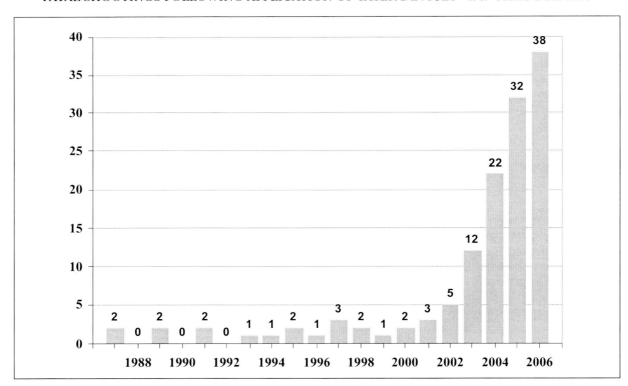

of impact weapons, chemical weapons, even human hands, which often inflict greater injuries than a TASER device.

SUMMARY

Many myths exist regarding the functions of the ADVANCED TASER M26 and the TASER X26 electronic control devices. Those myths generally center on the voltage the devices deliver and how those devices affect the muscles, the heart, and respiration. TASER devices deliver only enough voltage to trigger an action potential in the somatic nervous system, which, in turn, causes muscles to contract. The contractions sustain because of the repetitive charges. Most of the studies conducted on animals and human beings indicate that there is insufficient voltage and amperage in the TASER pulses to affect the sinus rhythm of the heart or interfere with the respiratory system, although a few studies do exist that demonstrate it is possible to stimulate the myocardium under certain laboratory conditions. No weapon is perfect, and that includes TASER electronic control devices. The use of any weapon implicates risk. Current studies conclude that the risks of unintended consequences of using a TASER device are very low.

KEY TERMS

Action Potential	Coulomb	Sensory Nervous System
Amperes	Joules	Somatic Nervous System
Autonomic Nervous System	Neuromuscular Junction	Tetanus
Central Nervous System	Ohm	Volts
Clonus	Peripheral Nervous System	Watts

Chapter 3

TASER CASE STUDIES

The cause is hidden; the effect is visible to all.
Ovid (Publius Ovidius Naso), *Metamorphoses, 4. 287*

Published media estimates of deaths supposedly related to the use of TASER devices have increased to more than 200, depending on the story and the source of the statistics. These estimates have generally come from one of three widely-quoted sources. Amnesty International published a report in 2004 that detailed the organization's concerns about using TASER electronic control devices. The 2004 Amnesty International report claimed that, following the application of a TASER device, 70 people had died between 2001 and 2004 (Amnesty International, 2004). In a subsequent report published 18 months later, Amnesty International claimed that the number of deaths in the United States and Canada had risen to more than 150 (Amnesty International, 2006). A 2006 news story, published in *The Arizona Republic*, identified 167 deaths in the United States and Canada from 1999 to the end of 2005 (Anglen, 2006).

Media reports published or broadcast shortly after a custody death often quote statistics from the Amnesty International and *Arizona Republic* reports, leading readers to believe that TASER devices were the cause of death. However, those three reports did not claim that the use of TASER devices caused the deaths. They merely reported that the unexpected deaths followed the use of a device, and they never established a causal relationship. Investigators usually need several days or weeks to determine the facts, complete the investigation, and determine whether a TASER pulse, or any other factor, played a role in an unexpected death. Consequently, although the numbers of deaths reported continue to rise, the public has little sense of the results of the investigations or of the coroners' findings.

Just because a death follows the use of an electronic control device does not mean that the device is the cause of the death. To determine the extent to which any factor may contribute to or cause a death, one must first know the details of each case. The Amnesty International and *Arizona Republic* reports give only a brief synopsis for each incident. They give few details partly because detailed information regarding custody deaths is so difficult to gather. First, no national database exists to collect information on custody deaths, so it is difficult to know with certainty how many deaths following the application of a TASER device have occurred. The Deaths in Custody Reporting Act of 2000 requires the States to report to the U.S. Justice Department any death of an arrested person or an intended arrestee that occurs during the process of arrest, a death in the custody of law enforcement officers, or a death as the result of lethal force by officers (Public Law 106–297, 114 Stat. 1045). Data collection began in 2003, but reporting is incomplete, and the Bureau of Justice Statistics, which is responsible for maintaining and analyzing the data, has not yet produced a report (Mumola, 2006). Some information is available from the States' custody death coordinators, but news reports are the principle way to identify cases for study.

Second, consistent and complete data collection is not possible under current records laws. Records laws throughout the United States, known variously as open records, public information, or freedom of information laws, vary greatly in defining what information is subject to disclosure and what information is exempt from disclosure. In some states, both police reports and autopsy reports are available. In some states, the law makes police reports available but exempts autopsy reports. In other states, the law exempts police reports but makes autopsy reports available. In some states, neither the police report nor the autopsy report is subject to disclosure.

Third, establishing the facts of each case is difficult. Even when police and autopsy reports are available, the circumstances leading to the death are often in dispute, as witnesses differ in their recollections about the incident. Additionally, medical evidence of the cause of death is not always clear. Different coroners and medical examiners may examine the same body and offer different opinions on the cause of death.

Following are 213 case reports from 1983 through 2005 in the United States. Each case involves an unexpected custody death following the use of a TASER. The information for these case studies comes from a compilation of wire service, television, newspaper, police, and autopsy reports. Because of the limitations in availability to the sources noted above, more cases may exist than this study identifies. Moreover, because some police and autopsy reports are exempt from disclosure, some narratives derive from news media accounts alone. In some deaths, the TASER device is neither the cause nor a contributing factor. In some deaths, the role of the device appears to be debatable. These case studies seek neither to justify the level of force used nor to determine whether officers used force appropriately. In some cases, the amount of force used might appear excessive, but that determination is not possible without more specific details than this study provides, and whether the amount of force used is excessive is a question separate from whether the TASER electronic control devices caused or contributed to the death.

In each case, the role of the TASER device as a cause of death or as a significant contributing factor is listed as excluded, doubtful, possible, confirmed, or undetermined.

- Excluded – The facts of the case and/or the pathological evidence demonstrate that the TASER device was neither the cause of nor a significant contributing factor to the death.
- Doubtful – The medical examiner or coroner cited the TASER device as a factor in the death, but the factual and pathological evidence fails to support a conclusion that, but for the use of the device, the death would not have occurred.
- Possible – The medical examiner or coroner cited or could not exclude the TASER device as a factor in the death, and the factual and pathological evidence tended to support a conclusion that, but for the use of the device, the death would not have occurred.
- Confirmed – The facts of the case and/or the pathological evidence demonstrate that the TASER device either was the cause of or was a significant contributing factor to the death.
- Undetermined – There is insufficient factual or pathological evidence from the available records to know whether the TASER device was the cause of or was a significant contributing factor to the death.

FIRST AND SECOND GENERATION TASER WEAPON CASES

The first 42 case studies, from Vincent Alvarez to David Torres Flores, represent deaths that followed the use of first generation and second generation TASER weapons, the TASER TF-76, the TASER-TRON, and the AIR TASER 34000, that occurred between 1983 and 1999.

VINCENT ALVAREZ

AGE: 27
RACE/GENDER: HISPANIC MALE
AGENCY: LOS ANGELES (CALIFORNIA) POLICE DEPARTMENT
DATE OF INCIDENT: AUGUST 10, 1983
CAUSE OF DEATH: CARDIAC DYSRHYTHMIE DUE TO ACUTE PHENCYCLIDINE TOXICITY
ROLE OF TASER DEVICE: EXCLUDED

The police report in the Alvarez case was exempt from disclosure under California law. However, media accounts reported that police officers observed Alvarez behaving strangely, shaking his body while holding onto a telephone pole. His actions led the officers to believe that Alvarez was under the influence of drugs. When the officers approached him, Alvarez was uncooperative, so the officers forced him to the ground, handcuffed him, bound his feet, and placed him in the back seat of a patrol car. Alvarez managed somehow to loosen his foot restraints. He began banging his head on the car window, and he demonstrated exceptional physical strength. Another officer, called to the scene because of his training in its use, fired a TASER device at Alvarez, striking him in the back. The officers subdued Alvarez and took him to the jail ward of a local hospital, where he later suffered a full cardiac arrest and died.

The medical examiner concluded that Alvarez died of cardiac dysrhythmia due to acute phencyclidine (PCP) toxicity. He ruled the death an accident (see County of Los Angeles (California) Coroner Report 1983–09964; *United Press International*, 1983; Palermo, 1986).

Although the available evidence did not specify how much time passed between the application of the TASER device and Alvarez's collapse, it was clear that officers had time to transport him to the jail ward of the hospital before he suffered the dysrhythmia. Consequently, because Alvarez did not collapse immediately following the use of the TASER pulse, indicating that the electrical current was not a cause of the dysrhythmia, and because the coroner attributed the cause of the dysrhythmia to PCP intoxication, the device is excluded as the cause of death and as a significant contributing factor in the Alvarez case.

RAUL GUEVARA, JR.

AGE: 30
RACE/GENDER: HISPANIC MALE
AGENCY: LOS ANGELES (CALIFORNIA) POLICE DEPARTMENT
DATE OF INCIDENT: JANUARY 22, 1984
CAUSE OF DEATH: HEART DISEASE AGGRAVATED BY EXERTION
ROLE OF TASER DEVICE: EXCLUDED

The police report in the Guevara case was exempt from disclosure under California law. However, media accounts reported that Guevara was in custody on charges of receiving stolen property when he began struggling with officers who were trying to retrieve a telephone receiver that he had torn from a pay phone in his cellblock. After a shock from a TASER device failed to subdue him, a jailer used a chokehold, placing his forearm across Guevara's throat. Guevara continued to struggle after four jailers wrestled him to the floor. A jailer again applied a chokehold while other jailers placed restraints on Guevara's wrists and ankles. As jailers were loading Guevara onto a gurney, they realized he had stopped breathing. Efforts to revive him at the jail and en route to a hospital failed.

Although the autopsy report was not available, media sources reported that the medical examiner determined that a preexisting heart problem, which the media reports did not define, aggravated by the effort Guevara exerted during the struggle, caused his death (see Palermo, 1986; Simon, 1986).

The available news media reports did not detail how much time passed between the application of the TASER device and Guevara's collapse. However, the accounts did indicate that, after the use of the device, Guevara struggled long enough to have four jailers wrestle him to the floor, place him in wrist and ankle restraints, and load him onto a gurney. Consequently, because Guevara did not collapse immediately following the application of the TASER pulse, indicating that the electrical charge did not significantly affect the rhythm of his heart, and because the coroner found sufficient evidence to attribute the death to a preexisting heart condition that was aggravated by Guevara's struggle, which had continued after the use of the TASER device, the device is excluded as the cause of death and as a significant contributing factor in the Guevara case.

LARRY DONNELL GARDNER

AGE: 32
RACE/GENDER: AFRICAN-AMERICAN MALE
AGENCY: BURKE COUNTY (GEORGIA) SHERIFF'S OFFICE
DATE OF INCIDENT: AUGUST 17, 1984
CAUSE OF DEATH: HYPERTHERMIA ASSOCIATED WITH SICKLE-CELL CRISIS
ROLE OF TASER DEVICE: EXCLUDED

A deputy saw Gardner walking down a street in Waynesboro, and he knew that Gardner had active warrants on charges of shoplifting and selling marijuana. When the deputy approached, Gardner ran. When deputies caught Gardner, he resisted. During the scuffle, a deputy applied a TASER device to Gardner. However, it was later determined that the device was inoperable and, probably, could not have delivered a shock to Gardner. The deputies took Gardner to the county jail, where he complained of being hot and short of breath. Deputies called for an ambulance, but Gardner collapsed and died. Because Gardner's death was a custody death, the Georgia Bureau of Investigation conducted the investigation.

A medical examiner for the Georgia Bureau of Investigation found that Gardner died of hyperthermia associated with sickle cell crisis, and he ruled the manner of death to be natural. A second autopsy, conducted at the family's request, found that Gardner died of severe sickle-cell crisis, and a coroner's jury ruled that Gardner died of a heart attack caused by sickle-cell trait (see *Associated Press*, 1984; *True Citizen*, 1984, August 22a; *True Citizen*, 1984, August 22b; Georgia Bureau of Investigation Report A84–1892).

Tests on the TASER device following Gardner's death showed that it was inoperative and was probably incapable of delivering a charge at the time the officer used it. Because the device was inoperative, because Gardner did not collapse immediately following the application of the TASER pulse, indicating that electrical current did not significantly affect the rhythm of his heart, and because two pathologists found evidence to attribute the death to sickle-cell, the device is excluded as the cause of death and as a significant contributing factor in the Gardner case.

CORNELIUS GARLAND SMITH

AGE: 35
RACE/GENDER: AFRICAN-AMERICAN MALE
AGENCY: LOS ANGELES (CALIFORNIA) POLICE DEPARTMENT
DATE OF INCIDENT: APRIL 11, 1985
CAUSE OF DEATH: CARDIAC ARRHYTHMIA
CONTRIBUTING FACTORS: SICK SINUS SYNDROME, MITRAL VALVE PROLAPSE, AND ELECTRICAL (TASER) STIMULATION WHILE UNDER THE INFLUENCE OF PHENCYCLIDINE
ROLE OF TASER DEVICE: EXCLUDED/DOUBTFUL

The police report in the Smith case was exempt from disclosure under California law, but media sources reported that officers found Smith writhing on the ground and screaming. When the officers approached, Smith jumped up, began break dancing and kicking the side of the patrol car. The officers shocked Smith one time with a TASER device. The officers took Smith to a local hospital. There, he suffered a cardiac arrest and died.

The medical examiner reported that Smith had suffered from a heart condition, sick sinus symptom, and needed a pacemaker, but had refused to get one. Additionally, Smith suffered from mitral valve prolapse, which, in rare circumstances, can lead to cardiac arrest resulting in sudden death. The medical examiner concluded that Smith died of cardiac arrhythmia, and he ruled the death a homicide. He noted significant contributing factors of sick sinus syndrome, mitral valve prolapse, and electrical stimulation (referring to the TASER device) while under the influence of phencyclidine (PCP) (see County of Los Angeles (California) Coroner

Report 1985–04921; *Los Angeles Times*, 1985, May 24; *San Jose Mercury News*, 1986).

The medical examiner explained that the TASER pulse was a contributing factor in Smith's death, saying that, because of preexisting medical problems, Smith did not survive the electrical stimulus of the TASER device. He did not attribute as a cause of death or as a contributing factor Smith's physical exertion either before or after officers arrived. Additionally, Smith did not suffer the cardiac arrhythmia until officers had taken Smith to a local hospital. Consequently, because Smith did not collapse immediately following the application of the TASER pulse, indicating that the electrical current did not significantly affect the rhythm of Smith's heart, and there was no evidence of electrical injury, the device is excluded as the cause of death in the Smith case. It is entirely possible that the TASER pulse contributed to the physiological stress Smith experienced that ultimately caused the arrhythmia. However, considering Smith's preexisting heart problems, the level of his exertion, and his PCP intoxication, it is questionable whether any other form of restraint would have produced a different result. Therefore, the role of the TASER device as a significant contributing factor is doubtful.

LANNIE STANLEY MCCOY

AGE: 35
RACE/GENDER: WHITE MALE
AGENCY: LOS ANGELES (CALIFORNIA) POLICE DEPARTMENT
DATE OF INCIDENT: OCTOBER 13, 1985
CAUSE OF DEATH: ACUTE COCAINE INTOXICATION
ROLE OF TASER DEVICE: EXCLUDED

Few details were available in the McCoy case because the police report was exempt from disclosure under California law, and the Los Angeles County Medical Examiner's Office could not produce an autopsy report. However, news media accounts reported that police responded to a call of a disturbance. When they arrived, they found McCoy stabbing the ground with a knife and a screwdriver. When McCoy ignored their orders to drop the knife and screwdriver, the officers shocked him three times with a TASER device. The officers subdued McCoy and took him to a local medical center, where he later collapsed while undergoing psychological evaluation at the medical center's psychiatric ward.

According to news reports, the medical examiner concluded that McCoy died of acute cocaine intoxication (see Los Angeles Times, 1985, August 30; Palermo, 1986). Although the news accounts do not specify how much time passed between the application of the TASER device and McCoy's collapse, enough time had passed for officers to get him to a hospital and for the staff to begin a psychological evaluation. McCoy did not collapse immediately following the applications of the TASER pulse, indicating that the electric current did not significantly affect the rhythm of McCoy's heart, and the medical examiner had found evidence to attribute the death to acute cocaine intoxication. Therefore, the device is excluded as the cause of death and as a significant contributing factor in the McCoy case.

JOSEPH RODRIGUEZ

AGE: 24
RACE/GENDER: HISPANIC MALE
AGENCY: SANTA CLARA (CALIFORNIA) COUNTY SHERIFF'S OFFICE
DATE OF INCIDENT: DECEMBER 27, 1985
CAUSE OF DEATH: PHENCYCLIDINE INTOXICATION
CONTRIBUTING FACTORS: BLUNT INJURIES AND STUN GUN INJURY
ROLE OF TASER DEVICE: EXCLUDED/DOUBTFUL

Police from the Santa Clara Police Department received a call to an elementary school on a report that suspicious people were trespassing on school grounds. When police arrived, five people fled, but one officer followed Rodriguez and watched him stash something in the bed of a pickup. Officers stopped Rodriguez, and in the truck they found a set of nunchakus, martial arts sticks connected by a short chain. The officers arrested Rodriguez and took him to the Santa Clara County jail. At the jail, Rodriguez was combative, rigid, and had slurred speech. Santa Clara County Sheriff's deputies reported that he was too wild for them to fingerprint or photograph. When deputies tried to put Rodriguez into a holding cell, he became violent and refused to enter. Deputies then used a TASER device to subdue him. When that failed, five deputies restrained Rodriguez. When they placed Rodriguez in a padded cell, he was kicking and screaming. Within nine minutes, the noise inside the cell had stopped. A deputy looked in on Rodriguez and found him lying on the floor unconscious. Medical rescue personnel took Rodriguez to a local hospital, where he died.

Laboratory results revealed phencyclidine (PCP) in Rodriguez's blood and cannabinoids in his urine. According to the coroner, Rodriguez died of PCP intoxication (see Santa Clara County (California) Sheriff's Office Report IR85–18511A; Santa Clara County (California) Medical Examiner Investigation Report C85–3109; Wright, 1985; Nakaso, 1986).

Rodriguez continued to kick and fight for several minutes following the application of the TASER pulse, indicating that the electric current did not significantly affect the rhythm of his heart. Consequently, because the coroner attributed the death to PCP intoxication, the device is excluded as the cause of death. The coroner did list blunt injuries and a stun gun injury, meaning the TASER device, as contributory, but he did not explain how the device contributed to Rodriguez's death. The TASER pulse did not cause an immediate collapse, and there was no evidence of electrical injury, only two sets of probe marks on Rodriguez's upper left back. Consequently, the role of the device as a significant contributing factor in Rodriguez's death is doubtful.

ANTHONY MANWELL WILLIAMS, III

AGE: 35
RACE/GENDER: AFRICAN-AMERICAN MALE
AGENCY: POMONA (CALIFORNIA) POLICE
  DEPARTMENT
DATE OF INCIDENT: MAY 9, 1986
CAUSE OF DEATH: CARDIAC ARREST FOLLOWING
  TASER DEVICE STIMULATION WHILE UNDER
  THE INFLUENCE OF COCAINE
ROLE OF TASER DEVICE: DOUBTFUL

Because the police report in the Williams case was exempt from disclosure under California law, and because news reports were sketchy, few facts were available in this incident. Media sources reported only that officers shocked Williams six to eight times with a TASER device in a hotel room after he had torn out plumbing fixtures and refused to submit to arrest.

According to news sources, the medical examiner concluded that Williams died due to cardiac arrest following TASER device stimulation while under the influence of cocaine. The medical examiner noted that, because Williams developed cardiac arrest immediately following application of the TASER device, and in the absence of other significant injuries, he assumed that the device caused the death. He ruled the death a homicide (see County of Los Angeles (California) Coroner Report 1986–06105; *Los Angeles Times*, 1986, July 10; Kelley, 1990, March 18).

In this case, the medical examiner provided no forensic evidence of how the TASER device or pulse caused the death. Instead, he relied on the temporal relationship between the shocks and the cardiac arrest, in the absence of other significant injuries, to assume a causal effect. However, he also concluded that Williams was under the influence of cocaine, and the elevated risk of cardiac dysrhythmia following the ingestion of sympathomimetic drugs, such as cocaine, was well known. The logic of the coroner's conclusion was clearly flawed, but no one has successfully challenged the ruling. Therefore, the role of the TASER device as the cause of death and as a significant contributing factor in the Williams case is doubtful.

ROBERT ZAPATA

AGE: 37
RACE/GENDER: HISPANIC MALE

AGENCY: LOS ANGELES (CALIFORNIA) POLICE
  DEPARTMENT
DATE OF INCIDENT: MAY 18, 1986
CAUSE OF DEATH: UNKNOWN
ROLE OF TASER DEVICE: UNDETERMINED

The police report in the Zapata case was exempt from disclosure under California law, but media accounts reported that Zapata's family called the police after Zapata tried to choke his mother, fought off his father, and began breaking furniture. When officers arrived, they found Zapata trying to swim in the debris on the floor. Zapata fought off several officers, exhibiting tremendous strength. One officer shocked him with a TASER device, but Zapata continued to fight. Zapata began to exhibit breathing difficulties, and he lost consciousness after two officers forced him to the floor and held their feet against his back. Paramedics called to the scene administered cardiopulmonary resuscitation, but they were unable to revive Zapata. He died within two hours (see *Los Angeles Times*, 1986, May 19).

The Los Angeles County Medical Examiner's Office was unable to locate an autopsy report, and media accounts did not report the cause of death. Because so little information is available regarding this incident, the role of the TASER device in the Zapata case is undetermined.

ROBERT (CHARLES) HERBERT BOBIER

AGE: 31
RACE/GENDER: WHITE MALE
AGENCY: LOS ANGELES (CALIFORNIA) POLICE
  DEPARTMENT
DATE OF INCIDENT: JUNE 2, 1986
DATE OF DEATH: JUNE 5, 1986
CAUSE OF DEATH: ACUTE COCAINE AND PHENCYCLIDINE INTOXICATION
ROLE OF TASER DEVICE: EXCLUDED

The police report in the Bobier case was exempt from disclosure under California law, but, according to news reports, witnesses told police that they saw Bobier banging his head against a curb before he climbed a fence and broke into a private school.

Police found him kicking and cursing an imaginary opponent in a darkened room inside the school. Bobier fought with officers, who shocked him four times with a TASER device, without effect. The officers then struck Bobier several times with their nightsticks. Bobier suddenly lost consciousness. The officers took Bobier to a local hospital, where he lapsed into a coma and died three days later.

Media accounts were not consistent regarding Bobier's first name. The *Daily News of Los Angeles* (Hart, 1986) reported the deceased's name as Charles Herbert Bobier, the *Los Angeles Times* (1986, July 4) reported his name as Robert Herbert Bobier, and the *San Jose Mercury News* (Nakaso, 1986) reported his name as Herbert Bobier.

The Los Angeles County Medical Examiner's Office was not able to produce the autopsy report, but news reports stated that Bobier's cause of death was acute cocaine and phencyclidine (PCP) intoxication (see Hart, 1986; *Los Angeles Times*, 1986, July 4; Nakaso, 1986).

The media accounts did not specify how much time passed between the last application of the TASER device and Bobier's collapse, but officers fought for a while, striking him several times him with batons, before he finally collapsed. The medical examiner found sufficient evidence to attribute the death to cocaine and PCP intoxication. Therefore, the TASER device is excluded as the cause of death and as a significant contributing factor in the Bobier case.

YALE LARY WILSON

AGE: 25
RACE/GENDER: AMERICAN INDIAN MALE
AGENCY: CALIFORNIA DEPARTMENT OF
  CORRECTIONS, VACAVILLE
DATE OF INCIDENT: OCTOBER 7, 1986
CAUSE OF DEATH: CARDIORESPIRATORY ARREST
  SECONDARY TO ASPHYXIA
ROLE OF TASER DEVICE: EXCLUDED

The Department of Corrections report on the Wilson case was exempt from disclosure under California law, but media accounts reported that

Wilson was a diagnosed psychotic with a history of hostile behavior. In 1986, the court convicted him of killing a drug dealer and sent him to the California Medical Facility at Vacaville for a psychiatric evaluation. When he refused to allow guards to handcuff him and move him to another cell, prison officials ordered him drugged. Wilson struggled with the guards and rolled under his bed to avoid the injection. Guards shot seven TASER cartridges at Wilson, although it is not known how many made contact, then rushed into his cell, overpowered him, wrapped his head with a bed sheet as a makeshift gag, and sedated him with Fluphenazine and Benztropine. Moments later, a prison nurse noticed that Wilson had stopped breathing. Twelve minutes after application of the TASER devices, a prison doctor also found that Wilson was not breathing. Efforts to revive him were unsuccessful.

The coroner concluded that Wilson died of cardiorespiratory arrest secondary to asphyxia. He also noted the presence of several burn-puncture wounds from the TASER devices, but he classified those as superficial (see Solano County (California) Coroner's Report 1086–1046–CCI–386; Wallace & Sward, 1994).

The media accounts do not specify how much time passed between the last application of the TASER device and Wilson's collapse, but it was long enough for guards to storm his cell, drag him from beneath his bunk, secure him, gag him, and administer an injection. Because Wilson did not collapse immediately following the application of the TASER pulse, indicating that the electric current did not significantly affect the rhythm of his heart, and because the medical examiner attributed the cardiac arrest to asphyxia, the device is excluded as the cause of death and as a significant contributing factor in the Wilson case.

MIGUEL CONTRERAS

AGE: 27
RACE/GENDER: HISPANIC MALE
AGENCY: LOS ANGELES (CALIFORNIA) POLICE DEPARTMENT
DATE OF INCIDENT: JULY 22, 1987
CAUSE OF DEATH: ACUTE PHENCYCLIDINE INTOXICATION
ROLE OF TASER DEVICE: EXCLUDED

The police report in the Contreras case was exempt from disclosure under California law, but news accounts reported that police spotted Contreras acting bizarrely, as if he were intoxicated on drugs. When the officer attempted to speak with him, Contreras was uncooperative. After backup officers arrived, Contreras lunged at one of the officers. He struggled with officers when they attempted to arrest him. The officers used a TASER device and several baton strikes to subdue him. Officers handcuffed Contreras, restrained his feet, and placed him in a police car. When they did, Contreras stopped breathing. Paramedics treated Contreras at the scene and took him to a local hospital, where he died.

The medical examiner ruled that Contreras died of acute phencyclidine (PCP) intoxication. He ruled the death an accident (see County of Los Angeles (California) Coroner Report 1987–06037; *Los Angeles Times*, 1987, June 22).

The media accounts did not specify how much time passed between the application of the TASER device and Contreras' collapse, but enough time passed to strike him several times with batons, subdue him, handcuff him, put him in leg restraints, and put him in the back seat of the police car. Therefore, because Contreras did not collapse immediately following application of the TASER pulse, and because the medical examiner found sufficient cause to attribute the death to PCP intoxication, the device is excluded as the cause of death and as a significant contributing factor in the Contreras case.

MARIO ANTONIO "TONY" GASTELUM

AGE: 24
RACE/GENDER: HISPANIC MALE
AGENCY: SAN DIEGO (CALIFORNIA) POLICE DEPARTMENT
DATE OF INCIDENT: NOVEMBER 2, 1987

DATE OF DEATH: NOVEMBER 4, 1987
CAUSE OF DEATH: CARDIORESPIRATORY FAILURE DUE TO COCAINE AND MORPHINE INTOXICATION
ROLE OF TASER DEVICE: EXCLUDED

The police report in the Gastelum case was exempt from disclosure under California law, but media sources claimed that a woman, who reported that Gastelum was under her house acting strangely, called officers to her residence. The officers arrived and found Gastelum under the house, dressed only in his underwear. Gastelum became combative and refused to come out. The officers shocked Gastelum one time with a TASER device, handcuffed him, and transported him to a local hospital. At the hospital, Gastelum again became violent, and officers shocked him again with a TASER device. After entering the emergency room, Gastelum suffered a heart attack, lapsed into a coma and died two days later.

The coroner ruled that Gastelum died of cardiorespiratory failure due to cocaine and morphine intoxication, but he did not classify Gastelum's death as an accident, homicide, or suicide. The autopsy report said there was insufficient medical evidence clearly to establish the interrelation of drug ingestion, the use of the TASER device and physical restraint, and how they related to cardiorespiratory failure (see Bernstein, 1987; County of San Diego (California) Coroner's Report 87–2253; Reza, 1987).

The media reports did not specify how much time lapsed between the last application of the TASER device and Gastelum's heart attack, but sufficient time passed for officers to subdue Gastelum, load him onto a gurney, and wheel him into the emergency room. Because Gastelum did not collapse immediately following the application of the TASER pulse, indicating that the electric current did not significantly affect his heart, and because the coroner found sufficient evidence to attribute the cardiorespiratory failure to cocaine and morphine intoxication, the device is excluded as the cause of death and as a significant contributing factor in the Gastelum case.

STEWART ALAN VIGIL

AGE: 29
RACE/GENDER: WHITE MALE
AGENCY: LOS ANGELES (CALIFORNIA) POLICE DEPARTMENT
DATE OF INCIDENT: DECEMBER 4, 1987
CAUSE OF DEATH: MULTIPLE BLUNT INJURIES AND METHAMPHETAMINE INTOXICATION
ROLE OF TASER DEVICE: EXCLUDED

The police report in the Vigil case was exempt from disclosure under California law, but media sources reported that police were notified of a man wandering in traffic. When officers arrived, they found Vigil standing in the middle of traffic, talking incoherently. They transported Vigil to the hospital for observation, but he began to struggle in the parking lot, so the officers called for a backup unit. Ultimately, a dozen officers arrived at the hospital, but they could not control Vigil, even after officers shocked him simultaneously with three TASER devices, and then struck him with batons. Several of the officers swarmed Vigil and knocked him to the ground, inflicting a head injury. Vigil was taken into the emergency room, where he died.

The coroner's office concluded that Vigil died of multiple blunt injuries and methamphetamine intoxication. He listed the mode of death as undetermined (see County of Los Angeles (California) Coroner Report 1987–11625; Rainey, 1987; Rainey, 1988).

Officers only subdued Vigil after knocking him to the ground and causing a head injury. The TASER pulse did not cause Vigil to collapse, indicating that the electric current did not significantly affect his heart. Because the TASER pulse did not cause an immediate collapse, and because the medical examiner attributed the death to blunt injuries and methamphetamine intoxication, the device is excluded as the cause of death and as a significant contributing factor in the Vigil case.

WILLIAM MCCALL

AGE: 39
RACE/GENDER: AFRICAN-AMERICAN MALE

AGENCY: LOS ANGELES (CALIFORNIA) POLICE DEPARTMENT
DATE OF INCIDENT: JANUARY 13, 1988
CAUSE OF DEATH: BLUNT FORCE TRAUMA TO THE NECK
ROLE OF TASER DEVICE: EXCLUDED

Little information is available on this incident because the police report in the McCall case was exempt from disclosure under California law, and the Los Angeles County Medical Examiner's Office was not able to produce the autopsy report. However, news reports indicated that a neighbor summoned officers, reporting that McCall had assaulted him and that McCall was acting as if he were under the influence of drugs. When officers arrived, McCall was shouting, "Shoot me! Shoot me!"

When McCall charged, one officer shocked him once with a TASER device, without effect. The struggle to subdue McCall took several minutes. Officers finally handcuffed McCall, hobbled his feet, and put him in the back seat of a patrol car, when he suddenly stopped breathing. While waiting for an ambulance, the officers administered cardiopulmonary resuscitation. Paramedics determined that McCall had suffered full cardiac arrest.

News accounts, quoting the medical examiner's report, said that McCall died of blunt force trauma to the neck. The medical examiner ruled that McCall's death was a homicide (see Kendall, 1988; *Los Angeles Times*, 1989, January 12).

McCall fought for several minutes following the application of the TASER pulse, indicating that the electric current did not significantly affect the rhythm of his heart. Because the TASER device did not cause an immediate collapse, and because the coroner attributed the death to a blunt force injury to McCall's neck, the device is excluded as the cause of death and as a significant contributing factor in the McCall case.

EDWARD BREEN

AGE: 38
RACE/GENDER: WHITE MALE
AGENCY: BELL-CUDAHY (CALIFORNIA) POLICE DEPARTMENT
DATE OF INCIDENT: SEPTEMBER 10, 1988
DATE OF DEATH: SEPTEMBER 11, 1988
CAUSE OF DEATH: COCAINE OVERDOSE
ROLE OF TASER DEVICE: EXCLUDED

The police report in the Breen case was exempt from disclosure under California law, but media accounts reported that officers from the Bell-Cudahy Police Department responded to a call of a man disturbing the peace. Upon arrival, they observed Breen running in the street amongst the traffic, gesturing at vehicles and shouting that people were in the trees with machine guns ready to kill him. The officers noted that Breen appeared to be under the influence of drugs. When they tried to take him into custody, Breen resisted. Officers used a TASER device to subdue Breen. They handcuffed and hobbled him, and then they noticed that he was having trouble breathing. Paramedics transported Breen to a local hospital, where he died.

The autopsy report was not available, but news sources reported that the coroner found Breen died of acute cocaine intoxication (see *Los Angeles Times*, 1988, October 14; *San Jose Mercury News*, 1988, September 12; Peavy, 2006, August 25).

Breen appeared to have collapsed immediately following the use of the TASER device, but, because the medical examiner attributed the death to a cocaine overdose and did not implicate the use of the TASER device, the device is excluded as the cause of death and as a significant contributing factor in the Breen case.

CHARLES EUGENE MILES

AGE: 37
RACE/GENDER: WHITE MALE
AGENCY: LOS ANGELES COUNTY (CALIFORNIA) SHERIFF'S OFFICE
DATE OF INCIDENT: SEPTEMBER 10, 1988
CAUSE OF DEATH: ACUTE PHENCYCLIDINE INTOXICATION COMPLICATED BY BLUNT FORCE TRAUMA AND RESTRAINING MANEUVERS
ROLE OF TASER DEVICE: EXCLUDED/DOUBTFUL

The Sheriff's Office report in the Miles case was exempt from disclosure under California law, but media accounts claimed that deputies found Miles standing nude in the street. When they stopped to investigate, Miles became combative, and a violent struggle ensued. The deputies twice used a TASER device to control Miles, but it was unsuccessful. When the device failed to bring Miles under control, the deputies struck him several times about the legs, side, and arms with batons, but that also had little effect. The deputies finally physically subdued, hogtied, and handcuffed him. The deputies took Miles to the jail, where they discovered he was not breathing. They then took Miles to a local hospital, where he died.

The coroner ruled that Miles died from a combination of phencyclidine (PCP) intoxication, the blows from officers, and the other actions taken to restrain him, which included the TASER device. He ruled the death a homicide (see County of Los Angeles (California) Coroner Report 1988–09032; *Los Angeles Times*, 1988, October 14; *San Jose Mercury News*, 1988, September 12).

The media accounts did not detail how much time passed between the applications of the TASER device and Miles' collapse, but enough time passed that the officers subdued him, hogtied him, and took him to the jail. Although the medical examiner attributed the death to PCP intoxication complicated by restraint maneuvers, he did not specify how the TASER device contributed to the death. Miles did not stop breathing for several minutes, indicating that the TASER pulse did not have a significant effect on the rhythm of his heart. It seems more likely that the baton strikes and the hogtying, which followed the failed use of the TASER device, were more likely significantly to contribute to the death. Consequently, the device is excluded as the cause of death, and its role as a significant contributing factor in Miles' death is doubtful.

JOSE TORRES

AGE: 31
RACE/GENDER: HISPANIC MALE
AGENCY: LOS ANGELES (CALIFORNIA) POLICE DEPARTMENT
DATE OF INCIDENT: FEBRUARY 14, 1989
CAUSE OF DEATH: UNKNOWN
ROLE OF TASER DEVICE: UNDETERMINED

Few details were available in the Torres case because the police report was exempt from disclosure under California law, and the Los Angeles County Coroner's Office could not locate an autopsy report. However, media reports said that Torres was visiting his family when he began acting irrationally, damaging property, and threatening family members. Fearing that he would injure a 15-month-old child, the family called police. When officers arrived, Torres became combative and attempted to kick them. One officer used a TASER devise to immobilize Torres so officers could apply handcuffs and leg restraints. Officers then called paramedics to treat Torres. A short time later, Torres stopped breathing. Paramedics took Torres to a local hospital, where he died (see *Los Angeles Times*, 1989, February 16). Because there is so little information on this case, the role of the TASER device in the Torres case is undetermined.

ANTHONY PUMA

AGE: 34
RACE/GENDER: MALE (RACE NOT DETERMINABLE FROM THE AVAILABLE RECORDS)
AGENCY: NEW YORK (NEW YORK) POLICE DEPARTMENT
DATE OF INCIDENT: MAY 22, 1989
CAUSE OF DEATH: UNKNOWN
ROLE OF TASER DEVICE: UNDETERMINED

Puma checked into a motor inn with his girlfriend. A few hours later, a desk clerk at the inn heard him yelling and called police. By the time officers arrived, Puma had left and was standing on a nearby service road next to a highway. He was agitated and flailing his arms, and he would not respond to police pleas to calm down. Puma attacked one officer who fired a TASER device at Puma, knocking him to the ground. Because of his condition, officers called for Emergency Medical Services. Paramedics treated Puma at the scene and took him to a local hospital, where he died about one hour later.

The girlfriend later told investigators that Puma had been drinking heavily and snorting cocaine, and that he started acting crazy and paranoid. Family members and friends told investigators that Puma had been diagnosed as manic-depressive and that he was taking lithium for the condition. He also reportedly had heart and liver problems (see Rivera, Holland, Queen, Jordan, Gelman, & Powell, 1989).

The police and autopsy reports were not available, and media sources do not report the results of an autopsy. Without additional information, the role of the TASER device in the Puma case is undetermined.

JEFFREY MICHAEL LEONTI

AGE: 37
RACE/GENDER: WHITE MALE
AGENCY: SANTA CLARA COUNTY (CALIFORNIA) SHERIFF'S OFFICE
DATE OF INCIDENT: MAY 24, 1989
CAUSE OF DEATH: ACUTE CARDIAC DEATH DUE TO APPLICATION OF STUN GUN
ROLE OF TASER DEVICE: POSSIBLE

The Sheriff's Office incident report in the Leonti case was exempt from disclosure under California law, but media sources reported that police arrested Leonti for exposing himself in front of an apartment building. The officers took Leonti, who did not resist arrest, to the county's main jail. Initially, he presented no problem for jail officials, but when they took him into a room to be strip-searched, he began to struggle with jailers. The jailers handcuffed Leonti and took him into another room with padded walls so a nurse could examine him. There, Leonti demonstrated exceptional strength and began kicking. Jailers then used a TASER device three times in rapid succession on Leonti. The first two shocks failed to subdue him, but he became unresponsive immediately with the third application, which jailers applied to the side of his chest.

The coroner found evidence of chronic intravenous drug abuse when he performed the autopsy. He found traces of marijuana metabolites, morphine and codeine in Leonti's urine. The coroner also reported that Leonti had a prescription for lithium carbonate, a chemical compound used as a mood stabilizer in psychiatric treatment of manic states and bipolar disorder, but there were no traces of it in his body. The coroner ruled the cause of death to be acute cardiac death due to application of a TASER device (see Bailey & Salner, 1989; County of Santa Clara (California) Coroner Report 89-144-003).

Leonti collapsed immediately following the application of the TASER device to the left side of his chest, and the medical examiner listed the cause of his cardiac death as the use of the device. However, Leonti suffered from several sudden death risk factors, including a history of intravenous drug use, mental illness, and ventricular hypertrophy, which mitigated against the TASER device as the sole source of Leonti's cardiac death. The coroner did not present forensic evidence to demonstrate that, but for the use of the TASER device, Leonti would have survived. Nevertheless, the available evidence supports that it is possible that a TASER pulse was the cause of death or was a significant contributing factor in the Leonti case.

DUANE JAY JOHNSON

AGE: 24
RACE/GENDER: AFRICAN-AMERICAN MALE
AGENCY: VENTURA (CALIFORNIA) POLICE DEPARTMENT
DATE OF INCIDENT: FEBRUARY 13, 1990
CAUSE OF DEATH: CARDIAC ARRHYTHMIA DUE TO CORONARY ARTERY ATHEROSCLEROSIS, MODERATE TO SEVERE; SCHIZOPHRENIC VIOLENT EXERTION; AND TASER GUN STIMULATION
ROLE OF TASER DEVICE: POSSIBLE

The police report in the Johnson case was exempt from disclosure under California law, but media sources reported that Johnson, who was in a hospital because of a near-fatal heart attack two weeks before, resisted hospital workers' efforts to transport him to the mental health ward. He began throwing himself into the walls and windows. Medical staff called the police for assistance with Johnson, a former athlete. Police shocked Johnson with a TASER device and another brand-named stun gun to force him to release his grip on a glass medical instrument during the initial struggle in the hospital room. A few minutes later, officers used the TASER device again to force him to release a hospital worker's leg. Officers then handcuffed Johnson's hands to the gurney railing, but he was still able to move his arms parallel to the rails. Officers applied the TASER device again when Johnson refused to allow officers to handcuff him while they moved him from the gurney into an isolation room. Johnson collapsed and died about 45 minutes after the incident began and between one and five minutes after the final TASER jolt. Police said they jolted Johnson four times, once with another brand-named stun gun and three times with a TASER device, but, according to the coroner's report, two mental health workers told county investigators that stun guns were used from seven to eleven times on the man during the 45 minutes before his death.

The coroner later ruled that shocks from the TASER device were a primary cause of Johnson's death. He concluded that officers had shocked Johnson nine times, but he noted that Johnson's fighting and struggling had put additional strain on his heart, which was already weakened by the previous severe atrial flutter and moderate-to-severe coronary artery atherosclerosis (see County of Ventura (California) Coroner Report 246–90; Kelly, 1990, March 13; Kelley, 1990, April 14).

Johnson did not collapse immediately following the last application of the TASER device, but he did quit breathing between one and five minutes later. Johnson had a heart already weakened from a previous heart attack, and the physiological stress he experienced from his exertion against the officers, the medical staff, and the restraints was too much for his heart to sustain. The TASER device was clearly part of that stress. Johnson also suffered at least two factors for sudden death that mitigated against declaring the TASER device as the sole cause of death, coronary artery atherosclerosis and schizophrenia. Nevertheless, because the application of the TASER pulse played a part in creating the physiological stress that Johnson experienced, its role as a cause of death and as a significant contributing factor in Johnson's death is possible.

DOUGLAS L. DANVILLE

AGE: 47
RACE/GENDER: WHITE MALE
AGENCY: LOS ANGELES COUNTY (CALIFORNIA) SHERIFF'S OFFICE
DATE OF INCIDENT: APRIL 21, 1991
CAUSE OF DEATH: VENTRICULAR FIBRILLATION CAUSED BY COCAINE INTOXICATION AND CARDIAC HYPERTROPHY
ROLE OF TASER DEVICE: EXCLUDED

The Sheriff's Office report in the Danville case was exempt from disclosure under California law, but media sources reported that deputies went to Danville's apartment after receiving a complaint of a man yelling, screaming, and banging something inside. When no one opened the door after they knocked and identified themselves, the deputies broke into the residence to see whether someone needed assistance. Danville was irrational and bleeding from self-inflicted wounds, so the deputies called for paramedics. Meanwhile, the deputies attempted to calm Danville, who had armed himself with a butcher knife and a piece of porcelain broken from a toilet. Danville would not quiet down, and he threw the piece of the toilet at the deputies. One deputy fired a TASER device to subdue Danville, detaining him without further incident. As paramedics were treating Danville for his self-inflicted wounds, he went into cardiac arrest. Paramedics revived Danville and took him to a local medical center, where he died.

The coroner ruled that Danville died from ventricular fibrillation caused by cocaine intoxication and cardiac hypertrophy. Nevertheless, the coroner ruled the mode of death to be homicide (see County of Los Angeles (California) Coroner Report 1991–03620; *Los Angeles Times*, 1991, April 22; Peavy, 2006, August 25).

Because Danville did not collapse immediately following the application of the TASER pulse, indicating that the electric current did not significantly affect the rhythm of his heart, and because the medical examiner attributed the cause of the ventricular fibrillation to cocaine intoxication and a preexisting heart condition, the device is excluded as the cause of death and as a significant contributing factor in the Danville case.

DOUGLAS CHARLES

AGE: 24
RACE/GENDER: AFRICAN-AMERICAN MALE
AGENCY: LOS ANGELES (CALIFORNIA) POLICE DEPARTMENT
DATE OF INCIDENT: JULY 1, 1991
CAUSE OF DEATH: COMBINED EFFECTS OF COCAINE AND ETHANOL
ROLE OF TASER DEVICE: EXCLUDED

The police report on the Charles case was exempt from disclosure under California law, but media sources reported that, according to witnesses, Charles injected himself with cocaine and began acting strangely and violently. An apartment manager phoned police after Charles came downstairs to the complex's pool and alarmed residents with his odd behavior. Scraps of carpeting that he held preoccupied him, and he picked nervously at weeds and a tree trunk. He kept muttering about people coming to get him. Charles forced his way into the manager's apartment, and then went upstairs to another unit, where he came crashing backward through the screen door, began breaking furniture and shouting, "They're going to kill me. They're going to kill me."

Charles also ingested either pills or a small packet that he pulled out of his pants pocket during the rampage according to a witness, who managed to hold Charles down with a chair until police arrived. When officers arrived, Charles resisted, and officers shocked him with a TASER device. They called for paramedics and placed Charles in a police car, where he began convulsing and went into cardiac arrest. Paramedics took Charles to a local hospital, where he was dead on arrival.

The coroner ruled that Charles died of the combined effects of cocaine and alcohol. He ruled the death an accident (see Berger, 1991; County of Los Angeles (California) Coroner Report 1991–05975).

Charles did not collapse immediately following application of the TASER pulse, indicating that the electric current did not significantly affect the rhythm of his heart. Because the TASER device did not cause an immediate collapse, and because the medical examiner attributed his death to the combined effects of alcohol and cocaine, the device is excluded as the cause of death and as a significant contributing factor in the Charles case.

MAX LEYZA GARCIA

AGE: 40
RACE/GENDER: HISPANIC MALE
AGENCY: FULLERTON (CALIFORNIA) POLICE DEPARTMENT
DATE OF INCIDENT: NOVEMBER 2, 1991
CAUSE OF DEATH: ACUTE PULMONARY EDEMA DUE TO ACUTE COCAINE INTOXICATION
ROLE OF TASER DEVICE: EXCLUDED

The police and autopsy reports were not available in this incident, but media and court sources reported that a security guard at a hotel called police because Garcia, a guest in the hotel, was causing a disturbance. The responding officers found Garcia in the corridor outside his room, where he had knocked over the furniture, ripped the drapes from the windows, and torn apart the beds. He appeared to be under the influence of a narcotic, and he was hallucinating. Garcia became combative when the officers tried to restrain his hands, struggling with the officers and screaming. Four officers finally got him to the floor and cuffed his hands behind his back, but he continued to struggle, spitting, biting, and kicking the officers. The officers tried to restrain his legs, first with a belt and then with plastic straps, but Garcia repeatedly broke free. One officer tried to subdue him with a TASER device, but it had no effect. There was a question whether the device was functioning. Eventually, the officers secured Garcia's legs with a hobble and attached it to the handcuffs. Several officers carried Garcia to a patrol car and placed him in the back seat. An officer and a police cadet drove Garcia to the police station. During the booking process, officers noticed that Garcia was having trouble breathing, so they called for medical assistance. Paramedics transported Garcia to a local hospital, where he died.

The autopsy report was not available, but court records indicate that the Orange County Coroner's office concluded Garcia died of acute pulmonary edema due to acute cocaine intoxication (see *Los Angeles Times*, 1991, November 4; *Garcia et al. v. City of Fullerton et al.*, 2002 Cal. App. Unpub. LEXIS 8245).

Garcia continued to struggle with officers for several minutes following the application of the TASER device, and there was indication that the device was malfunctioning. In a subsequent lawsuit, Garcia's family claimed that Garcia died from the hobble restraints and the lack of medical attention. They made no allegation against the TASER electronic control device. Because the TASER pulse did not have an immediate effect and the medical examiner attributed the death to the effects of cocaine, the device is excluded as the cause of death and as a significant contributing factor in the Garcia case.

DONNIE RAY WARD

AGE: 38
RACE/GENDER: AFRICAN-AMERICAN MALE
AGENCY: CALIFORNIA DEPARTMENT OF
 CORRECTIONS, DEUEL
DATE OF INCIDENT: DECEMBER 9, 1991
CAUSE OF DEATH: CARDIAC ARRHYTHMIA DUE TO
 METABOLIC COLLAPSE DUE TO PROLONGED
 STATE OF MENTAL AND PHYSICAL EXHAUSTION
ROLE OF TASER DEVICE: EXCLUDED

The Department of Corrections report in the Ward case was exempt from disclosure under California law, but media accounts reported that Ward, a former mental patient, arrived at Deuel State Prison in November 1991. Although he had been taking antipsychotic drugs for several years to control his mental illness, Ward refused to take his medication at Deuel. Ward began behaving oddly December 7. Guards moved him to a cell in the prison infirmary where he began exhibiting psychotic behavior, suffering hallucinations, babbling incoherently, and intermittently yelling and pacing around his cell with a jerky, unstable gait. On December 9, prison officials decided to remove Ward from the cell and give him a tranquilizer, but he resisted. Guards fired a TASER device six times at Ward, but none had any effect. Guards then swarmed into his cell in helmets, face shields, and armor. Shortly after, Ward collapsed. Efforts to resuscitate him were unsuccessful, and he died at a local hospital an hour later.

The coroner's office concluded that Ward's death was accidental. The pathologist listed the cause of death as irregular heartbeat and metabolic collapse combined with a prolonged state of mental and physical exhaustion. In the report, the coroner noted old abrasions and bruises, along with new abrasions, but he noted that none of those cuts or bruises was serious enough to cause death. He also excluded the TASER pulses as a cause of death (see San Joaquin County (California) Coroner's Report 658-91; Gordon, 1992; Wallace, 1992; Wallace & Sward, 1994).

How many times the TASER device delivered a charge to Ward is unknown. However, Ward continued to struggle following the last application of the TASER pulses, indicating that the electric current did not significantly affect the rhythm of his heart. Additionally, the coroner attributed the cause of the cardiac arrhythmia to metabolic collapse due to Ward's struggle and his mental illness. Consequently, the device is excluded as the cause of death and as a significant contributing factor in the Ward case.

JAMES RICARD

AGE: 32
RACE/GENDER: HISPANIC MALE
AGENCY: LOS ANGELES COUNTY (CALIFORNIA)
 SHERIFF'S OFFICE
DATE OF INCIDENT: JULY 18, 1992
CAUSE OF DEATH: SEQUELAE OF PHENCYCLIDINE
 INTAKE AND NONSPECIFIC MYOCARDIAL FIBRO-
 SIS
ROLE OF TASER DEVICE: EXCLUDED

The Sheriff's Office report in the Ricard case was exempt from disclosure under California law, but media sources reported that Ricard had told deputies he was under the influence of phencyclidine when they arrested him on two outstanding warrants and suspicion of vandalism. After officers took him to the jail and handcuffed him to a bench in a hallway awaiting transportation to a psychiatric ward, Ricard began swinging a chain that was attached to a wall at deputies and other inmates. To subdue him, deputies first tried to put additional restraints on him. When that failed, a jail sergeant shot Ricard twice with a TASER device. Officers placed Ricard on a gurney, took him to a ward for mental observation, and under supervision of the medical staff, secured him to the bed in ankle restraints. Ricard stopped breathing shortly afterward, and a jail physician could not revive him. Paramedics transported Ricard to a local hospital, where he died an hour later.

The Los Angeles County Medical Examiner's Office was unable to locate an autopsy report, but media sources reported that the coroner concluded Ricard died of the sequelae of phencyclidine (PCP) intake and nonspecific myocardial fibrosis (see Ford, 1992; Peavy, 2006, August 25).

The media sources were not specific in how much time passed between the application of the TASER device and Ricard's collapse, but it was long enough for jailers to place Ricard on a gurney, take him to a jail ward, and place him in restraints on a bed. Ricard did not collapse immediately, indicating that the TASER pulse did not significantly affect the rhythm of his heart. Because the TASER device did not cause an immediate collapse, and because the coroner attributed the death to the intake of PCP and a preexisting heart condition, the device is excluded as the cause of death and as a significant contributing factor in the Ricard case.

CLARICE A. YOUNGER

AGE: 62
RACE/GENDER: FEMALE (RACE NOT DETERMINABLE FROM THE AVAILABLE RECORDS)
AGENCY: PRINCE GEORGE'S COUNTY (MARYLAND) SHERIFF'S OFFICE
DATE OF INCIDENT: AUGUST 20, 1992

CAUSE OF DEATH: BLUNT FORCE INJURIES
CONTRIBUTING FACTORS: OSTEOPOROSIS AND THE USE OF A TASER DEVICE
ROLE OF TASER DEVICE: EXCLUDED

The Prince George's County Sheriff's Office did not respond to a request for records in the Younger case, but media sources reported that deputies went to Younger's apartment to serve an emergency psychiatric order obtained by her son. When they knocked on her door, she approached the deputies waving a butcher knife. One of the deputies twice fired a TASER device at Younger, but it was ineffective. The deputy then shot Younger with a 37mm rubber bullet, striking her in the chest. The deputies transported Younger to a local hospital, where she died.

The Virginia State Medical Examiner's Office did not respond to a request for records either, but media sources reported that the coroner ruled Younger died from blunt force injuries from the TASER device and the rubber bullets. Younger suffered from osteoporosis. The rubber bullet broke three of her ribs, puncturing her heart with a bone splinter. The coroner noted that use of the TASER device was a contributing factor in Younger's death, but he did not explain how it contributed (see Duggan, 1992; Keary, 1992).

Younger continued to resist after officers used the TASER device, indicating that the TASER did not significantly affect the rhythm of her heart. The officers fired rubber bullets only after the device had failed to subdue Younger. Because the TASER pulse did not cause an immediate collapse, and because the blunt force injury from the rubber bullet was clearly the cause of the broken ribs puncturing the heart, the TASER is excluded as the cause of death and as a significant contributing factor in the Younger case.

DAVID MARTINEZ

AGE: 27
RACE/GENDER: HISPANIC MALE
AGENCY: LOS ANGELES (CALIFORNIA) POLICE DEPARTMENT

DATE OF INCIDENT: SEPTEMBER 14, 1992
CAUSE OF DEATH: METHAMPHETAMINE INTOXICATION
ROLE OF TASER DEVICE: EXCLUDED

The police report in the Martinez case was exempt from disclosure under California law, but news accounts reported that an apartment manager summoned police to Martinez's apartment. Martinez was upset over marital problems, and he had been rampaging through his apartment, breaking windows and destroying furniture. Friends told officers they believed that Martinez intended to commit suicide. When officers arrived, they found that Martinez had barricaded the front door. After officers forced their way in, Martinez tried to force them out. When that did not work, he lay on the floor and began sawing at his own throat with a large piece of broken glass. To subdue Martinez, officers fired four TASER cartridges at Martinez, striking him with three, and they struck him once in the chest with a baton. Paramedics, who the officers had called to the scene, said that Martinez was experiencing heart problems. They took Martinez to a local hospital where he died about 80 minutes later.

The coroner ruled that Martinez died of methamphetamine intoxication, and he ruled the death an accident (see County of Los Angeles (California) Coroner Report 1992-08402; Curtiss, 1992).

Martinez continued to resist following application of the TASER device, indicating that the electric current did not significantly affect the rhythm of his heart. Because the TASER pulse did not cause an immediate collapse, and because the coroner attributed the death to methamphetamine intoxication, the device is excluded as the cause of death and as a significant contributing factor in the Martinez case.

MICHAEL JAMES BRYANT

AGE: 37
RACE/GENDER: AFRICAN-AMERICAN MALE
AGENCY: LOS ANGELES (CALIFORNIA) POLICE DEPARTMENT

DATE OF INCIDENT: MARCH 8, 1993
CAUSE OF DEATH: ACUTE COCAINE INTOXICATION AND ASPHYXIATION FROM RESTRAINT PROCEDURES
CONTRIBUTING FACTORS: OBESITY, A FATTY LIVER, AND SCARRING OF HEART TISSUE
ROLE OF TASER DEVICE: EXCLUDED

The police reports regarding the Bryant case were exempt from disclosure under California law, but media sources reported that Bryant flagged down a San Marino police officer. The officer said he believed Bryant was under the influence of drugs. When the officer tried to stop him from driving, Bryant sped away. A chase through three jurisdictions ended when Bryant fell into a swimming pool. Los Angeles police officers shocked Bryant with a TASER device. Bryant continued to struggle with officers, so they hogtied him and placed him in the back of a police car, where he stopped breathing.

The coroner's report was not available, but news accounts reported that the coroner's report showed Bryant died because of acute cocaine intoxication and asphyxiation from restraint procedures. Bryant had a potentially lethal level of cocaine in his blood when he died. Nevertheless, the coroner ruled the death a homicide. The coroner listed contributing factors in Bryant's death as obesity, a fatty liver and scarring of heart tissue (see Newton, 1993; *San Jose Mercury News*, 1993, April 25).

Bryant continued to resist officers following the application of the TASER device, indicating that the electric current did not significantly affect the rhythm of his heart. Because the TASER pulse did not cause an immediate collapse, and because the coroner attributed the death to cocaine intoxication and asphyxiation, the device is excluded as the cause of death and as a significant contributing factor in the Bryant case.

VITAL MONTILLA

AGE: 28
RACE/GENDER: AFRICAN-AMERICAN MALE
AGENCY: NEW YORK (NEW YORK) POLICE DEPARTMENT

DATE OF INCIDENT: DECEMBER 2, 1993
CAUSE OF DEATH: ACUTE COCAINE TOXICITY
ROLE OF TASER DEVICE: EXCLUDED

The police report on this incident was not available, but media sources reported that Montilla had been smoking marijuana and crack cocaine and performing a voodoo ritual when he suddenly took off all of his clothes and ran into the hallway of his apartment. When police arrived, Montilla struggled with the officers, who used chemical spray and two jolts from a TASER in an attempt to subdue Montilla, but those efforts were ineffective. Eventually, thirteen officers arrived to subdue Montilla. Once they secured him, Montilla collapsed. Paramedics took him to a local hospital, where he died.

The autopsy report was exempt from disclosure under New York law. However, news reports quote the coroner's office as saying that Montilla died of acute cocaine toxicity, and that his death was an accident (see James, 1993; *New York Times*, 1993).

Media accounts are not specific on how long Montilla continued to struggle following the application of the TASER device, but the fact that it took thirteen officers to subdue him after use of the TASER device indicates that the electric current had no significant affect on the rhythm of his heart. Because the TASER pulse did not cause an immediate collapse, and because the medical examiner attributed the death to cocaine toxicity, the device is excluded as the cause of death and as a significant contributing factor in the Montilla case.

DANIEL SCOTT GIZOWSKI

AGE: 25
RACE/GENDER: WHITE MALE
AGENCY: LOS ANGELES COUNTY (CALIFORNIA) SHERIFF'S OFFICE
DATE OF INCIDENT: JANUARY 6, 1994
CAUSE OF DEATH: METHAMPHETAMINE INTOXICATION AND ASPHYXIA
ROLE OF TASER DEVICE: EXCLUDED

The Sheriff's Office report on the Gizowski case was exempt from disclosure under California law, but news sources reported that deputies received a call from a woman who said Gizowski was acting strangely. When they arrived, the deputies found him on a living room couch yelling incoherently. When the deputies entered the room, Gizowski immediately jumped from the couch and attacked them. They used chemical spray and two shocks from a TASER device on Gizowski, but those efforts had no effect. After the deputies physically restrained Gizowski, they noticed that he was having trouble breathing. Paramedics took Gizowski to a local hospital, where he died.

The coroner later ruled that Gizowski died of methamphetamine intoxication and asphyxia. He ruled the death a homicide (see County of Los Angeles (California) Coroner Report 1994–00214; *Los Angeles Times*, 1994, January 8; Peavy, 2006, August 25).

The media accounts were not specific in reporting how long Gizowski struggled with officers following the last application of the TASER device, but he did continue to struggle, indicating that the electric current did not significantly affect the rhythm of his heart. Because the TASER pulse did not cause an immediate collapse, and because the medical examiner attributed the cause of death to methamphetamine intoxication and asphyxia, the device is excluded as the cause of death and as a significant contributing factor in the Gizowski case.

RICHARD WAYNE HARRIS

AGE: 32
RACE/GENDER: WHITE MALE
AGENCY: LOS ANGELES (CALIFORNIA) POLICE DEPARTMENT
DATE OF INCIDENT: APRIL 16, 1994
CAUSE OF DEATH: UNKNOWN
ROLE OF TASER DEVICE: UNDETERMINED

The police report in this incident was not available, but media sources reported that officers responded to an east Hollywood apartment after receiving a report of a violent man. When they arrived, police found Harris, bleeding, ripping apart bathroom fixtures and holding a pair of scissors. When he lunged at the officers, they used chemical spray and two shocks from a TASER device to take him into custody. After they subdued him, police noticed that Harris was having difficulty breathing, and they called for paramedics. The paramedics took Harris to a local hospital, where he died about an hour later (see *Los Angeles Times*, 1994, April 18; *Press-Telegram*, 1994; Maislin, 2007).

The autopsy report was not available, and media sources do not report the results of an autopsy. Without additional information, the role of the TASER device in this case is undetermined.

LEEGRAND GRIFFIN

AGE: 39
RACE/GENDER: AFRICAN-AMERICAN MALE
AGENCY: CINCINNATI (OHIO) POLICE DEPARTMENT
DATE OF INCIDENT: JUNE 5, 1994
CAUSE OF DEATH: ACUTE COCAINE TOXICITY
ROLE OF TASER DEVICE: EXCLUDED

Griffin exited his home, naked and brandishing a knife. He stabbed himself in the abdomen and then accosted several citizens. He broke into a home and confronted an elderly woman. Witnesses confronted Griffin and forced him into a kitchen closet until police arrived. When the responding officers opened the door to the closet, Griffin lunged at them, now holding a pan. One officer fired a TASER device, striking Griffin in the waist. The charge had no effect. Several officers wrestled Griffin to the floor while administering chemical spray, which had no effect. The officers put handcuffs on Griffin, but he continued to kick at officers. Officers had called for medical assistance, but while they were holding Griffin down on the floor, he suddenly stopped breathing. Paramedics took Griffin to a local hospital, where he died.

The coroner ruled that Griffin had died from acute cocaine intoxication. He ruled the death an accident (see Cincinnati (Ohio) Police Homicide Unit, 1994; Hamilton County (Ohio) Coroner's Report 122314; Lecky & Jackson, 1996).

Griffin continued to resist officers after application of the TASER device, indicating that the electric current did not significantly affect the rhythm of his heart. Because Griffin did not collapse immediately following the application of the TASER pulse, and because the coroner attributed the death to cocaine, the device is excluded as the cause of death and as a significant contributing factor in the Griffin case.

BRUCE KLOBUCHAR

AGE: 25
RACE/GENDER: MALE (RACE NOT DETERMINABLE WITH THE AVAILABLE RECORDS)
AGENCY: LOS ANGELES (CALIFORNIA) POLICE DEPARTMENT
DATE OF INCIDENT: AUGUST 18, 1995
CAUSE OF DEATH: INTOXICATION ON MULTIPLE DRUGS AND RESTRAINT ASPHYXIA
ROLE OF TASER DEVICE: EXCLUDED

The police report in this incident was exempt from disclosure under California law, but media sources reported that police responded to a complaint about a violent male. An apartment manager and the man's roommate met police when they arrived and told them that Klobuchar had gone crazy. When officers entered the apartment, they found Klobuchar in the bathroom, covered with blood, sweating profusely and holding a pair of scissors. He had knocked over the toilet, ripped the sink from the wall, and smashed the mirror. When he lunged and swung a wooden pole at the officers, an officer sprayed him with chemical spray. Klobuchar was still violent, so a second officer stunned him with a TASER device. As the officers worked to handcuff him, Klobuchar suddenly sat up and threw off four officers. A second shock from the TASER device subdued him enough for officers to put on handcuffs and bind his feet. Minutes later, as Klobuchar lay on his side, officers noticed he had stopped breathing. He died a short time later.

Although the autopsy report was not available, media sources reported that the coroner determined Klobuchar died from intoxication of multiple drugs, which the sources did not identify, and restraint asphyxia (see *City News Service*, 1997).

Klobuchar continued to struggle with officers after the application of the TASER device, indicating that the electric current did not immediately affect the rhythm of his heart. Because the TASER pulse had no immediate effect and the coroner attributed the death to restraint asphyxia and drug intoxication, the device is excluded as the cause of death and as a significant contributing factor in the Klobuchar case.

BYRON WILLIAMS

AGE: 36
RACE/GENDER: AFRICAN-AMERICAN MALE
AGENCY: LOS ANGELES (CALIFORNIA) POLICE DEPARTMENT
DATE OF INCIDENT: JANUARY 5, 1996
CAUSE OF DEATH: ACUTE COCAINE INTOXICATION
ROLE OF TASER DEVICE: EXCLUDED

Little information is available on the Williams case because the police report was exempt from disclosure under California law, and news accounts were limited. One newspaper account reports that officers responded to a call about unspecified trouble in an alley, and they found Williams tearing at a mattress. Officers tried to talk with him, but he continued to act in a disturbed and aggressive manner. Officers attempted to subdue him by firing a TASER electronic control device. Shortly afterward, Williams appeared to be in medical distress. An ambulance took him to a local hospital, where he died.

The coroner found that Williams died of acute cocaine intoxication, and he ruled the death an accident (see County of Los Angeles (California) Coroner Report 1996–00138; *Los Angeles Times*, 1996, January 6).

Media accounts of this incident are not specific in describing how much time passed between the application of the TASER device and when officers noticed that Williams was in distress. However, because the coroner attributed the death to cocaine and did not mention the TASER device, the device is excluded as the cause of death and as a significant contributing factor in the Williams case.

SCOTT JARON NORBERG

AGE: 33
RACE/GENDER: WHITE MALE
AGENCY: MARICOPA COUNTY (ARIZONA) SHERIFF'S OFFICE
DATE OF INCIDENT: JUNE 1, 1996
CAUSE OF DEATH: POSITIONAL ASPHYXIA
CONTRIBUTING FACTORS: CARDIOMEGALY AND CORONARY ATHEROSCLEORSIS
ROLE OF TASER DEVICE: EXCLUDED

The Sheriff's Office report on this case was not available, but media sources reported that Norberg, who had been hospitalized for dehydration and heat exhaustion that day, was jailed on Friday night for assaulting a Mesa police officer. He began behaving oddly in jail, telling people that he thought he was going to die. He was uncooperative during processing Saturday afternoon. Detention officers were trying to put Norberg into a restraining chair when he began fighting with them and spitting on them. Norberg also displayed extraordinary strength while fighting with detention officers, and showed no signs that a TASER device that officers used affected him. Following the applications of the device, five officers struggled with Norberg for about 11 minutes before subduing him. They restrained his arms behind his back, and wrapped a towel around his face to prevent him from spitting. While in the restraining chair, an officer held his head down against his chest. A few minutes later, officers noticed that Norberg was not breathing. He died later at a local hospital.

The medical examiner noted traces of methamphetamine and propoxyphene in Norberg's body, but he concluded that Norberg died of positional asphyxia. He concluded that an enlarged heart and coronary artery disease contributed to his death. He also noted that the presence of methamphetamine and propoxyphene might have caused Norberg's bizarre behavior, paranoia, and cardiotoxicity. The medical examiner noted that autopsy results showed no evidence of neck damage, broken bones, or internal injuries. Norberg's parents later filed a wrongful death lawsuit. The parties agreed to a pretrial settlement of $8.25 million. The suit alleged excessive force on the part of deputies, but did not include allegations that the TASER device was involved in Norberg's death (see *Associated Press*,

1996, June 6; *Associated Press*, 1996, October 17; Maricopa County (Arizona) Medical Examiner's Report 96–01510; Kammer, 1999).

Norberg continued to struggle with jailers following the application of the TASER device, indicating that the electric current did not immediately affect the rhythm of his heart. Because the TASER pulse did not have an immediate effect and the medical examiner attributed the death to restraint asphyxia, the devise is excluded as the cause of death and as a significant contributing factor in the Norberg case.

JAMES QUENTIN PARKINSON

AGE: 25
RACE/GENDER: WHITE MALE
AGENCY: FAIRFIELD (CALIFORNIA) POLICE DEPARTMENT
DATE OF INCIDENT: JUNE 8, 1996
CAUSE OF DEATH: SUDDEN CARDIAC DEATH DURING VIOLENT STRUGGLE DUE TO SCHIZOPHRENIA WITH EXCITED DELIRIUM
CONTRIBUTING FACTORS: CLOZAPINE INTOXICATION
ROLE OF TASER DEVICE: EXCLUDED

Parkinson, who had a history of mental illness, walked away from an adult psychiatric ward. He began running through the streets, breaking the windows of several homes while tearing off his clothes and yelling that he was looking for Jesus Christ. He then broke into a house and assaulted a woman, ran down the street, and began pummeling a light pole with his fists. When officers arrived, Parkinson charged an officer three times. The officer sprayed Parkinson with chemical spray each time, but the chemical spray had little effect. When Parkinson began running around a swimming pool, another officer shot him with a TASERTRON-conducted energy weapon. The probes struck Parkinson in the back and buttocks. Parkinson hit the ground, rolled over, and started to masturbate. He then jumped up and began running again. Finally, one of the officers tackled Parkinson, and three officers held him down to handcuff him. Parkinson began to suffer a seizure. The officers rolled him on his side, and the seizure stopped. When paramedics arrived, they strapped Parkinson face-up on a gurney. Parkinson somehow managed to turn over onto his stomach and seemed to calm down. Paramedics were preparing to treat him when they realized he was not breathing anymore, and his heart had stopped. The paramedics took Parkinson to a local hospital, where he died shortly after arrival.

The coroner listed the cause of death as sudden cardiac death during violent struggle due to schizophrenia with excited delirium. He stated he believed that Parkinson suffered a cardiac arrhythmia brought on by the struggle. He also noted that Parkinson was taking Clozapine, an antipsychotic drug, and that Parkinson's blood level of Clozapine was significantly above the therapeutic range, but slightly below the ranges reported in fatal overdoses. Nevertheless, he concluded that an overdose of Clozapine could have produced many of Parkinson's symptoms; including arrhythmia and death (see Shioya, 1996; Solano County (California) Coroner's Report 696–1071–CCI–160; Davis, 1998).

Parkinson continued to struggle for several minutes following application of the TASER device, indicating that the electric current did not significantly affect the rhythm of his heart. Because Parkinson did not collapse immediately following the use of the TASER device, and because the coroner attributed the death to excited delirium and an overdose of Clozapine, the device is excluded as the cause of death and as a significant contributing factor in the Parkinson case.

KIMBERLY LASHON WATKINS-OLIVER

AGE: 29
RACE/GENDER: AFRICAN-AMERICAN FEMALE
AGENCY: LOS ANGELES COUNTY (CALIFORNIA) SHERIFF'S OFFICE
DATE OF INCIDENT: JULY 20, 1996
CAUSE OF DEATH: ACUTE COCAINE INTOXICATION
ROLE OF TASER DEVICE: EXCLUDED

Information was limited in this case because the Sheriff's Office report in the Watkins-Oliver case was exempt from disclosure under California law, the Los Angeles County Medical Examiner's Office could not locate an autopsy report, and news accounts were sketchy. Watkins-Oliver had reportedly been driving under the influence of drugs. She crashed her car into a wall and a parked car. When deputies arrived to investigate, she resisted arrest. A deputy shocked her at least once with a TASER device. Once deputies had her subdued, they had her taken to a local hospital, where she died.

News reports indicate the coroner ruled that Watkins-Oliver died of acute cocaine intoxication (see *Los Angeles Times*, 1996, July 22; Peavy, 2006, November 15).

The media accounts are not specific regarding when Watkins-Oliver collapsed relative to the application of the TASER device. However, because the coroner attributed the death to cocaine and did not mention the TASER pulses, the device is excluded as the cause of death and as a significant contributing factor in the Watkins-Oliver case.

ANDREW HUNT, JR.

AGE: 38
RACE/GENDER: WHITE MALE
AGENCY: POMONA (CALIFORNIA) POLICE
   DEPARTMENT
DATE OF INCIDENT: DECEMBER 27, 1996
CAUSE OF DEATH: EXCITED DELIRIUM ASSOCIATED
   WITH COCAINE ABUSE
ROLE OF TASER DEVICE: EXCLUDED

The police report in the Hunt case was exempt from disclosure under California law, and there were no news accounts detailing this incident available. However, the medical examiner's investigation report says that Hunt was involved in an altercation with Pomona police officers at his apartment. The report does not describe the altercation, but it does mention that officers used a TASER device to subdue Hunt. They transported Hunt to the local jail, where they later found him to be unresponsive. Paramedics transported Hunt to a local hospital, where he died.

The autopsy lists Hunt's cause of death as excited delirium associated with cocaine abuse. The medical examiner concluded that the metabolic and thermal effects of excited delirium associated with cocaine abuse produced Hunt's death. He added that the contribution, if any, produced by infliction of a TASER device was problematic (see County of Los Angeles (California) Coroner's Report 1996–09471).

Hunt did not collapse immediately following the use of the TASER device, indicating that the electric current had no significant affect on the rhythm of his heart. Consequently, because the medical examiner attributed the death to excited delirium due to cocaine, and because the medical examiner stated that the TASER device played no significant role, the device is excluded as the cause of death and as a significant contributing factor in the Hunt case.

GARNER ROOSEVELT HICKS, JR.

AGE: 25
RACE/GENDER: AFRICAN-AMERICAN MALE
AGENCY: SANTA ANA (CALIFORNIA) POLICE
   DEPARTMENT
DATE OF INCIDENT: JULY 6, 1997
CAUSE OF DEATH: ACUTE COCAINE AND
   COAETHYLENE INTOXICATION DUE TO COMBINED
   EFFECT OF COCAINE AND ETHANOL
ROLE OF TASER DEVICE: EXCLUDED

The police report in this incident was exempt from disclosure under California law, but media sources reported that Hicks was acting crazy and threatening to cut his wrists on the balcony of a residence. Before police arrived, Hicks barricaded himself inside a friend's condominium by putting a sofa next to the door. He threatened to kill himself, and he sliced his wrist with a piece of glass. Officers watched through a window as the man put a phone cord around his neck in an apparent attempt to choke himself. Police decided to end the standoff when they saw that Hicks had set fire to the kitchen. The officers were worried that a fire would spread to other condominiums if firefighters could not get inside to extinguish it. Officers fired a TASER device to subdue the individual and take him into custody. While he was handcuffed, he continued to struggle. Officers put him on a gurney to take him to a hospital, but he stopped breathing shortly afterward. Paramedics took him to a local hospital, where he died.

The coroner reported that Hicks died from acute cocaine and cocaethylene intoxication. He did not mention the TASER device as a cause or as a significant contributing factor (see Orange County (California) Coroner's Report 97–04401–EY; Rams, 1997).

Hicks continued to struggle with officers after the application of the TASER device, indicating that the electric current did not significantly affect the rhythm of his heart. Because Hicks continued to resist following the application of the TASER pulse, and because the medical examiner attributed the death to cocaine and alcohol intoxication, the device is excluded as the cause of death and as a significant contributing factor in the Hicks case.

MARK ANDREW BROWN

AGE: 43
RACE/GENDER: WHITE MALE
AGENCY: LOS ANGELES COUNTY (CALIFORNIA) SHERIFF'S OFFICE
DATE OF INCIDENT: MARCH 9, 1998
DATE OF DEATH: MARCH 10, 1998
CAUSE OF DEATH: CARDIOPULMONARY ARREST
CONTRIBUTING FACTORS: AIRWAY OBSTRUCTION DUE TO FOREIGN OBJECT AND RESTRAINT DUE TO COCAINE INTOXICATION
ROLE OF TASER DEVICE: EXCLUDED

The police report on the Brown case was exempt from disclosure under California law, but media sources reported that deputies were sent to a home after a woman reported that her son was drunk, tearing up the house and behaving violently. Four deputies and two sergeants entered the home and found Brown nude and behaving bizarrely. The deputies said Brown was cooperative at first, but then he became combative. Due to Brown's aggression, a sergeant deployed a TASER device. The prongs struck Brown, who dropped to the floor but continued to struggle and kick. Deputies restrained him with a nylon device that restrained his arms and legs. When the deputies got Brown to the car, they noticed he was having trouble breathing. Paramedics called to the scene pulled three wads of magazine pages from Brown's throat, which he apparently swallowed prior to the arrival of deputies. He returned to normal breathing. The paramedics took him to a local hospital, but he died the next day.

The Los Angeles County Medical Examiner's office was unable to provide a copy of the autopsy report, but media sources reported that the medical examiner listed the cause of Brown's death as cardiopulmonary arrest. The autopsy report also indicated significant associated factors of an airway obstruction related to a foreign object and restraint due to cocaine intoxication (see *City News Service*, 1998; Peavy, 2006, November 15).

Brown continued to struggle with deputies after the application of the TASER device, indicating that the electric current did not significantly affect the rhythm of his heart. Because Brown continued to resist following the application of the TASER pulse, and because the medical examiner attributed the death to cocaine and an airway obstruction, the device is excluded as the cause of death and as a significant contributing factor in the Brown case.

MICHAEL LABMEIER

AGE: 48
RACE/GENDER: MALE (RACE NOT DETERMINABLE WITH THE AVAILABLE RECORDS)
AGENCY: KENTON COUNTY (KENTUCKY) POLICE DEPARTMENT
DATE OF INCIDENT: JANUARY 29, 1999
CAUSE OF DEATH: IRREGULAR HEARTBEAT CAUSED BY AN ADRENALINE RUSH
CONTRIBUTING FACTORS: OBESITY
ROLE OF TASER DEVICE: UNDETERMINED

The Sheriff's Office report was not available in this incident, but media sources reported that a jail doctor visited Labmeier in an isolation cell and determined he needed a mental evaluation. Labmeier was agitated, refused to come out of his cell, and was clothed in only his underwear. After a 15-minute wait, jailers called for paramedics. Labmeier was banging himself off the walls and working himself into a frenzy. The jail administrator sprayed Labmeier once with chemical spray. Two different reports conflicted in whether jailers used a TASER device, one report saying a device was not used and the other saying it was. After jailers had Labmeier handcuffed, he began kicking at officers. Five deputies took him to the floor and shackled his legs and hands. Jail staff checked for a pulse, but paramedics found none 30 seconds later. Labmeier was pronounced dead at the scene.

The autopsy report was not available, but media sources reported that the coroner said Labmeier might have lost his pulse because of an irregular heartbeat caused by an adrenaline rush while he was being restrained. He said there was evidence of heart disease in a major artery, but the reports did not define that disease further. The medical examiner also said that Labmeier's weight—he was 6-feet tall and weighed 302 pounds—was likely a contributing factor (see Driehaus, 1999; Schaefer, 1999).

There is a question whether a TASER device was actually used in this case. If jailers did not use the device, then clearly the device is excluded as a cause of death. If jailers did use a TASER device, when they used it is not known, and whether it had an effect is not known. The medical examiner attributed the death to an irregular heartbeat caused by the physical exertion and the preexisting coronary artery disease. Consequently, without more information, the role of the TASER device in the Labmeier case is undetermined.

DAVID TORRES FLORES

AGE: 37
RACE/GENDER: HISPANIC MALE
AGENCY: FAIRFIELD (CALIFORNIA) POLICE
  DEPARTMENT
DATE OF INCIDENT: SEPTEMBER 28, 1999
CAUSE OF DEATH: SUDDEN DEATH DURING POLICE
  RESTRAINT DUE TO AGITATED DELIRIUM DUE TO
  ACUTE COCAINE AND METHAMPHETAMINE
  INTOXICATION
CONTRIBUTING FACTORS: CHRONIC COCAINE
  ABUSE

ROLE OF TASER DEVICE: EXCLUDED

Flores' ex-wife called the police to report that Flores was at her residence. She said that Flores was paranoid and acting strangely. She said Flores believed that the police were after him and that people were coming through the walls. She also told the police that Flores' behavior was drug-related. A few minutes later, 9-1-1 received a hang-up call. When the dispatcher called the number, Flores' ex-wife answered and asked for an ambulance. She said that Flores, who had locked himself in the bathroom, was making strange noises and tearing up things. She was afraid that he was either ill or injured. When officers arrived, Flores resisted the officers' attempts to get him out of the bathroom. At one point, he picked up the lid of a toilet and threatened officers with it. Officers used chemical spray and a TASER device in an attempt to subdue Flores, but both were ineffective. Officers finally wrestled Flores to the ground and secured him with handcuffs. Flores had a seizure, so the officers requested medical personnel, who had staged just outside the apartment, to assist. The paramedics took Flores to a local hospital, where he died.

The coroner concluded that Flores died of sudden death during police restraint due to agitated delirium due to acute cocaine and methamphetamine intoxication. Toxicology results revealed that Flores had toxic levels of cocaine and methamphetamine in his body. At autopsy, the medical examiner discovered cocaine in Flores' gastric contents (see Fairfield (California) Police Report 99–13233; Goodyear, 1999; *San Francisco Chronicle*, 1999; Solano County (California) Coroner's Report 999–1720–CCI–261; Burkeman & Borger, 2001; Anglen, 2006).

Flores continued to struggle with police after the application of the TASER device, indicating that the electric current had no significant affect on the rhythm of his heart. Because the TASER pulse did not cause an immediate collapse, and because the medical examiner attributed the death to agitated delirium due to cocaine and methamphetamine, the device is excluded as the cause of death and as a significant contributing factor in the Flores case.

THIRD AND FOURTH GENERATION TASER WEAPON CASES

The 171 case studies, from Enrique Juarez Ochoa to David L. Moss, Jr., represent deaths that followed the use of third generation and fourth generation TASER weapons, the ADVANCED TASER M26 and the TASER X26, that occurred between 2000 and 2005. Because so many of the investigations of 97 cases that occurred in 2006 are not complete, those case studies are not included in this chapter. Appendix A, Unexpected Deaths Following Application of TASER Devices–2006, contains a listing of those cases.

ENRIQUE JUAREZ OCHOA

AGE: 34
RACE/GENDER: HISPANIC MALE
AGENCY: BAKERSFIELD (CALIFORNIA) POLICE DEPARTMENT
DATE OF INCIDENT: MAY 14, 2000
CAUSE OF DEATH: DISSEMINATED INTRAVASCULAR COAGULATION DUE TO BLUNT IMPACT TRAUMA WHILE IN A HYPER-EXCITABLE STATE AND COCAINE TOXICITY
ROLE OF TASER DEVICE: EXCLUDED

The police report in the Ochoa case was exempt from disclosure under California law, but media sources reported that Ochoa died after his mother called police for help because he was acting frightened and confused. Bakersfield police responded, determined that Ochoa was high on drugs. The officers twice applied a TASER device to subdue him. The officers placed Ochoa on the ground in physical restraints for 15 to 20 minutes before they transported him to a local hospital for evaluation. When he arrived at the hospital, Ochoa's face and body were badly swollen; injuries that the officers said were self-inflicted. About 15 minutes later, officers noticed that Ochoa had stopped moving.

The autopsy report was not available, but media sources reported that the coroner ruled Ochoa had died from disseminated intravascular coagulation due to blunt impact trauma while in a hyperexcitable state and cocaine toxicity (see *Fresno Bee*, 2001; Anglen, 2006).

Ochoa did not collapse until at least 30 minutes following the last application of the TASER device, indicating that the electrical current did not significantly affect the rhythm of his heart. Because the TASER pulse did not cause an immediate collapse, and because the coroner attributed the death to intravascular coagulation and cocaine, the device is excluded as the cause of death and as a significant contributing factor in the Ochoa case.

MARK LORENZO BURKETT

AGE: 18
RACE/GENDER: AFRICAN-AMERICAN MALE
AGENCY: ALACHUA COUNTY (FLORIDA) SHERIFF'S OFFICE
DATE OF INCIDENT: JUNE 13, 2001
DATE OF DEATH: JUNE 17, 2001
CAUSE OF DEATH: ACUTE EXHAUSTIVE MANIA
CONTRIBUTING FACTORS: BLOOD ALCOHOL CONTENT OF 0.11 FOUR HOURS AFTER ARREST
ROLE OF TASER DEVICE: EXCLUDED

Burkett, who suffered from paranoid schizophrenia, had suddenly become agitated and aggressive after a football scholarship offer was withdrawn, so his mother called 9-1-1 for assistance taking him to the hospital. He had been walking around the house swinging a hammer when officers arrived, and he injured several officers during the struggle to detain him. A court issued an order for blood tests, and while jail personnel were attempting to collect a sample, Burkett attacked the officers. When jailers shocked him with a TASER device, he suddenly collapsed and became unresponsive. Officers took Burkett to the hospital, where he remained on life support until he died four days later.

The autopsy report listed Burkett's cause of death as acute exhaustive mania, meaning that he worked himself into a frenzy that caused him to suffer a cardiac arrest. The medical examiner noted that Burkett's condition resembled excited delirium, except that toxicology reports revealed no traces of cocaine, steroids, or methamphetamine. However, the toxicology tests did detect traces of Diphenhydramine, an over-the-counter antihistamine and sedative, and Phenytoin, a commonly used anti-epileptic medication. The coroner noted that mania in psychiatric patients can lead to death, and he reported Burkett's family history of paranoid schizophrenia. However, a pathologist hired by the Burkett family found that Burkett died of diffuse traumatic axonal injury, one of the most common and devastating types of brain injury, and, in part, from blunt trauma on Burkett's heart (see Alachua County (Florida) Sheriff's Report 01–007808; Office of the Medical Examiner, Eighth District (Florida) Report ME 01–203; Amnesty International, 2004; Barton, 2005; Anglen, 2006).

Burkett did collapse immediately following the use of the TASER device, but two pathologists attributed the death to other causes, one to exhaustive mania, the other to diffuse traumatic axonal injury. Neither implicated the TASER pulse. Consequently, the device is excluded as the cause of death and as a significant contributing factor in the Burkett case.

STEVEN VASQUEZ

AGE: 40
RACE/GENDER: HISPANIC MALE
AGENCY: CORAL SPRINGS (FLORIDA) POLICE DEPARTMENT
DATE OF INCIDENT: DECEMBER 15, 2001
DATE OF DEATH: DECEMBER 21, 2001
CAUSE OF DEATH: COMBINED DRUG TOXICITY
ROLE OF TASER DEVICE: EXCLUDED

The police report in the Vasquez case was not available, but media sources reported that on December 15, 2001, officers arrested Vasquez for disorderly conduct and resisting arrest. The officers shocked Vasquez with a TASER device during an altercation as they were attempting to escort him out of a bar. The officers took Vasquez to the local jail, and he was later released.

On December 19, 2001, Vasquez was attending a party with his son. The son told officers that Vasquez was intoxicated, so the son tried to drive him home. On the way, Vasquez had his son stop at a friend's house so Vasquez could purchase Xanax. When they got home, Vasquez stayed in the truck and fell asleep. Vasquez's wife found him in the truck the next morning and called for assistance when she could not arouse him. Fire/Rescue personnel arrived, but they could not revive Vasquez, and he was pronounced dead on the scene.

The medical examiner discovered that Vasquez was taking hydrocodone, a prescribed painkiller, for a previous back injury. Toxicology tests also revealed the presence of alcohol, Carisoprodol, a muscle relaxant, Meprobamate, an anti-anxiety drug, and Alprazolam, also known as Xanax, which is commonly used to treat anxiety disorders and as an adjunctive treatment for depression. The medical examiner concluded that Vasquez died of combined drug toxicity, and she ruled the death an accident. The coroner added that the TASER device was not a contributing factor in Vasquez's death (see Broward County (Florida) Medical Examiner's Report 01–1664; Amnesty International, 2004; Barton, 2005; Anglen, 2006).

The Vasquez case is the first case in the 2004 Amnesty International report questioning the safety of TASER technology. What that report does not say is that Vasquez died, without previous symptoms or complaints, five days after the application of the TASER device. Because Vasquez did not collapse for five days following application of the TASER pulse, and because the medical examiner attributed Vasquez's death to drug toxicity, the device is excluded as the cause of death and as a significant contributing factor in the Vaquez case.

MARVIN HENDRIX

AGE: 27
RACE/GENDER: AFRICAN-AMERICAN MALE
AGENCY: HAMILTON (OHIO) POLICE DEPARTMENT
DATE OF INCIDENT: DECEMBER 17, 2001
CAUSE OF DEATH: COCAINE TOXICITY
CONTRIBUTING FACTORS: EXCITED DELIRIUM
ROLE OF TASER DEVICE: EXCLUDED

Paramedics received a call for emergency assistance at Hendrix's home. When they arrived, they found that Hendrix had vomited a pink fluid into the toilet and in other areas. When the paramedics approached Hendrix, he charged at them. The paramedics called for police assistance saying that Hendrix was out of control. When police arrived, Hendrix was fighting with paramedics in the house. An officer applied a TASER device to Hendrix's shoulder, but the shock was not effective. The officer then shocked Hendrix a second time for 10 seconds to handcuff him. Hendrix lost consciousness approximately two minutes later. Paramedics took Hendrix to a local hospital, where he died. During the investigation, investigators learned that Hendrix had swallowed a plastic bag containing crack cocaine to conceal it from officers. The bag had burst, releasing toxic levels of cocaine into his system.

The medical examiner listed the cause of death as cocaine abuse, and he ruled the death accidental. He reported that the exact role of the TASER device in Hendrix's death was unknown, but he listed no evidence implicating the device (see *Associated Press*, 2001; Hamilton (Ohio) Police Report 2001–071684; Hamilton (Ohio) Police Report 2001–071691; Hamilton (Ohio) Police Report 2001–071692; *Business Wire*, 2002; Amnesty International, 2004; Anglen, 2006).

Hendrix lost consciousness about two minutes following the last application of the TASER electronic control device. Because Hendrix did not collapse immediately after the application of the TASER pulse, and because the medical examiner attributed the death to cocaine, the device is excluded as a cause of death and as a significant contributing factor in the Hendrix death.

VINCENT DELOSTIA

AGE: 31
RACE/GENDER: WHITE MALE
AGENCY: HOLLYWOOD (FLORIDA) POLICE
   DEPARTMENT
DATE OF INCIDENT: JANUARY 27, 2002
CAUSE OF DEATH: COCAINE TOXICITY

CONTRIBUTING FACTORS: HISTORY OF DRUG
   ABUSE AND PSYCHOSIS
ROLE OF TASER DEVICE: EXCLUDED

Delostia was running around in traffic. He then ran into the lobby of a hotel and refused to leave. When the police arrived, they found Delostia half-dressed, banging on the front door of the hotel, bleeding from a cut on the back, sweating and grunting. He laid down and kicked at officers. Police were trying to take him into custody for a mental health evaluation when they shocked him with a TASER device, which seemed to have no effect. Delostia pulled out one probe and broke a wire. Officers rolled him onto his stomach, placed handcuffs around his arms, and placed hobbles on his legs. Shortly thereafter, Delostia stopped breathing. Paramedics transported Delostia to a local hospital, where he died.

A medical examiner said the cause of death was cocaine toxicity, and she ruled the death to be accidental. She also noted that Delostia had a history of psychosis, drug abuse, and asthma. She reported that Delostia exhibited multiple signs of excited delirium, including agitation, bizarre behavior, and perspiration. She also noted the presence in Delostia's system of the antidepressant and panic disorder medication Prozac, the antiepileptic drug Dilantin, and Temazepam, an anticonvulsant. She excluded the TASER device as a contributing factor (see Broward County (Florida) Medical Examiner Report 02–0117; *Business Wire*, 2002; Hollywood (Florida) Police Report 02–015065; Amnesty International, 2004; Barton, 2005; Anglen, 2006).

Delostia continued to struggle with officers following application of the TASER device, indicating that the electric current did not significantly affect the rhythm of his heart. Because the TASER pulse showed no immediate effect and the medical examiner attributed the death to cocaine, the device is excluded as the cause of death and as a significant contributing factor in the Delostia case.

ANTHONY SPENCER

AGE: 35
RACE/GENDER: MALE (RACE NOT DETERMINABLE

FROM THE AVAILABLE RECORDS)
AGENCY: PHILADELPHIA (PENNSYLVANIA) POLICE
  DEPARTMENT
DATE OF INCIDENT: FEBRUARY 13, 2002
CAUSE OF DEATH: COCAINE INTOXICATION
ROLE OF TASER DEVICE: EXCLUDED

Information regarding Spencer's death was limited because the City of Philadelphia was not able to produce a police report, and the autopsy report was exempt from disclosure under Pennsylvania law. However, news sources indicated that police, responding to a domestic disturbance call, found Spencer outside his house intoxicated, naked, and brandishing a folding knife with an open 7-inch blade. As the officers approached, Spencer ran into the house. He went upstairs then came back down as officers tried to speak with him. Spencer then rushed past the officers and ran outside. The officers used chemical spray and a TASER device to subdue Spencer. Spencer was still resisting when officers and paramedics loaded him into a patrol wagon for transport to a local hospital. He was conscious and belligerent all the way to the hospital, but he died about 30 minutes later.

News reports quote Philadelphia officials as saying that tests found Spencer's death was due to cocaine intoxication, and that shocks from a TASER device were not a contributing factor (see Boyer, 2002; Gibbons, 2002; Nolan, 2002; Amnesty International, 2004; Anglen, 2006).

Spencer continued to resist officers and paramedics for several minutes following application of the TASER device, indicating that the electric current had no significant affect in his heart. Because the TASER pulse did not have an immediate effect and the medical examiner attributed the death to cocaine, the device is excluded as the cause of death and as a significant contributing factor in the Spencer case.

HENRY WILLIAM CANADY

AGE: 46
RACE/GENDER: AFRICAN-AMERICAN MALE
AGENCY: NASSAU COUNTY (FLORIDA) SHERIFF'S
  OFFICE
DATE OF INCIDENT: MARCH 27, 2002
CAUSE OF DEATH: COCAINE TOXICITY
CONTRIBUTING FACTORS: CORONARY ATHEROSCLEROSIS AND THE STRESS OF THE STRUGGLE
  WITH POLICE
ROLE OF TASER DEVICE: EXCLUDED/DOUBTFUL

Deputies were attempting to arrest Canady on a warrant related to an undercover drug buy. A deputy had Canady walk over to his car and put his hands on the vehicle. At first, Canady complied. However, Canady removed his hands from the vehicle and ran down the street toward his residence. A deputy fired his TASER device at Canady, but it had no effect. A second deputy also fired a TASER device, but again, it had no effect. Canady dove through the back door of the house, picked up a headboard from a bed, and shielded himself with it. A deputy then drive-stunned Canady with the TASER device. Canady dropped the headboard, but continued to struggle. The deputies wrestled with Canady until they finally got handcuffs on him. Even then, Canady resisted and tried to crawl away. Deputies once again applied the device to Canady, but he continued to struggle. Moments later, Canady suddenly became calm. The deputies turned him on his side when he suddenly quit breathing. The deputies called for an ambulance, but Canady died at the scene.

A medical examiner's investigation determined Canady's death was due to cocaine toxicity. He stated that coronary atherosclerosis was a significant condition contributing to the death because all three major coronary arteries were blocked greater than 70 percent. The medical examiner also noted that the death occurred during or shortly after a physical altercation, and the exertion and stress may have contributed to the death. He ruled the death an accident (see Kinner, 2002; Nassau County (Florida) Medical Examiner's Report 02–0423; Nassau County (Florida) Sheriff's Office Report 2002–01492; Treen, 2002; Amnesty International, 2004; Anglen, 2006).

Canady continued to struggle with officers following the last application of the TASER device, indicating that the electric current had no immediate effect on the rhythm of his heart. Because the

TASER pulse did not have an immediate effect and the medical examiner attributed the death to cocaine, the device is excluded as the cause of death. The TASER pulses undoubtedly contributed to Canady's stress level during the struggle, and the medical examiner, in part, attributed the stress of his was arrest as a contributing factor. However, based on his severe coronary artery blockage, it is questionable whether, but for the application of the TASER device, Canady would have survived, or that he would have survived had the police used some other form of restraint. Therefore, the role of the device as a significant contributing factor in Canady's death is doubtful.

RICHARD JOSEPH BARALLA

AGE: 36
RACE/GENDER: WHITE MALE
AGENCY: PUEBLO (COLORADO) POLICE DEPARTMENT
DATE OF INCIDENT: MAY 17, 2002
CAUSE OF DEATH: SUDDEN CARDIAC ARREST DURING EXCITED DELIRIUM NECESSITATING RESTRAINT
CONTRIBUTING FACTORS: CHRONIC THROIDITIS
ROLE OF TASER DEVICE: EXCLUDED

Police arrested Baralla after they saw him walking down the middle of a street exhibiting strange behavior. He refused to obey several commands to get out of the street, and he would not answer the officers' questions. When officers tried to prevent Baralla from jumping in front of traffic, he resisted. Officers sprayed him with chemical spray and tried to handcuff him. He resisted their efforts to secure him, but they finally were able to get two pairs of cuffs linked together behind his back. Baralla then refused to stand up so the officers could get him into a car. An officer drive-stunned Baralla on the left pectoral muscle to get him to quit resisting, but the shock had little effect. The officer then fired the probes into Baralla's back. Baralla tensed up, but the TASER pulses did not have the desired effect. Officers then put Baralla on his stomach and hobbled his legs. When they rolled Baralla onto his side, the officers noticed that he was not breathing. The officers called for an ambulance and removed the restraints. Fire/Rescue personnel took Baralla to a local hospital, where he died.

The initial autopsy report was inconclusive. A second report, requested by the Federal Bureau of Investigation, concluded that Baralla's death was the result of cardiac arrest during a state of excited delirium that necessitated restraint. The coroner noted a history of drug abuse, but toxicology reports showed no alcohol or drugs in Baralla's system. The medical examiner noted two sets of marks consistent with the application of a TASER device, but he also noted that the injuries were not sufficiently severe to have contributed to Baralla's death (see Pueblo County (Colorado) Coroner Report 02A-MD08; Pueblo (Colorado) Police Report 02–10782; St. Mary-Corwin Regional Medical Center (Pueblo, Colorado) Autopsy Report P–02070; Huntley, 2003, Oct. 1; Amnesty International, 2004; Anglen, 2006).

Baralla continued to struggle with officers following the final application of the TASER device, indicating that the electric current did not immediately affect the rhythm of his heart. Because the TASER pulses did not have an immediate effect and the medical examiner attributed the cardiac arrest to excited delirium, the device is excluded as the cause of death and as a significant contributing factor in the Baralla case.

EDDIE RENE ALVARADO

AGE: 36
RACE/GENDER: HISPANIC MALE
AGENCY: LOS ANGELES (CALIFORNIA) POLICE DEPARTMENT
DATE OF INCIDENT: JUNE 10, 2002
CAUSE OF DEATH: SEQUELAE OF METHAMPHETAMINE INTOXICATION AND COCAINE USE
CONTRIBUTING FACTORS: SEVERE RESPIRATORY CONGESTION, RESTRAINT, AND TASER USE
ROLE OF TASER DEVICE: DOUBTFUL

Little information was available regarding this incident because the police report was exempt from disclosure under California law and news accounts are sketchy. However, news reports and the medical examiner's investigation report say that paramedics responded to Alvarado's residence to find him violent. They were unable to approach until he collapsed and began to have a seizure. The paramedics called police for assistance. The police handcuffed Alvarado so the paramedics could prepare him for transport. Alvarado again erupted in violence, and the officers shocked him five times with a TASER device to gain control of him. The paramedics checked Alvarado immediately after the officers secured him, and they found him to have declining vital signs. Alvarado suffered a cardiac arrest, and paramedics performed CPR en route to a local hospital. Medical personnel could not revive Alvarado, and he died.

The coroner reported that Alvarado died from the sequelae of methamphetamine and cocaine use while being restrained. The coroner added that he could not rule out the TASER electronic control device as a cause of death. He reported that Alvarado had no history of heart disease, and he noted a temporal relationship between the application of the device and Alvarado's heart attack (see Amnesty International, 2004; County of Los Angeles (California) Coroner's Report 2002–04388; Anglen, 2006).

After the final application of the TASER device, when the paramedics first checked Alvarado, he had a heartbeat. It was shortly afterward that he suffered the cardiac arrest. Although the coroner noted the temporal relationship between the application of a TASER device and Alvarado's declining vital signs, there was no pathological evidence that the device caused the decline. Alvarado was already having seizures before officers arrived, and whatever was causing the seizures could have been the cause for Alvarado's collapse. It is possible that Alvarado's vital signs declined for reasons other than the application of a TASER device, such as the ingestion of sympathomimetic drugs like cocaine and methamphetamine, and that the officers stopped using the device when Alvarado became too weak to resist. Therefore, the role of the TASER device as the cause of death and as a significant contributing factor in the Alvarado case is doubtful.

NICHOLAS AGUILAR

AGE: 39
RACE/GENDER: HISPANIC MALE
AGENCY: PHOENIX (ARIZONA) POLICE DEPARTMENT
DATE OF INCIDENT: JUNE 13, 2002
CAUSE OF DEATH: METHAMPHETAMINE INTOXICATION
CONTRIBUTING FACTORS: ATHEROSCLEROTIC CARDIOVASCULAR DISEASE
ROLE OF TASER DEVICE: EXCLUDED

Aguilar had returned to the home he shared with his girlfriend around 6:00 P.M. While he was there, he had several beers with his brother. He left to take his brother to work, and then he returned to work on his brother's van. Aguilar's girlfriend later told investigators that, when he returned, Aguilar acted very drunk, and he kept trying to pick a fight with her, breaking a vase in the process. After he went out to the garage, his girlfriend reported hearing Aguilar drop his tools, and she heard him yelling, as though he were arguing with someone, but no one was there. A few minutes later, she observed him lying in the yard, struggling to get up. She assisted him into the house where he began breaking things. She tried to calm him, but she could not. She had Aguilar's son call 9-1-1.

When officers arrived, they found Aguilar outside in the driveway. As they approached him, he turned and started toward his girlfriend, who was now standing at the front door. When he would not comply with the officers commands to stop, one officer applied a TASER device. Aguilar went down, but got back up again. The officer applied a second charge. Aguilar went down again and pulled out the prongs. The officer then sprayed chemical spray in Aguilar's face, but that had no effect. Aguilar grabbed the officer in a bear hug, so the officer struck him several times in the face. Another officer struck Aguilar with a flashlight several times on his legs and arms. Aguilar continued to struggle, and it took six officers more than three minutes to subdue him.

Officers called for paramedics because Aguilar was disoriented, muddy, and bloody. When paramedics arrived, they described Aguilar as still kicking and resisting. While they were treating him, Aguilar stopped breathing. Paramedics transported Aguilar to a local hospital, where he died.

Toxicology reports indicated that Aguilar had amphetamines and methamphetamine in his system. The medical examiner concluded that Aguilar died of methamphetamine intoxication. He also reported that atherosclerotic cardiovascular disease was a contributory factor (see Maricopa County (Arizona) Medical Examiner's Report 02–01878; Phoenix (Arizona) Police Report 2002–21082428; Phoenix (Arizona) Police Report 2002–21082428-A; Amnesty International, 2004; Anglen, 2006).

Because Aguilar continued to struggle with officers following the application of the TASER device, indicating that the electric current did not affect the rhythm of his heart, and because the medical examiner attributed the death to methamphetamine intoxication and cardiovascular disease, the device is excluded as the cause of death and as a significant contributing factor in the Aguilar case.

JASON NICHOLS

AGE: 21
RACE/GENDER: AFRICAN-AMERICAN MALE
AGENCY: OKLAHOMA CITY (OKLAHOMA) POLICE DEPARTMENT
DATE OF INCIDENT: JUNE 15, 2002
CAUSE OF DEATH: HEAD TRAUMA
CONTRIBUTING FACTORS: MULTIPLE ABRASIONS, MULTIPLE CONTUSIONS
ROLE OF TASER DEVICE: EXCLUDED

Nichols was involved in a family fight. When the police arrived, he struggled with officers, who were attempting to arrest him for assaulting family members. The officers shocked him with a TASER device, subdued him, and took him to a local hospital because of the various wounds he received in the fight with his family. He suffered a cardiac arrest about 13 minutes after his arrival at the hospital.

The autopsy revealed multiple injuries, the most serious being three lacerations to the back of the head. Each of those wounds contributed to a large, underlying soft tissue hematoma and an underlying superficial skull fracture. The medical examiner concluded that it was extremely unlikely that the TASER device or restraint asphyxia had a direct part in the death because Nichols had presented to the hospital with spontaneous breathing, a weak pulse, and a normal blood pressure. Drug tests were negative for all but a slight trace of marijuana. The medical examiner ruled that Nichols death was a result of head trauma, and he ruled the death a homicide (see Office of the Chief Medical Examiner (Oklahoma) Report 0203435; Oklahoma City (Oklahoma) Police Report 02–056072; Amnesty International, 2004; Anglen, 2006).

The application of the TASER device had no immediate effect on Nichols because his heart did not arrest until after he had arrived at the hospital. Because the TASER pulse did not have an immediate effect, the medical examiner attributed the death to severe head injury, and the medical examiner noted that it was highly unlikely that the TASER device had a direct part in Nichols' death. The device is excluded as the cause of death and as a significant contributing factor in the Nichols case.

FERMIN RINCON

AGE: 24
RACE/GENDER: HISPANIC MALE
AGENCY: FONTANA (CALIFORNIA) POLICE DEPARTMENT
DATE OF INCIDENT: JUNE 27, 2002
CAUSE OF DEATH: ACUTE CARDIAC ARRHYTHMIA DUE TO METHAMPHETAMINE USE
CONTRIBUTING FACTORS: PROLONGED METHAMPHETAMINE USE
ROLE OF TASER DEVICE: EXCLUDED

The police report in the Rincon case is exempt from disclosure under California law, but media sources reported that Rincon died after a struggle with five police officers at a business complex. Officers reportedly shocked Rincon three times with a TASER device and placed him in a chokehold to subdue him after he fell from a roof.

The autopsy report was not available: however, news reports claimed that the coroner attributed the cause of Rincon's death to acute cardiac arrhythmia due to methamphetamine use. The coroner's report also documented three fractured ribs and an assortment of superficial injuries from the TASER device and other causes (see Amnesty International, 2004; Anglen, 2006).

The media reports were not specific in reporting how much time passed between the last application of the TASER device and Rincon's collapse, but he did continue to resist officers for at least a short time. Because the TASER pulse had no immediate effect and the coroner attributed the cardiac arrhythmia to methamphetamine, the device is excluded as the cause of death and as a significant contributing factor in the Rincon case.

CLEVER CRAIG, JR.

AGE: 46
RACE/GENDER: AFRICAN-AMERICAN MALE
AGENCY: MOBILE (ALABAMA) POLICE
 DEPARTMENT
DATE OF INCIDENT: JUNE 28, 2002
CAUSE OF DEATH: CARDIAC DYSRHYTHMIA DURING EPISODE OF EXCITED DELIRIUM AND FOLLOWING ELECTRICAL SHOCK FROM TASER DEVICE WHILE RESISITING ARREST
ROLE OF TASER DEVICE: EXCLUDED

The police report in the Craig case was not available, but media sources reported that relatives called 9-1-1 saying Craig was acting strangely. Police found the 6-foot, 200-pound Craig, a diagnosed paranoid schizophrenic with a heart problem, holding a barbell. When he refused to drop it, officers shocked him twice with a TASER device in about 40 seconds. Craig continued to struggle for five minutes after being shocked. When officers handcuffed him, he was unresponsive.

The medical examiner reported that Craig died of a cardiac dysrhythmia during an episode of excited delirium and following electrical shock from a TASER device while resisting arrest. He did not state in his report how the TASER device or the resisting arrest contributed to Craig's dysrhythmia. However, the report did record that Craig suffered from cardiomegaly and fusions in the aortic and mitral valves (see Alabama Department of Forensic Sciences Report of Autopsy 01A–02MB05430; *Associated Press*, 2002; *Associated Press*, 2004, August 6; Amnesty International, 2004; Anglen, 2006).

Craig continued to resist officers for about five minutes following the last application of the TASER device, indicating that the electric current did not immediately affect the rhythm of his heart. The TASER pulse did not have an immediate effect, and the medical examiner provided only a temporal statement regarding the use of the device. He offered no pathological evidence or explanation of how the TASER pulse may have contributed to Craig's death more than his preexisting cardiomegaly. Consequently, the device is excluded as the cause and as a significant contributing factor to Craig's death.

GORDON RANDALL JONES

AGE: 37
RACE/GENDER: WHITE MALE
AGENCY: ORANGE COUNTY (FLORIDA) SHERIFF'S
 OFFICE
DATE OF INCIDENT: JULY 19, 2002
CAUSE OF DEATH: POSITIONAL ASPHYXIA SECONDARY TO APPLICATION OF RESTRAINTS IN SETTING OF ACUTE COCAINE INTOXICATION
ROLE OF TASER DEVICE: DOUBTFUL

Jones was drunk in a hotel lobby where he was a paying guest. He had ingested cocaine earlier that day, and he was acting strangely when Orange County Sheriff's deputies arrived in response to a call from the hotel manager. When deputies ordered Jones to leave, he dumped his clothes from a duffle bag and refused to leave. When a deputy attempted to place Jones under arrest, he struggled with them. One deputy shocked Jones at least eight times with a TASER device, but each time the deputies approached, Jones continued to kick and resist their efforts to handcuff him. One deputy applied a pulse continuously for several cycles so that other deputies could handcuff Jones while he was incapacitated. Once they had secured Jones, he walked with the deputies to the front of the hotel and to a waiting ambulance, where he was restrained face down on a stretcher. On the way to the hospital, Jones screamed at the paramedic crew and tried to bite and spit on them. He quit breathing while en route. Paramedics and emergency room personnel could not revive him.

The autopsy report was not available, but media sources reported that the county medical examiner determined that Jones suffocated, and he listed TASER shocks and cocaine intoxication as contributing factors. The TASER shocks, he said, interfered with the muscles Jones needed to breathe, making him already short of breath when deputies eventually handcuffed him while he was lying on his chest. As a result, Jones' blood was oxygen starved. Orange County officials hired another pathologist to render a second opinion. That pathologist found that Jones' death resulted primarily from cocaine intoxication, with being restrained face down reduced to a contributing cause. He claimed that the multiple TASER shocks did not contribute to Jones' death (see Orange County (Florida) Sheriff's Office Report 02–071994; Amnesty International, 2004; Barton, 2005; Anglen, 2006).

Clearly, following the final application of the TASER device, Jones was breathing as he continued to resist arrest and as he walked to the ambulance. He was showing no signs of being unable to breathe when paramedics loaded him into the ambulance, and he continued to scream and resist paramedics while en route to the hospital. Consequently, because the TASER pulses did not have an immediate effect, because Jones did not stop breathing until he had been restrained while on his stomach in the ambulance, and because a second pathologist concluded that the TASER device was not a factor, the role of the device as a cause of death and as a significant contributing factor in Jones' death is doubtful.

JOHNNY LOZOYA

AGE: 24
RACE/GENDER: HISPANIC MALE
AGENCY: GARDENA (CALIFORNIA) POLICE
    DEPARTMENT
DATE OF INCIDENT: JULY 19, 2002
CAUSE OF DEATH: HYPOXIC ENCEPHALOPATHY
    FOLLOWING CARDIOPULMONARY ARREST

CONTRIBUTING FACTORS: SEQUELAE OF COCAINE
    INTOXICATION NEEDING RESTRAINT
ROLE OF TASER DEVICE: DOUBTFUL

Residents reported seeing Lozoya running on the roof of a convalescent home. A few minutes later, police received reports that he was running in and out of traffic and jumping on a parked car a few blocks away. Officers found Lozoya in the street, foaming at the mouth, and having an apparent seizure. They called for paramedics, who arrived and transported Lozoya to a local hospital. At the hospital, as medical staff attempted to put soft restraints on Lozoya, he started thrashing his arm about. Medical staff could not secure him, and neither could an officer. Another officer drive-stunned Lozoya one time to get him to stop flailing his arm, but the TASER charge had no effect. Officers eventually pulled Lozoya's arm down and held him while medical staff secured him with soft restraints. Following the restraint, Lozoya's condition worsened, and he died a few minutes later.

The autopsy report shows that Lozoya died of hypoxic encephalopathy, cardiac arrest, and cocaine intoxication, but the medical examiner reported that he could not exclude the TASER device as causing the damage to the heart. He left the manner of death as undetermined (see County of Los Angeles (California) Coroner's Report 2002–05448; Gardena (California) Police Report 02–6515; Amnesty International, 2004; Anglen, 2006).

The medical examiner's inability to exclude the TASER pulses as causing damage to Lozoya's heart cannot be construed to imply that the TASER device did damage Lozoya's heart. It means nothing more than the medical examiner does not have enough evidence to know what caused the heart damage. It is equally likely that the ingestion of sympathomimetic drugs, like cocaine, caused the heart damage. Because Lozoya continued to struggle following the application of the TASER device, and because the medical examiner could not attribute the death to the TASER pulses, the role of the device in Lozoya's death is doubtful.

FREDERICK STEVEN WEBBER

AGE: 44

RACE/GENDER: WHITE MALE
AGENCY: VOLUSIA COUNTY (FLORIDA) SHERIFF'S OFFICE
DATE OF INCIDENT: SEPTEMBER 1, 2002
CAUSE OF DEATH: CARDIAC ARRHYTHMIA DUE TO COCAINE-INDUCED AGITATED DELIRIUM
ROLE OF TASER DEVICE: EXCLUDED

Webber was at a campground when he got into a fight with another camper. Webber's friend came over to break up the fight. He separated the men, but as he was holding Webber down, a woman walked up and kicked Webber in the head. The friend said that Webber's eyes rolled back in his head, and he quit breathing. The friend started cardiopulmonary resuscitation, and Webber started breathing again. Once Webber resuscitated, he became combative again.

As a deputy arrived, the friend was still trying to hold Webber to keep him from fighting. When the friend released him, Webber charged the deputy, who shocked him with a TASER device. Webber continued to fight and ignored the deputy's commands to stop resisting. After a few minutes, another deputy arrived and assisted the first deputy in handcuffing Webber, who was now on his stomach with his hands behind his back. As soon as the deputies had Webber secured, they rolled him over and found that he was not breathing. Paramedics, who had staged nearby, could not revive Webber, and he died at the scene.

The autopsy report was not available, but media sources reported that the coroner ruled Webber's death was an accident caused by cardiac arrhythmia attributed to cocaine-induced agitated delirium with restraint (see Volusia County (Florida) Sheriff's Office Report 02–28810; Barton, 2005; Anglen, 2006).

Webber continued to fight for several minutes following the last application of the TASER device, indicating that the electric current did not have an immediate affect on the rhythm of his heart. Because the TASER pulse had no immediate effect and the medical examiner attributed the cardiac arrhythmia to cocaine-induced agitated delirium, the device is excluded as the cause of death and as a significant contributing factor in the Webber case.

STEPHEN L. EDWARDS

AGE: 59
RACE/GENDER: WHITE MALE
AGENCY: OLYMPIA (WASHINGTON) POLICE DEPARTMENT
DATE OF INCIDENT: NOVEMBER 7, 2002
CAUSE OF DEATH: CARDIAC ARRHYTHMIA
CONTRIBUTING FACTORS: HYPERTENSION, DIABETES AND OBESITY
ROLE OF TASER DEVICE: EXCLUDED

Edwards fought with a store security officer who was trying to get him back into the store. The security officer had observed that Edwards possessed a handgun, and, when the first police officer arrived, the security officer told him that the subject had a handgun. Edwards was trying to get into his van, and he would not obey the officer's commands. The officer then used a TASER device, striking Edwards in the back, and applying four charges. Edwards went to the ground, but continued to try to get to the handgun that he had dropped beneath his van. Other officers arrived and assisted in subduing Edwards, who was handcuffed and on his stomach. When officers turned him over, they realized that he was not breathing.

The autopsy report was exempt from disclosure under Washington law, but media sources reported that the coroner concluded Edwards had died of cardiac arrhythmia due to diabetes and obesity. He added that the TASER device was not a factor (see Olympia (Washington) Police Report 2002–9614; Thurston County (Washington) Sheriff's Office Report 02–45433–12; Amnesty International, 2004; Anglen, 2006).

Edwards continued to resist officers following the last application of the TASER device, indicating that the electric current did not have a significant affect on the rhythm of his heart. Because the TASER pulse did not have an immediate effect and the medical examiner attributed the cause of the arrhythmia to hypertension, diabetes, and obesity, the device is excluded as the cause of death and as a significant contributing factor in the Edwards case.

RONALD EDWARD WRIGHT

AGE: 35
RACE/GENDER: WHITE MALE
AGENCY: ARLINGTON (TEXAS) POLICE DEPARTMENT
DATE OF INCIDENT: DECEMBER 31, 2002
CAUSE OF DEATH: BLUNT TRAUMA
ROLE OF TASER DEVICE: EXCLUDED

Police received several 9-1-1 calls from witnesses who saw Wright running in and out of traffic. When police arrived, he was sitting on a retaining wall atop a highway overpass. Officers and a police negotiator tried to persuade him not to jump. Throughout the incident, Wright repeatedly climbed on and off the railing. Toward the end, he straddled the wall. Wright became agitated after he saw a television news van, and police decided to use a TASER device after negotiations broke down. The shock was not effective, lasting only a moment. After the shock stopped, Wright jumped from the overpass and was declared dead on the scene.

The autopsy report was not available, but media sources reported that the medical examiner ruled Wright's death was a suicide, and he found that Wright died of blunt trauma from the fall (see Arlington (Texas) Police Report 02–97548; Eiserer, 2003). The TASER device is, therefore, excluded as the cause of death and as a significant contributing factor in the Wright case.

CHRISTOPHER SMITH

AGE: 31
RACE/GENDER: WHITE MALE
AGENCY: ALBUQUERQUE (NEW MEXICO) POLICE DEPARTMENT
DATE OF INCIDENT: MARCH 15, 2003
DATE OF DEATH: MARCH 16, 2003
CAUSE OF DEATH: DRUG AND ETHANOL INTOXICATION
ROLE OF TASER DEVICE: UNDETERMINED

A witness flagged down officers and directed them to Smith. The officers found him jumping on parked cars and breaking windows. Smith had cuts on his arms from the breaking glass, so officers called for an ambulance. Smith resisted arrest and fought with the officers, who used chemical spray, a baton, and two charges from a TASER device to subdue him. He lost consciousness and died in the ambulance en route to the hospital.

The autopsy report was not available, but media sources reported that hospital officials determined that Smith had a blood alcohol content of 0.065. They also told police that they found evidence of cocaine, methamphetamine, and marijuana in Smith's system (see Albuquerque (New Mexico) Police Report 03–53426; *Albuquerque Journal*, 2003; Hovey, 2003; Amnesty International, 2004; Anglen, 2006). With so little information on the Smith case, the role of the TASER device in his death is undetermined.

COREY CALVIN CLARK

AGE: 33
RACE/GENDER: WHITE MALE
AGENCY: AMARILLO (TEXAS) POLICE DEPARTMENT
DATE OF INCIDENT: APRIL 16, 2003
CAUSE OF DEATH: ACUTE COCAINE INTOXICATION DUE TO CHRONIC COCAINE ABUSE
CONTRIBUTING FACTORS: GUNSHOT WOUNDS
ROLE OF TASER DEVICE: EXCLUDED

Officers attempted to stop a van that Clark was driving because he was wanted for a string of car thefts. Clark led officers on a chase, which ended when he wrecked the van. When officers ordered Clark out of his car, he refused and reached behind the seat for a knife. When an officer saw Clark with a knife, the officer shot him three times, striking him in the chest and arm. However, Clark continued to reach around in the vehicle as if he were looking for something. The officer activated his TASER device at least three times while trying to get Clark out of the vehicle and secured. When the officers finally succeeded in pulling Clark from the van and handcuffing him, he asked the officer to turn off the TASER device. Clark then began losing consciousness. Paramedics transported Clark to a local hospital, where he died.

According to the autopsy report, Clark died of acute cocaine intoxication due to chronic cocaine abuse. The medical examiner noted that the bullet wounds did not cause the death. He also noted that Clark had a previous history of emergency room visits complaining of chest pains that were the result of cocaine intoxication (see Abbey, 2003; Amarillo (Texas) Police Report 2003–00036037; Texas Tech University Health Sciences Center Autopsy Examination Report FA03–0280; Yates, 2005).

The TASER pulse did not have an immediate effect on Clark, indicating the electric current did not significantly affect the rhythm of his heart. Because the TASER device did not have an immediate effect and the medical examiner attributed the death to cocaine, the device is excluded as the cause of death and as a significant contributing factor in the Clark case.

JOSHUA ALVA HOLLANDER

AGE: 22
RACE/GENDER: WHITE MALE
AGENCY: SAN DIEGO (CALIFORNIA) POLICE DEPARTMENT
DATE OF INCIDENT: MAY 8, 2003
DATE OF DEATH: MAY 10, 2003
CAUSE OF DEATH: ANOXIC/ISCHEMIC ENCEPHALOPATHY DUE TO CARDIAC ARREST DUE TO INCISED WOUNDS OF WRISTS AND RIGHT HAND
CONTRIBUTING FACTORS: UNDER INFLUENCE OR METHAMPHETAMINE, HYPOXIC BRAIN DAMAGE
ROLE OF TASER DEVICE: EXCLUDED

The police report in the Hollander case was exempt from disclosure under California law, but news accounts reported that Hollander entered the apartment of his ex-girlfriend, stabbed her to death, and then slashed his own wrists. Police found him in the bathroom. Despite his wounds, he struggled with police who used a carotid restraint and shocked him twice with a TASER electronic control device. The officers carried him to the courtyard of the apartment where he was talking incoherently. After the officers restrained him, paramedics began treating Hollander, who answered their questions and spoke with them for about 30 minutes. Hollander went into cardiac arrest while in the ambulance en route to a trauma hospital, and he died two days later.

The autopsy report lists the manner of death as suicide. The medical examiner reported that Hollander died of anoxic/ischemic encephalopathy due to cardiac arrest due to the self-inflicted wounds to his wrists. He also noted the presence of amphetamine and methamphetamine in Hollander's system. The medical examiner added that the carotid restraint and the TASER device did not contribute to Hollander's death (see San Diego County (California) Medical Examiner's Report 03–00963; Amnesty International, 2004; Anglen, 2006).

Hollander was conscious and speaking with medical rescue personnel for 30 minutes following the last application of the TASER pulses, indicating that the electric current did not immediately affect the rhythm of his heart. Because the TASER device had no immediate effect and the medical examiner attributed the cardiac arrest to Hollander's self-inflicted wounds, the device is excluded as a cause of death and as a significant contributing factor in the Hollander case.

TIMOTHY ROY SLEET

AGE: 44
RACE/GENDER: AFRICAN-AMERICAN MALE
AGENCY: SPRINGFIELD (MISSOURI) POLICE DEPARTMENT
DATE OF INCIDENT: JUNE 9, 2003
CAUSE OF DEATH: RESTRAINT ASPHYXIA SECONDARY TO PRONE RESTRAINT SYNDROME
CONTRIBUTING FACTORS: PSYCHOSIS DUE TO PCP INTOXICATION
ROLE OF TASER DEVICE: EXCLUDED

Police responded to a 9-1-1 call from a child saying her father was killing her mother. Officers found that Sleet and his girlfriend had been involved in a disturbance. The girlfriend stabbed Sleet in the right shoulder with a kitchen knife. Sleet took the knife away from her and stabbed her to death. When officers arrived, Sleet refused to obey commands. He kept telling the officers that he was God. The officers used a TASER device, eight rounds from a beanbag shotgun, batons, and chemical spray in their attempt to subdue Sleet, but none of those efforts was effective. Finally, the officers piled on top of Sleet to subdue him. Sleet stopped breathing, lost consciousness, and died at the scene.

The medical examiner observed that Sleet's stab wound had struck no vital organs, and the wound was not the cause of his death. He determined that Sleet had died of restraint asphyxia secondary to prone restraint syndrome. He also reported that Sleet was in a state of psychosis due to phencyclidine (PCP) intoxication (see Green County (Missouri) Medical Examiner Report A03–097; Springfield (Missouri) Police Report 03–27280; Anglen, 2006).

Sleet continued to struggle with officers following the application of the TASER pulse, indicating that the electric current did not significantly affect the rhythm of his heart. Because the TASER device did not have an immediate effect and the medical examiner attributed the death to asphyxia, the device is excluded as the cause of death and as a significant contributing factor in the Sleet case.

DAVID SEAN LEWANDOWSKI

AGE: 26
RACE/GENDER: WHITE MALE
AGENCY: ESCAMBIA COUNTY (FLORIDA) SHERIFF'S OFFICE
DATE OF INCIDENT: JUNE 26, 2003
CAUSE OF DEATH: SHARP FORCE INJURY OF FOREARM AND ACUTE PSYCHOSIS
ROLE OF TASER DEVICE: EXCLUDED

Lewandowski went into a convenience store where his sister was a clerk. He swept a bread rack clean of its loaves, sexually harassed an elderly woman, and muttered about how hydrogen and atom bombs were going to kill everyone. A witness called deputies while another witness chased Lewandowski from the store. Lewandowski ran down the street, stripping off his clothes while he ran. He ran in and out of traffic clad only in his underwear, which was falling down. He alternated between screaming and making obscene gestures at passing vehicles to kneeling in the middle of the road, hands cupped and outstretched, apparently in prayer. Two deputies arrived, one a K-9 officer, as Lewandowski ran across the street and down a hill into a clay pit. Lewandowski exited the pit, ran to a nearby house, and ran into the front door, trying to get in. When that did not work, he shattered a window and cut his arm severely on the glass. When the deputies caught up to Lewandowski, they ordered him to get on the ground. When he would not comply, one officer shocked him with a TASER device, but it had no effect. Lewandowski pulled out the probes and ran back into the pit. When the deputies again confronted Lewandowski in the pit, one officer shot him, striking him in the arm. The deputies eventually subdued Lewandowski and transported him to a local hospital, where he died.

The medical examiner concluded that Lewandowski died from loss of blood from the cut on his forearm that he received when he broke the window. Neither the gunshot wound to his upper arm nor the TASER device were contributing factors (see Florida Department of Law Enforcement Report PE–27–0018; Norman, 2003; Office of the Medical Examiner, First District (Florida) Report MLA03–446; Smith, 2004).

Lewandowski continued to struggle with officers following the application of the TASER pulse, indicating that the electric current did not significantly affect the rhythm of his heart. Because the TASER device did not have an immediate effect and the medical examiner attributed the death to loss of blood from the cut on his arm, the device is excluded as the cause of death and as a significant contributing factor in the Lewandowski case.

TROY DALE NOWELL

AGE: 51
RACE/GENDER: WHITE MALE
AGENCY: AMARILLO (TEXAS) POLICE DEPARTMENT
DATE OF INCIDENT: August 4, 2003

CAUSE OF DEATH: CARDIOPULMONARY ARREST DURING VIOLENT PHYSICAL STRUGGLE WITH RESTRAINT IN AN INDIVIDUAL WITH PROBABLE ENDOGENOUS MENTAL DISEASE
CONTRIBUTING FACTORS: ARTERIOSCLEROTIC AND HYPERTENSIVE HEART DISEASE
ROLE OF TASER DEVICE: EXCLUDED

Police said Nowell, who had a history of psychiatric disorder, left his home one morning, and police subsequently escorted him to a local psychiatric center. Personnel at the center released him later that day. Shortly thereafter, Nowell arrived at a local hospital seeking treatment for blisters on his feet. Medical personnel treated and released him. Later that same day, Nowell appeared at a carpenter's union hall requesting a drink of water. He began making strange statements that the police were after him and that his wife had his son killed. He then followed two elderly women and a man out of the hall, jumped into their vehicle, and assaulted them. The man placed Nowell in a headlock and held him until police could arrive. When police arrived, Nowell refused to obey the officers' commands. When Nowell rushed the officers, one officer struck him repeatedly with a baton. Another officer tried to drive-stun him with a TASER device, but the drivestuns were not effective. He then deployed the probes. Once officers had Nowell secured, they called for an ambulance. Before the ambulance could arrive, Nowell stopped breathing.

The medical examiner reported the cause of death to be cardiopulmonary arrest during a violent struggle with restraint in an individual with probable endogenous mental disease. He listed a contributory cause of death as arteriosclerotic and hypertensive heart disease (see Amarillo (Texas) Police Report 2003-00080149; Texas Tech University Health Sciences Center Autopsy Examination Report FA03-0590; Amnesty International, 2004; Anglen, 2006; Yates, 2006).

Nowell continued to struggle with police following the application of the TASER pulse, indicating that the electric current did not significantly affect the rhythm of his heart. Because the TASER device did not have an immediate effect, because the medical examiner attributed the cardiopulmonary arrest to the stress of the struggle, and because the medical examiner listed arteriosclerotic and hypertensive heart disease instead of the TASER device as contributing factors, the device is excluded as the cause of death and as a significant contributing factor in the Nowell case.

JOHN LEE THOMPSON

AGE: 45
RACE/GENDER: AFRICAN-AMERICAN MALE
AGENCY: CARROLLTON TOWNSHIP (MICHIGAN) POLICE DEPARTMENT
ZILWAUKEE (MICHIGAN) POLICE DEPARTMENT
SAGINAW COUNTY (MICHIGAN) SHERIFF'S OFFICE
DATE OF INCIDENT: AUGUST 6, 2003
DATE OF DEATH: AUGUST 8, 2003
CAUSE OF DEATH: METABOLIC ACIDOSIS
CONTRIBUTING FACTORS: ARTERIOSCHLEROTIC HEART DISEASE
ROLE OF TASER DEVICE: EXCLUDED

Thompson became violent while playing a card game with friends, who called the police. When officers arrived, Thompson threw a lamp at them. Officers shocked Thompson several times with a TASER device. They took him to jail where he continued to struggle with officers. Later, in an isolation cell, Thompson became unresponsive. He was taken to a hospital, where he died two days later.

A medical examiner said that Thompson's death was not a result of physical force, and he listed the mode of death as unknown. He concluded that Thompson died of metabolic acidosis with complications. The medical examiner also found that a contributing factor was arteriosclerotic heart disease. He added that the TASER device did not produce a lethal effect on Thompson. Later, in a product liability lawsuit filed on behalf of Thompson's daughter, the trial judge dismissed the suit ruling that the plaintiff failed to prove the TASER pulses had caused Thompson's metabolic acidosis (see Saginaw County (Michigan) Medical Examiner's

Report S03–106; Saginaw County (Michigan) Sheriff's Office Report 173–0004792–03; Anglen, 2006; *Associated Press*, 2006, June 8).

Thompson continued to struggle with officers following the application of the TASER pulses, indicating that the electric current did not significantly affect the rhythm of his heart. Because the TASER device did not have an immediate effect, the medical examiner attributed the death metabolic acidosis, the medical examiner excluded the TASER device as a contributing cause, and a trial court ruled that there was not sufficient evidence to prove that the TASER pulses caused the acidosis, the device is excluded as the cause of death and as a significant contributing factor in the Thompson case.

WALTER CURTIS BURKS, JR.

AGE: 36
RACE/GENDER: AFRICAN-AMERICAN MALE
AGENCY: MINNEAPOLIS (MINNESOTA) POLICE DEPARTMENT
DATE OF INCIDENT: AUGUST 7, 2003
DATE OF DEATH: AUGUST 8, 2003
CAUSE OF DEATH: EXCITED DELIRIUM WITH COCAINE USE
CONTRIBUTING FACTORS: HEART DISEASE, PULMONARY EMPHYSEMA
ROLE OF TASER DEVICE: EXCLUDED

Neither the police incident report nor the autopsy reports regarding Burks were available under Minnesota law. However, news reports give the following account of Burks' death. Burks had gained release from a Minnesota prison on Monday, August 5, 2003, having been in prison about one year. On Wednesday, August 7, just before midnight, Minneapolis police received a call of a disturbance at a convenience store. When officers arrived, they found Burks, who was uncooperative and assaultive. The officers used chemical spray and two shocks from a TASER device to subdue Burks. Because of his behavior, officers decided to take him to a crisis center at a local hospital rather than to jail or to a detoxification center. Burks collapsed and died shortly before 1 A.M. August 8, before the crisis center could admit him.

Media sources reported that the coroner listed Burks' cause of death as excited delirium associated with cocaine use. He also listed heart disease, which the media did not define, and pulmonary emphysema as contributing causes (see Chanen, 2003; Duchschere, 2003; Amnesty International, 2004; McKinney, 2004).

Burks did not collapse until about one hour following the application of the TASER pulses, indicating that the electric current did not significantly affect the rhythm of his heart. Because the TASER device did not have an immediate effect and the coroner attributed the death to cocaine-induced excited delirium, the device is excluded as the cause of death and as significant contributing factor in the Burks case.

GORDON BENJAMIN RAUCH

AGE: 39
RACE/GENDER: WHITE MALE
AGENCY: SACRAMENTO COUNTY (CALIFORNIA) SHERIFF'S OFFICE
DATE OF INCIDENT: AUGUST 17, 2003
CAUSE OF DEATH: SUDDEN CARDIAC DEATH DURING STRUGGLE WITH POLICE DUE TO EXCITED DELIRIUM DUE TO BIPOLAR PSYCHIATRIC DISORDER
CONTRIBUTING FACTORS: HYPERTENSIVE CARDIOVASCULAR DISEASE, MORBID OBESITY

The Sheriff's Office report in the Rauch case was exempt from disclosure under California law, but media sources reported that Rauch had a history of violent behavior. Rauch's father called police to report that his son was threatening to kill him. The younger Rauch had been awake for three days, and he had been exhibiting bizarre behavior. He had attended the state fair with his parents, but they had to leave because the younger Rauch would not stop shouting obscenities at people. When they returned home, Rauch went for a swim in a nearby lake, still wearing his clothes. He then went back to the house and began breaking things. He smashed a television and hit his father in the face with a broom handle. When deputies arrived, they found Rauch dressed in a leather coat, sweating profusely. The bedroom in which they found Rauch was strewn with broken glass. When one of the officers attempted to speak with him, Rauch hit him. Two officers fired TASER devices at Rauch, but they had no effect. Officer forced Rauch to the ground and handcuffed him. A few minutes later, the officers noticed that Rauch was not breathing. They initiated CPR and called for medical assistance. Fire department rescue personnel took Rauch to a local hospital, where he died about one hour later.

The coroner found that Rauch suffered several minor abrasions and contusions, but none that were life threatening, and he noted that Rauch was displaying symptoms of hyperthermia and acute psychosis. He reported that Rauch's psychiatric disorder and medical problems associated with his morbid obesity were the underlying cause of his death. The only drug in Rauch's system was Loxapine, a typical antipsychotic medication used primarily in the treatment of schizophrenia. The coroner ruled that Rauch died of excited delirium (see Sacramento County (California) Coroner Report 03–03978; Hume, 2004; Anglen, 2006).

The media accounts do not specify how much time passed between the last application of the TASER device and the time Rauch collapsed, but he continued to struggle with officers after application of the TASER pulses, indicating that the electric current did not significantly affect the rhythm of his heart. Because the TASER device did not have an immediate effect, the coroner attributed the death to excited delirium, and the coroner cited hypertensive cardiovascular disease and morbid obesity as contributing factors, but not the TASER device, the device is excluded as the cause of death and as a significant contributing factor in the Rauch case.

RAY CHARLES AUSTIN

AGE: 25
RACE/GENDER: AFRICAN-AMERICAN MALE
AGENCY: GWINNETT COUNTY (GEORGIA) SHERIFF'S OFFICE
DATE OF INCIDENT: SEPTEMBER 24, 2003
DATE OF DEATH: SEPTEMBER 26, 2003
CAUSE OF DEATH: HEART ATTACK, CAUSE NOT CLEARLY DETERMINED

ROLE OF TASER DEVICE: EXCLUDED

The Sheriff's Office reports on this incident were not available, but media sources reported that deputies went to Austin's residence to arrest him for a probation violation on September 3, 2003. The deputies reported a bizarre scene after Austin, who had a history of mental illness, let them into his apartment. He would not respond to their commands, and he kept referring to the urine that he was cooking in the kitchen. The officers found about 20 knives on the kitchen counter and a hammer in the living room. Austin began walking toward the kitchen and disregarded orders to stop. A deputy drive stunned Austin on the right arm with a TASER device, but it had no effect. Officers were able to secure Austin and transport him to jail.

Austin was a disciplinary problem while in jail. On at least two occasions, deputies used TASER devices to control him. On September 24, 2003, while incarcerated and awaiting trial on a parole violation, Austin exhibited disruptive behavior, beating on things in his room. Deputies approached Austin to secure him so the nurse could administer an injection to calm him. When the nurse entered his room, Austin began to resist. He scuffled with deputies, biting off a portion of one deputy's ear. Deputies shocked Austin three times with a TASER device and struck him with fists to subdue him. Once officers had secured Austin, they placed him in a restraining chair, and the nurse administered psychotropic drugs. He lost consciousness and was transported to a local hospital. He died two days later.

The autopsy report was not available, but news sources reported that the autopsy revealed Austin died of a heart attack of undetermined cause. The coroner reported that the physical restraint might have impaired his breathing (see Mungin, 2003; Amnesty International, 2004; Mungin, 2004, September 9; Anglen, 2006).

Austin continued to struggle with jailers following the last application of the TASER pulses, indicating that the electric current did not significantly affect the rhythm of his heart. He did not collapse until after jailers had placed him in the restraint chair and the jail nurse had administered an injection to sedate him. Because the TASER device did not have an immediate effect, and because the coroner attributed the death to factors unknown, speculating that the physical restraint may have contributed but not speculating on the TASER device, the device is excluded as the cause of death and as a contributing factor in Austin's death.

GLENN RICHARD LEYBA

AGE: 37
RACE/GENDER: WHITE MALE
AGENCY: GLENDALE (COLORADO) POLICE DEPARTMENT
DATE OF INCIDENT: SEPTEMBER 29, 2003
CAUSE OF DEATH: CARDIAC ARREST DURING COCAINE-INDUCED AGITATED DELIRIUM NECESSITATING RESTRAINT
CONTRIBUTING FACTORS: PULMONARY EDEMA, MILD ATHEROSCLEROSIS
ROLE OF TASER DEVICE: EXCLUDED

The police report for the Leyba case was exempt from disclosure under Colorado law, but media sources reported that a witness called 9-1-1 and told dispatchers that Leyba was going crazy. Firefighters who arrived at the scene, called police and told them that Leyba was out of control. Leyba was screaming, flailing, and suffering seizure-like symptoms when police arrived at his condominium. When Leyba refused medical treatment, a police officer shocked him with a TASER device. Police said he was on the ground, kicking and thrashing at officers, who shocked Leyba at least five times. When he stopped breathing, paramedics tried to resuscitate Leyba, but he was dead on arrival at a local hospital.

The coroner concluded that the TASER device did not contribute to Leyba's death. The autopsy report listed the cause of death as cardiac arrest during cocaine-induced agitated delirium (see Arapahoe County (Colorado) Coroner's Report 03A278; *Associated Press*, 2003; Huntley, 2003, October 1; Huntley, 2003, October 28; Amnesty International, 2004; Anglen, 2006).

Leyba did collapse immediately following the application of the TASER pulse. However, the coroner found sufficient forensic evidence to attribute the cardiac arrest to cocaine-induced excited delirium. He cited pulmonary edema and atherosclerosis as contributing factors, but not the TASER device. Consequently, the device is excluded as the cause of death and as a significant contributing factor.

ROMAN GALLIUS PIERSON

AGE: 40
RACE/GENDER: WHITE MALE
AGENCY: BREA (CALIFORNIA) POLICE DEPARTMENT
DATE OF INCIDENT: OCTOBER 7, 2003
CAUSE OF DEATH: CARDIAC ARREST DUE TO ACUTE METHAMPHETAMINE INTOXICATION
CONTRIBUTING FACTORS: HISTORY OF CORONARY ARTERY DISEASE
ROLE OF TASER DEVICE: EXCLUDED

The police report in the Pierson case is exempt from disclosure under California law, but media sources reported that police received a report of a reckless driver, who they later identified as Pierson. As officers responded, they got additional reports that a man dressed in a black leather jacket, black pants and wearing a helmet was running in and out of traffic. The man, Pierson, scaled a fence and ran through the parking lot of a supermarket. Pierson then entered a service station, where witnesses said he was acting strangely and complaining that he was hot and thirsty. He ran out of the station and across the parking lot to the supermarket, then ran out and returned to the station, again complaining that he was hot. Pierson broke into an ice machine in front of the supermarket and rubbed ice on his face. When officers arrived, Pierson ignored orders to lie down and took a fighting stance. An officer shocked him once with a TASER electronic control device, and then shocked him again when he began grappling with police. The officers handcuffed Pierson, but after about a minute, they noticed that he had stopped breathing. Paramedics treated Pierson and took him to a local hospital, where he died.

The autopsy report was not available, but media sources reported that the coroner listed Pierson's cause of death as cardiac arrest due to acute methamphetamine intoxication and that he noted Pierson's history of coronary artery disease (see *City News Service*, 2003; Amnesty International, 2004; Anglen 2006).

Because the TASER pulses did not have an immediate effect and the coroner discovered enough forensic evidence to attribute the cardiac arrest to methamphetamine intoxication, the TASER device is excluded as the cause of death and as a significant contributing factor in the Pierson case.

DENNIS D. HAMMOND

AGE: 31
RACE/GENDER: WHITE MALE
AGENCY: OKLAHOMA CITY (OKLAHOMA) POLICE DEPARTMENT
DATE OF INCIDENT: OCTOBER 11, 2003
CAUSE OF DEATH: ACUTE METHAMPHETAMINE INTOXICATION
ROLE OF TASER DEVICE: EXCLUDED

Hammond was walking down the street screaming at the sky, and witnesses called 9-1-1. When police arrived, they found him perched on a brick mailbox, and when they approached, Hammond screamed at them. The officers saw that Hammond was bleeding from wounds on his feet, so they called for medical assistance. When assistance arrived, Hammond refused to get off the mailbox, and he resisted the officers' attempts to get him down. During the struggle that ensued, officers shot Hammond three times with a beanbag shotgun and five times with a TASER device. After the officers handcuffed Hammond, paramedics began bandaging wounds to his calves and soles of his feet. As they treated him, Hammond stopped breathing. Paramedics transported him to a local hospital. Upon arrival, Hammond had a pulse, but he was not breathing on his own. Medical staff placed him on a ventilator, but about five minutes later, he died.

The medical examiner found several abrasions, cuts, and contusions, none of which were life threatening. The postmortem toxicology showed a high level of methamphetamine in Hammond's system. The medical examiner reported that Hammond's mental and behavioral agitation was likely a result of acute, agitated delirium due to methamphetamine toxicity. He said that the beanbags and TASER shocks, while significant, did not have an immediate role in Hammond's death (see Office of the Chief Medical Examiner (Oklahoma) Report 0305864; Oklahoma City (Oklahoma) Police Report 03–102727; Amnesty International, 2004; Anglen, 2006).

Hammond did not collapse immediately following the application of the TASER pulses, indicating that the electric current did not significantly affect the rhythm of his heart. Because the TASER shocks did not have an immediate effect and because the coroner attributed the death to methamphetamine intoxication, the TASER device is excluded as the cause of death and as a significant contributing factor in the Hammond case.

LOUIS N. MORRIS, JR.

AGE: 50
RACE/GENDER: WHITE MALE
AGENCY: ORANGE COUNTY (FLORIDA) SHERIFF'S OFFICE
DATE OF INCIDENT: OCTOBER 21, 2003
CAUSE OF DEATH: COCAINE-INDUCED EXCITED DELIRIUM, CARDIAC ARRHYTHMIA DUE TO RESTRAINT
CONTRIBUTING FACTORS: ATHEROSCHLEROTIC CORONARY ARTERY DISEASE
ROLE OF TASER DEVICE: EXCLUDED

Morris was driving a van erratically through the parking lot of a supermarket. When store security officers approached, Morris said a passenger in the van needed medical attention, but nobody else was in the van. He went into the supermarket and started yelling that someone was breaking into his van, prompting a call to the police. He accused several customers in the store of being responsible for the burglary. When officers arrived, Morris fled to a nearby convenience store. Trying to take him into custody for a mental health evaluation, officers shocked Morris twice with a TASER device. After police handcuffed him and hobbled his legs, Morris began banging his head on the ground. When officers turned him over, they saw he was in distress. The officers called for paramedics who treated Morris at the scene and transported him to a local hospital, where he died.

The autopsy report was not available, but media sources reported that the coroner listed the cause of death as cocaine-induced excited delirium and cardiac arrhythmia due to restraint. The coroner also noted that Morris had atherosclerotic coronary artery disease that contributed to his death (see Orange County (Florida) Sheriff's Office Report 03-096603; Amnesty International, 2004; Barton, 2005; Anglen, 2006).

Morris continued to struggle after the final application of the TASER pulses, indicating that the electric current did not immediately affect the rhythm of his heart. Because the TASER shocks did not have an immediate effect, the coroner attributed the cardiac arrhythmia to cocaine-induced excited delirium, and the coroner cited atherosclerotic coronary artery disease, but not the TASER device, as a contributing factor, the device is excluded as the cause of death and as a significant contributing factor in the Morris case.

JAMES LEE BORDEN

AGE: 47
RACE/GENDER: WHITE MALE
AGENCY: MONROE COUNTY (INDIANA) SHERIFF'S OFFICE
DATE OF INCIDENT: NOVEMBER 6, 2003
CAUSE OF DEATH: CARDIAC DYSRHYTHMIA SECONDARY TO HYPERTROPHIC CARDIOMYOPATHY, PHARMACOLOGIC INTOXICATION, AND ELECTRIC SHOCK
CONTRIBUTING FACTORS: PREEXISTING HEART CONDITION, DIABETES
ROLE OF TASER DEVICE: EXCLUDED

The police and autopsy reports on this incident were not available, but media sources reported that on the eve of his father's funeral, deputies of the Lawrence County Sheriff's Office arrested Borden for a house arrest violation. Reportedly, he had been causing problems all day at the hotel where he was staying. The deputies had paramedics check Borden before they took him to the Monroe County jail. At the jail, officers reported that Borden had been disoriented and thrashing around in the squad car, and he would not follow commands of jailers. A jailer shocked Borden at least three times with a TASER device because he was resisting jailers' efforts to book him. Three jailers held Borden down while they pulled off his shoes and prepared him to go into a cell. Jailers noticed that Borden had stopped breathing, and they began CPR. Paramedics arrived and transported Borden to a local hospital, where he died. The Indiana State Police conducted the investigation.

News sources reported that the original autopsy report listed the cause of death as cardiac dysrhythmia secondary to hypertrophic cardiomyopathy and electric shock from the TASER device. A forensic pathologist, hired by a jailer who was facing criminal charges in this incident, disputed the original autopsy report. The pathologist concluded that, because the TASER probes had struck Borden in the lower abdomen and buttocks, that the TASER effect could not have penetrated the fatty tissues sufficiently to cause dysrhythmia. He concluded that Borden's death was directly attributable to his underlying cardiovascular pathology, specifically, cardiac dysrhythmia secondary to hypertrophic cardiomyopathy and pharmacological intoxication.

During a later product liability civil trial, the original coroner admitted that, at the time of the autopsy, he knew nothing of TASER technology. He also admitted that the TASER shock did not directly affect Borden's heart. He stated that other factors, only tangentially included in the original autopsy report, contributed to Borden's death: the stress of his father's death, the stress of his arrest, obesity, a lethal overdose of ephedrine, and an overdose of Promethazine.

The jailer who shocked Borden was charged with two counts of felony battery, including battery while armed with a deadly weapon. He later pleaded guilty to a felony charge of criminal recklessness and received a one and a half-year suspended sentence (see Amnesty International, 2004; *Associated*

Press, 2004, July 23; *PR Newswire US*, 2004; *PR Newswire US*, 2005; Anglen, 2006; Nolan, 2006).

Because the first medical examiner reconsidered his original diagnosis and acknowledged that the TASER shock had not directly affected Borden's heart, and he ultimately concluded that the physiological stress of the restraint, the stress of his father's death, his obesity, and overdoses of ephedrine and Promethazine were contributing factors to Borden's death, and because the second pathologist concluded that the TASER shocks could not have been the cause of death, the TASER device is excluded as the cause of death and as a significant contributing factor in the Borden case.

MICHAEL SHARP JOHNSON

AGE: 32
RACE/GENDER: AFRICAN-AMERICAN MALE
AGENCY: OKLAHOMA CITY (OKLAHOMA) POLICE DEPARTMENT
DATE OF INCIDENT: NOVEMBER 9, 2003
DATE OF DEATH: NOVEMBER 10, 2003
CAUSE OF DEATH: ACUTE CONGESTIVE HEART FAILURE DUE TO COCAINE-INDUCED CARDIAC ARREST
CONTRIBUTING FACTORS: DRUG ABUSE (COCAINE); AGITATED DELIRIUM, PHYSICAL EXERTION
ROLE OF TASER DEVICE: EXCLUDED

Officers responded to a burglary call and found Johnson sitting in a chair, looking exhausted. When Johnson did not respond to their commands, officers shocked him with a TASER device. Johnson began struggling with officers after being shocked. Officers shocked Johnson at least five times. About two minutes later, Johnson stopped breathing and went into cardiac arrest. Emergency medical personnel took Johnson to a local hospital. When he arrived, he was in full cardiopulmonary arrest. Medical staff placed him on a ventilator and moved him to the intensive care unit with multisystem failure. He died about 22 hours later. A drug screen at that hospital was positive for cocaine and marijuana.

The medical examiner reported that the autopsy revealed acute congestive heart failure with diffuse cerebral edema, acute pulmonary edema, and early aspiration pneumonia. The toxicology results indicated a high level of cocaine and cocaine metabolites in Johnson's system. The medical examiner ruled that Johnson' death was the result of acute congestive heart failure due to cocaine-induced cardiac arrest. He reported that the sudden cardiopulmonary arrest during or shortly after a struggle and restraint procedures was not likely caused by restraint or trauma, but it was likely caused by the effects of cocaine abuse (see Office of the Chief Medical Examiner (Oklahoma) Report 0306462; Oklahoma City (Oklahoma) Police Report 03–11117; Amnesty International, 2004; Anglen, 2006).

Because the TASER shocks did not have an immediate effect and the medical examiner attributed the cardiac arrest to cocaine intoxication, the TASER device is excluded as the cause of death and as a contributing factor in the Johnson case.

KERRY KEVIN O'BRIEN

AGE: 31
RACE/GENDER: WHITE MALE
AGENCY: PEMBROKE PINES (FLORIDA) POLICE DEPARTMENT
DATE OF INCIDENT: NOVEMBER 11, 2003
CAUSE OF DEATH: POSITIONAL ASPHYXIA
CONTRIBUTING FACTORS: EXCITED DELIRIUM, HISTORY OF ASTHMA AND DEPRESSION
ROLE OF TASER DEVICE: EXCLUDED

The police received a call complaining of O'Brien running through an intersection, yelling and pounding on cars. When officers arrived, they found O'Brien lying on the sidewalk. As they approached, O'Brien got up and lunged at one of the officers, who shocked him one time with a TASER device. The shocks had no effect, and O'Brien ran off with officers in pursuit. When the officers caught O'Brien, they pushed him to the ground, handcuffed him, and hogtied his legs. Rescue personnel, who the officers had called to the scene, administered Haldol, an antipsychotic drug.

When the rescue personnel lifted O'Brien off the ground to put him on the stretcher, they realized that he was not breathing.

The autopsy report was not available, but media sources reported that the coroner determined O'Brien died from being hogtied, reporting that he was a victim of restraint asphyxia. The coroner ruled the death as accidental. The autopsy report noted that O'Brien was obese and had a history of psychotic behavior. The coroner also concluded that the TASER pulses did not contribute to O'Brien's death (see Pembroke Pines (Florida) Police Report 2003–092431; Amnesty International, 2004; Barton, 2005; Anglen, 2006).

O'Brien was able to run off after the application of the TASER shocks, and he continued to struggle with officers when they tackled him. Because the TASER device did not have an immediate effect and the medical examiner attributed the death to asphyxiation, the device is excluded as the cause of death and as a contributing factor in the O'Brien case.

LEWIS SANKS KING

AGE: 39
RACE/GENDER: AFRICAN-AMERICAN MALE
AGENCY: ST. JOHNS COUNTY (FLORIDA) SHERIFF'S OFFICE
DATE OF INCIDENT: DECEMBER 9, 2003
CAUSE OF DEATH: HYPERTROPHIC AND ISCHEMIC CARDIOMYOPATHY
ROLE OF TASER DEVICE: EXCLUDED

Deputies made a routine traffic stop on King for a defective brake light. King gave a deputy his license and registration, but as the deputy went to his vehicle to write a citation, King fled in his vehicle, dragging another deputy down the street. The deputies pursued King's vehicle until it crashed, and King ran off into the woods. The deputies located King a few minutes later, lying in the grass near the edge of the woods. When deputies approached, King got up, but he refused to obey their commands. One deputy fired a TASER device at King, who then fell to the ground. King still would not comply with orders, so another deputy fired a TASER device. However, a probe pulled off, so the charge failed. The deputy replaced the cartridge, but the second cartridge would not fire. Deputies finally subdued King by bodily force, holding him to the ground, handcuffing him and putting him in hobbles. Deputies would later find a small packet of cocaine in King's vehicle. Due to the wreck and the application of the TASER devices, deputies called for an ambulance. Rescue personnel transported King to a local hospital, where he died about one hour later.

The autopsy report was not available, but media sources reported that preliminary autopsy findings showed King had severe heart disease and a massively enlarged heart. He suffered cardiac arrest during restraint. The medical examiner also considered the possibility that King died of restraint asphyxia. In the final autopsy report, according to the Sheriff's Office incident report, the medical examiner stated the cause of death as hypertrophic and ischemic cardiomyopathy (see St. John's County (Florida) Sheriff's Office Report 03–343014; Sundlin, 2003, December 12; Sundlin, 2003, December 24; Amnesty International, 2004; Barton, 2005; Anglen, 2006).

After the application of the TASER shocks, King continued to struggle with deputies while they held him down, handcuffed him, and applied leg restraints. When medical personnel arrived, he still had a normal heartbeat. He did not collapse until he got to the hospital. King did not collapse immediately following the application of the TASER pulses, indicating the electric current had no immediate affect on the rhythm of his enlarged heart. Because the TASER device did not have an immediate effect and the medical examiner attributed the death to preexisting heart disease, the device is excluded as the cause of death and as a contributing factor in the King case.

CURTIS LAMAR LAWSON

AGE: 40
RACE/GENDER: AFRICAN-AMERICAN MALE
AGENCY: HOUSTON COUNTY (GEORGIA) SHERIFF'S OFFICE
DATE OF INCIDENT: DECEMBER 9, 2003
CAUSE OF DEATH: COMPLICATIONS OF ACUTE

COCAINE TOXICITY
CONTRIBUTING FACTORS: CARDIOMEGALY
ROLE OF TASER DEVICE: EXCLUDED

Lawson confronted a woman as she was pumping gas. Barefoot and clad only in a T-shirt and shorts, he tried to make the woman leave with him, but she ran to another convenience store. Lawson chased her into the store, where employees attempted to keep Lawson from grabbing her. One of the clerks called 9-1-1, and Warner Robins police were dispatched. Meanwhile, a Peach County Sheriff's deputy, on routine patrol, pulled into the parking lot. When Lawson saw the deputy, he ran across the street. A few minutes later 9-1-1 got a call from a man saying that he was in a room of a hotel across the street from the store. It is unclear whether Lawson or someone else made the call. The man asked the operator to "get me some police now."

When police arrived at the hotel, Lawson was outside the room. He ran back inside and refused to come out. When officers forced open the door, Lawson ran to the bathroom. Unable to restrain the suspect with chemical spray, Peach County deputies called for assistance from the Houston County Sheriff's Office for the use of a TASER electronic control device. A Houston County deputy arrived with a TASER device, and he fired it at Lawson when Lawson charged him. The first attempt missed, so the deputy made a second attempt, which shocked and stunned Lawson, though he continued to fight with the officers. After about 15 minutes, while officers were waiting for a transport van, Lawson suddenly ceased fighting, leading the officers to ask if he was okay. When Lawson did not respond, officers checked his pulse and discovered he was not breathing. The officers started CPR and called for EMS, but they could not revive Lawson. Paramedics transported Lawson to a local hospital, where he died.

The Georgia Bureau of Investigation, which conducted the investigation, concluded that Lawson died from complications of acute cocaine toxicity. The autopsy report added that cardiomegaly was a contributing factor (see *Cordele Dispatch*, 2003; Georgia Bureau of Investigation Record of Medical Examiner Report 2003–4005152; Amnesty International, 2004; *Associated Press*, 2004, March 8; Mungin & Bentley, 2004; Anglen, 2006).

Lawson continued to struggle for fifteen minutes following the application of the TASER shocks, indicating that the electric current did not significantly affect his heart. Because the TASER pulses did not have an immediate effect, the medical examiner attributed the death to cocaine toxicity, and the medical examiner listed Lawson's cardiomegaly, but not the TASER device, as a contributing factor, the device is excluded as the cause of death and as a contributing factor in the Lawson case.

DAVID GLOWCZENSKI

AGE: 35
RACE/GENDER: WHITE MALE
AGENCY: SOUTHAMPTON VILLAGE (NEW YORK) POLICE DEPARTMENT
DATE OF INCIDENT: FEBRUARY 4, 2004
CAUSE OF DEATH: EXHAUSTIVE MANIA DUE TO SCHIZOPHRENIA
ROLE OF TASER DEVICE: UNDETERMINED

Neither the Southampton Village Police Department nor the Suffolk County Medical Examiner's Office responded to requests for records regarding this incident. However, media sources reported that Glowczenski began hearing voices, started screaming incoherently, and stalked out of his home. His family called 9-1-1 to ask for help. Officers found Glowczenski and attempted to calm him, but Glowczenski resisted and threw an officer to the ground. A struggle ensued in which the officers used chemical mace and shocked Glowczenski with a TASER device nine times. The officers eventually handcuffed Glowczenski, but, a short time later, he stopped breathing. Officers called for an ambulance and attempted to revive him. Paramedics transported Glowczenski to a local hospital, where he died.

News accounts reported that a preliminary autopsy was unable to determine the cause of death, but the medical examiner later said that Glowczenski died from exhaustive mania due to schizophrenia. Another pathologist, hired by Glowczenski's family, claimed that the cause of death was excessive use of force by the police. That pathologist noted that Glowczenski was bleeding from his right testicle and his back, and his knees were rubbed raw (see Amnesty International, 2004; Healy, 2004; Tavernise, 2004; Williams, 2004; Anglen, 2006).

The media accounts do not specify how much time passed between the application of the TASER shocks and when officers noticed that Glowczenski had stopped breathing. The medical examiner for Suffolk County attributed the death to exhaustive mania, and, although the specific cause of death from the second pathologist was not available through media reports, the family did file a lawsuit against the Southampton Village Police Department and TASER International, indicating that they believed the TASER device was either the cause of death or was a significant contributing factor. With no more information than is available in the media accounts, the role of the TASER device in the Glowczenski case is undetermined.

RAYMOND SIEGLER

AGE: 40
RACE/GENDER: WHITE MALE
AGENCY: MINNEAPOLIS (MINNESOTA) POLICE DEPARTMENT
DATE OF INCIDENT: FEBRUARY 6, 2004
DATE OF DEATH: FEBRUARY 13, 2004
CAUSE OF DEATH: CARDIAC ARREST
CONTRIBUTING FACTORS: HYPERTENSION AND HEART DISEASE
ROLE OF TASER DEVICE: EXCLUDED

Siegler, who lived in a group home for the mentally ill, was schizophrenic and slightly overweight. His family said that he had stopped taking his medication, and he had become argumentative. While celebrating his engagement, Siegler consumed some alcohol and created a disturbance by threatening people at the group home. When officers arrived, they shot a TASER cartridge at Siegler, hitting him in the neck and chest. He collapsed shortly afterward and went into cardiac arrest. He never regained consciousness, and he died seven days later.

Under Minnesota law, autopsy reports are not available for 30 years. However, news accounts reported that the medical examiner listed a number of causes of death, including preexisting conditions of hypertension and heart disease. According to those news reports, nowhere in the autopsy report did the coroner mention the TASER device or electric shock as a cause or as a contributing factor in Siegler's death (see Amnesty International, 2004; Chanen, 2004; McKinney, 2004; Minneapolis (Minnesota) Police Report MP-04-028424; Anglen, 2006).

Media accounts and that part of the police report that was public record do not detail how much time passed between the application of the TASER shocks and the time that Siegler collapsed. However, Siegler had preexisting heart disease, and the medical examiner did not mention the TASER device. Consequently, the device is excluded as the cause of death and as a contributing factor in the Siegler case.

WILLIAM D. LOMAX, JR.

AGE: 26
RACE/GENDER: AFRICAN-AMERICAN MALE
AGENCY: LAS VEGAS METROPOLITAN (NEVADA) POLICE DEPARTMENT
DATE OF INCIDENT: FEBRUARY 20, 2004
CAUSE OF DEATH: CARDIAC ARREST RESULTING FROM RESTRAINING PROCEDURES WHILE INTOXICATED
CONTRIBUTING FACTORS: TASER DEVICE, OBESITY, EARLY STAGES OF PNEUMONIA, AND PHENCYCLIDINE (PCP) INTOXICATION
ROLE OF TASER DEVICE: DOUBTFUL/POSSIBLE

The police and autopsy reports in this incident were not available under Nevada law, but media sources reported that Lomax was acting erratically when security guards at an apartment complex asked whether he needed medical assistance. The guards tried to subdue Lomax because they believed he might hurt himself or others, but they could not handcuff him because he was too strong and combative. When paramedics arrive, Lomax began fighting with them. When an officer arrived, he repeatedly warned Lomax that he was going to stun Lomax with a TASER device if he did not stop fighting. When Lomax grabbed one of the security guards, the officer shocked Lomax on the neck. After the shock wore off, Lomax became combative again. Following a second shock, the security officers handcuffed Lomax, but he continued to struggle. During the next 20 minutes, the officer shocked Lomax seven times, for between two seconds and eight seconds each time. The officers called for an ambulance for Lomax, but he stopped breathing and his heart stopped during transport to a local hospital. Paramedics revived him, and he was breathing on his own when they arrived at the hospital. Lomax, who spent some of the ensuing 20 hours hooked to ventilators, died the next day.

Media sources reported that the medical examiner who performed an autopsy on Lomax listed the cause of death as cardiac arrest during restraint procedures. He said the TASER device was a contributing factor in the restraint used by the police and the guards, but he added that phencyclidine (PCP) intoxication and the fact Lomax had early stages of pneumonia also contributed to the cardiac arrest. The coroner could not say whether Lomax would have died had the officer not used the TASER device, and he could not say that the device caused Lomax's death. A coroner's inquest jury ruled that the cause of death was a combination of drugs, restraining force and the use of the TASER electronic control device (see Amnesty International, 2004; Curreri, 2004; Geary, 2004, June 26; Anglen, 2006).

Lomax continued to struggle following the applications of the TASER shocks and did not develop difficulty breathing until he was in the ambulance en route to the hospital. His continued struggle indicates that the TASER pulses did not immediately affect the rhythm of his heart. Because the TASER pulses did not have an immediate effect and the medical examiner could not say that, but for the use of the TASER device, Lomax would have survived, the role of the device as the cause of death in the Lomax case is doubtful. However, the medical examiner included the TASER device as one of the forms of restraint as a significant contributing factor in Lomax's death. The TASER shocks could have added to the overall physiological stress that Lomax experienced during his violent struggle against police restraint. Therefore, the role of the device as a significant contributing factor is possible. Whether other forms of restraint would have produced a different outcome is unknown.

CURT LEE ROSENTANGLE

AGE: 44
RACE/GENDER: WHITE MALE
AGENCY: KITDAP COUNTY (WASHINGTON) SHERIFF'S OFFICE
DATE OF INCIDENT: FEBRUARY 21, 2004
CAUSE OF DEATH: EXCITED DELIRIUM WITH CARDIAC ARRHYTHMIA DUE TO ACUTE COCAINE INTOXICATION
CONTRIBUTING FACTORS: SIGNS OF SUBSTANCE INDUCED PSYCHOSIS
ROLE OF TASER DEVICE: EXCLUDED

The police and autopsy reports on this incident were exempt from disclosure under Washington law. However, media sources reported that deputies responded to calls claiming that Rosentangle was acting erratically, breaking lights in an apartment complex, pounding on doors, and trying to force himself inside apartments. The first deputy who responded to the scene found Rosentangle outside of an apartment building wearing only jeans, and he was covered in blood. She tried to get him to go inside his apartment. Rosentangle became aggressive toward her after trying to kick down an apartment door. The deputy ordered Rosentangle to stay on the ground, but he tried to get up, and she shocked him with a TASER device. The internal log showed that the deputy fired the device four times. Three other deputies arrived and struggled with Rosentangle for about 90 seconds before finally putting him in handcuffs. Once officers got Rosentangle to the ground, he went limp. Sometime later, deputies noticed that Rosentangle had trouble breathing as he lay handcuffed on the ground. They turned Rosentangle on his side, released the handcuffs, and called for medical assistance. Paramedics transported Rosentangle to a local hospital, where he died of heart failure.

According to media sources, the coroner concluded that Rosentangle died of excited delirium due to acute cocaine intoxication, and she added that the TASER device did not contribute to his death (see Amnesty International, 2004; Le, 2004; *Associated Press*, 2005, March 22; Anglen, 2006).

Rosentangle continued to struggle after the application of the TASER pulses, indicating that the electric current did not significantly affect the rhythm of his heart. Because the TASER shocks did not have an immediate effect and the medical examiner attributed the death to cocaine-induced cardiac arrhythmia, the TASER device is excluded as the cause of death and as a contributing factor in the Rosentangle case.

TERRY L. WILLIAMS

AGE: 45
RACE/GENDER: AFRICAN-AMERICAN MALE
AGENCY: MADISON (ILLINOIS) POLICE
  DEPARTMENT
DATE OF INCIDENT: MARCH 28, 2004
DATE OF DEATH: MARCH 29, 2004
CAUSE OF DEATH: SUDDEN DEATH ASSOCIATED
  WITH MARKED AGITATION AND PHYSICAL
  RESTRAINT
ROLE OF TASER DEVICE: EXCLUDED

A woman, who complained that her niece's boyfriend had beaten her niece, contacted police. When an officer arrived, he found the girlfriend suffering from injuries to her head, fingers and ribs, and he found Williams asleep in his bed. Williams woke up as the officer was trying to handcuff him, and he began to struggle with the officer, finally lying down on the floor to resist arrest. The officer used his TASER device on Williams' leg when Williams refused to stand up or to put on his clothes. After the officer placed him in a patrol car, Williams began trying to knock out the side window by repeatedly hitting it with his head. The officer used the TASER device on Williams again, striking him this time in the chest. A second officer was driving Williams to the police station when Williams suddenly stopped talking and laid down on the back seat. After dragging Williams into the station, officers realized that Williams was completely unresponsive, and they called an ambulance. Williams died about 40 minutes later at a local hospital. The Illinois State Police Public Integrity Unit conducted the investigation. Williams' girlfriend told investigators that Williams had drunk two-fifths of whiskey in the hours before his arrest.

The coroner concluded that Williams' death was the type of sudden death that occurs in individuals who are markedly agitated and are physically restrained. He ruled out the TASER device as a cause of death. A Madison County coroner's jury ruled Williams' death an accident. Although there were no illicit drugs in Williams' system, he had a postmortem blood alcohol content of 0.213 (see Amnesty International, 2004; Howard, 2004, March 30; Howard, 2004, August 5; Madison County (Illinois) Coroner's Report 04–0471; Anglen, 2006).

Williams continued to struggle with officers following the application of the TASER pulses, and he did not go quiet until an officer was transporting him to the jail, indicating that the electric current did not immediately affect the rhythm of his heart. Because the TASER shocks did not have an immediate effect and the medical examiner attributed the death to Williams' agitation and the restraint, the TASER device is excluded as the cause of death and as a contributing factor in the Williams case.

PHILLIP LEBLANC

AGE: 36
RACE/GENDER: HISPANIC MALE
AGENCY: LOS ANGELES (CALIFORNIA) POLICE
  DEPARTMENT
DATE OF INCIDENT: APRIL 1, 2004
CAUSE OF DEATH: EXCITED DELIRIUM DUE TO
  COCAINE INTOXICATION
ROLE OF TASER DEVICE: EXCLUDED

The police report in this incident was exempt from disclosure under California law, however, media sources reported that a security guard observed LeBlanc acting strangely. The guard, after persuading LeBlanc to sit by the curb, calmed him for a moment. When the guard put one handcuff on

LeBlanc, he began fighting, so the guard attached the other cuff to a fence and called for police and paramedics. When paramedics arrived, they noted that LeBlanc was aggressive and combative. Officers used two shocks from a TASER device to subdue LeBlanc. He continued to resist while they handcuffed him. LeBlanc suddenly became unresponsive and stopped breathing. Paramedics took LeBlanc to a local hospital, where he died.

The coroner's report listed LeBlanc's cause of death as excited delirium due to cocaine intoxication, and he ruled the death to be an accident (see County of Los Angeles (California) Coroner Report 2004–02590; Anglen, 2006).

LeBlanc continued to resist officers following the application of the TASER pulses, indicating that the electric current did not immediately affect the rhythm of his heart. Because the TASER shocks did not have an immediate effect, the medical examiner attributed the death to cocaine-induced excited delirium. The TASER device is excluded as the cause of death and as a contributing factor in the LeBlanc case.

MELVIN SAMUEL

AGE: 28
RACE/GENDER: AFRICAN-AMERICAN MALE
AGENCY: HOUSTON COUNTY (GEORGIA) SHERIFF'S OFFICE
DATE OF INCIDENT: APRIL 16, 2004
CAUSE OF DEATH: POSITIONAL ASPHYXIA COMPLICATED BY OBESITY AND SICKLE CELL CRISIS
CONTRIBUTING FACTORS: PULMONARY INTERSITIAL PNEUMONITIS
ROLE OF TASER DEVICE: EXCLUDED

The police report on this incident was not available, but media sources reported that Samuel called the Savannah police to report a burglary. Officers subsequently arrested him on a warrant for failing to pay $700 in traffic fines, and they took him to the Houston County jail. Jail officials said he was uncooperative, and jailers shocked Samuel three times with a TASER device while moving him out of a holding cell. About 10 minutes later, Samuel became unresponsive and died. The Georgia State Bureau of Investigation conducted the investigation.

An autopsy found that Samuel asphyxiated after officers hogtied him and laid him on his stomach. Samuel's obesity and sickle cell disease contributed to the death. The coroner found traces of tetrahydrocannabinol (THC) metabolites in Samuel's body, but no other drugs. The coroner reported that a contributing factor to hypoxia might have been pulmonary interstitial pneumonitis. A Georgia Bureau of Investigation autopsy report said the TASER device was not a factor in the death (see Amnesty International, 2004; *Atlanta Journal-Constitution*, 2004; Georgia Bureau of Investigation Record of Medical Examiner Report 2004–4001604; Anglen, 2006).

Samuel continued to struggle with jailers following the application of the TASER shocks, indicating that the electric current did not immediately affect the rhythm of his heart. Because the TASER pulses did not have an immediate effect and the medical examiner attributed the death to restraint asphyxia, the device is excluded as the cause of death and as a contributing factor in the Samuel case.

ROBERT HAROLD ALLEN

AGE: 45
RACE/GENDER: AFRICAN-AMERICAN MALE
AGENCY: LITTLE ROCK (ARKANSAS) POLICE DEPARTMENT
DATE OF INCIDENT: APRIL 17, 2004
CAUSE OF DEATH: ARTERIOSCLEROTIC CARDIOVASCULAR DISEASE
CONTRIBUTING FACTORS: ACUTE COCAINE INTOXICATION, PHYSICAL EXERTION, SUPERFICIAL INJURIES
ROLE OF TASER DEVICE: EXCLUDED

Allen struggled with ambulance workers and officers when they tried to take him to the hospital for treatment after stitches for a wound on his leg opened up. He had lost his other leg a few years earlier because of a shotgun injury, and he was using an electric wheelchair. Concerned about the stitches, a friend, whom Allen was visiting, called for an ambulance. Allen fought with emergency workers and officers, throwing books and glasses. An officer shocked Allen four times with a TASER device and arrested him. An ambulance took Allen to a local hospital, where he died.

The medical examiner noted a long history of Allen's severe medical problems. Allen suffered moderate to severe coronary artery arteriosclerosis, and he had sustained a sizeable myocardial infarct sometime in the past. He also suffered from type II diabetes mellitus, hyperthyroidism, hepatitis, and bipolar disorder. He noted that Allen continued to struggle for several minutes after the last application of the TASER shocks, even after he arrived at the hospital and the emergency room staff had given him a sedative, Lorazepam. Toxicology reports indicated a relatively high level of cocaine and a low level of Darvocet in Allen's blood. The medical examiner concluded that Allen's death was a complicated situation involving severe underlying medical disease, cocaine, and the use of force. However, he emphasized that there was no evidence to suggest that the TASER device had any direct, adverse effect on his cardiac system, including production of an arrhythmia. Media accounts report that the Allen family contracted a private pathologist who concluded that the TASER device had caused Allen's heart attack. That report was not available (see Arkansas State Crime Laboratory Medical Examiner Division Report ME–322–04; Hand, 2004; Little Rock (Arkansas) Police Department Information Report 2004–45639; Little Rock (Arkansas) Police Department Information Report 2004–45667; Frazier, 2005; Heard & Hillen, 2006).

Allen continued to struggle against emergency medical workers and the police for several minutes after the final application of the TASER pulses, indicating that the electric current did not immediately affect the rhythm of his seriously diseased heart. Because the TASER shocks did not have an immediate effect, because the first medical examiner specifically ruled out the TASER device as a significant contributing factor to Allen's heart attack, and because the medical examiner attributed the death to arteriosclerotic cardiovascular disease, cocaine intoxication, and his physical exertion, the device is excluded as the cause and as a significant contributing factor in Allen's death.

ANTHONY ALFREDO DIAZ

AGE: 29
RACE/GENDER: HISPANIC MALE
AGENCY: ORANGE COUNTY (FLORIDA) SHERIFF'S OFFICE
DATE OF INCIDENT: APRIL 18, 2004
CAUSE OF DEATH: LYSERGIC ACID DIETHYLAMIDE-INDUCED PSYCHOSIS WITH HYPERTHERMIA
CONTRIBUTING FACTORS: STRUGGLE WITH THE POLICE AND APPLICATION OF THE TASER DEVICE
ROLE OF TASER DEVICE: EXCLUDED/UNDETERMINED

The incident report and the autopsy report in this incident were not available, but media sources reported that deputies responded to a 9-1-1 call about a man running naked in the street. Diaz's brother also called 9-1-1 and reported that someone had slipped LSD into Diaz's drink. He told dispatchers that Diaz was going crazy. When deputies arrived, they found Diaz running around in an intersection stripping off his clothes. When the deputies approached him, Diaz threatened to kill them. Deputies sprayed Diaz with chemical spray and then shocked him at least twice with a TASER device. After deputies handcuffed Diaz, they noticed he was having trouble breathing, and they called an ambulance. Paramedics took him to a local hospital, where he died. Deputies said Diaz's brother told them he had allegedly been smoking marijuana, which someone might have laced with something, causing the strange behavior.

News accounts say that the medical examiner found Diaz's death was an accident caused by LSD-induced psychosis with hyperthermia, and that significant contributing conditions were the struggle with police and the application of the TASER device (see Amnesty International, 2004; Barton, 2005; Anglen, 2006).

The media accounts were not specific regarding the time that passed between the application of the TASER pulses and when Diaz developed his problem breathing, but, because the medical examiner attributed the death to LSD-induced psychosis, the TASER device is excluded as the cause of death in the Diaz case. Without more information, however, the role of the device as a significant contributing factor cannot be determined.

ERIC WOLLE

AGE: 45
RACE/GENDER: WHITE MALE
AGENCY: MONTGOMERY COUNTY (MARYLAND) POLICE DEPARTMENT
DATE OF INCIDENT: APRIL 27, 2004
CAUSE OF DEATH: CARDIAC ARRHYTHMIA IN SETTING OF ACUTE PSYCHOSIS AND PHYSICAL RESTRAINT
CONTRIBUTING FACTORS: ACUTE ALCOHOL INTOXICATION AND FOCAL MYOCARDIAL FIBROSIS
ROLE OF TASER DEVICE: EXCLUDED

Wolle, a diagnosed schizophrenic, had long believed that the CIA and FBI wanted to take him away. A car carrying Chinese delivery food parked in front of his house. When the driver left the lights on and the engine running while making his delivery next door, Wolle believed that the police had come to kill him. Family members said Wolle pushed his mother and ran out of the house. Wolle's mother ran next door and told the neighbors, who called the police. When officers arrived, they found Wolle hiding in the bushes behind the house. Wolle had a machete in his belt, and he was screaming that the police would never take him alive. When he refused orders to show his hands, an officer used a TASER device. The first shock was ineffective, but the second shock took Wolle to his knees. One officer grabbed the machete while six other officers struggled physically to subdue Wolle. During the struggle, officers used the TASER device two more times. Even after officers handcuffed Wolle, he continued to kick and struggle. While officers were waiting for a cage car to transport Wolle, he suddenly stopped breathing. Officers started CPR and called for medical assistance. Paramedics transported Wolle to a local hospital, but he was dead upon arrival.

The autopsy report was not available, but media sources reported that the medical examiner ruled Wolle had died from cardiac arrhythmia in the setting of acute psychosis during restraint. He also noted that acute alcohol intoxication and focal myocardial fibrosis contributed to Wolle's death. Wolle had bipolar disorder and schizophrenia, but he had refused to take his medications (see Amnesty International, 2004; Montgomery County (Maryland) Police Report M04–022711; Snyder, 2004; Anglen, 2006).

Wolle continued to struggle with officers for several minutes following the last application of the TASER shocks, indicating that the electric current did not immediately affect the rhythm of his heart. Because the TASER pulses did not have an immediate effect and the medical examiner attributed the cardiac arrhythmia to other causes, the TASER device is excluded as the cause of death and as a contributing factor in the Wolle case.

HENRY JOHN LATTARULO

AGE: 40
RACE/GENDER: WHITE MALE
AGENCY: HILLSBOROUGH COUNTY (FLORIDA) SHERIFF'S OFFICE
DATE OF INCIDENT: MAY 22, 2004
CAUSE OF DEATH: COCAINE-INDUCED AGITATED DELIRIUM
ROLE OF TASER DEVICE: EXCLUDED

Sheriff's deputies were called to a trailer park on a report that a man was trying to stab people with a screwdriver and a pair of scissors. They found Lattarulo fighting with a friend. When the first deputy got out of her patrol car, Lattarulo attacked her, so she shocked him with a TASER device and struck him several times with her baton, neither of which stopped him. Another deputy arrived and fired a second TASER device, but Lattarulo pulled the probes out of his chest, so it had no effect. Four more deputies arrived and physically restrained Lattarulo. After they placed him in handcuffs and leg restraints, Lattarulo stopped breathing.

Toxicological tests detected the presence of cocaine and opiates in Lattarulo's body. The medical examiner determined that Lattarulo's death was caused by cocaine-induced excited delirium, and he stated that the TASER shocks were not a factor (see Amnesty International, 2004; Colavecchio-Van Sickler, 2004; Hillsborough County (Florida) Medical Examiner Report 04–03066 A; Hillsborough County (Florida) Sheriff's Office Report 2004–50360; Barton, 2005; Anglen, 2006).

Lattarulo continued to struggle with deputies for several minutes following the last application of the TASER pulses, indicating that the electric current did not immediately affect the rhythm of his heart. Because the TASER device did not have an immediate effect and the medical examiner attributed the death to cocaine-induced excited delirium, the device is excluded as the cause of death and as a contributing factor in the Lattarulo case.

FREDERICK JEROME WILLIAMS

AGE: 31
RACE/GENDER: AFRICAN-AMERICAN MALE
AGENCY: GWINNETT COUNTY (GEORGIA) SHERIFF'S OFFICE
DATE OF INCIDENT: MAY 27, 2004
DATE OF DEATH: MAY 29, 2004
CAUSE OF DEATH: HYPOXIC ENCEPHALOPATHY DUE TO CARDIOPULMONARY ARREST OF UNCERTAIN ETIOLOGY
ROLE OF TASER DEVICE: EXCLUDED

Williams, who had a history of epilepsy, was acting strangely when police officers responded to a domestic violence call at his house. His wife told officers that Williams had not been taking his medication. Officers arrested Williams and took him to the Gwinnett County jail. Williams was combative during the transport. At the jail, jailers shocked him five times with a TASER device during a violent struggle and placed him in a restraint chair. The jailers then noticed that Williams lost consciousness. A jail nurse checked Williams and found that he had a pulse and was breathing. However, he would not respond. Jailers began CPR and attached a portable defibrillator, but the defibrillator showed not to give a shock. Paramedics transported Williams to a local hospital, where he died two days later.

A coroner said Williams died of hypoxic encephalopathy due to cardiopulmonary arrest of uncertain etiology. However, the coroner said there was no evidence that shocks from a TASER device caused or contributed to Williams' death (see Amnesty International, 2004; Gwinnett County (Georgia) Medical Examiner's Office Report 04G–0402; Gwinnett County (Georgia) Sheriff's Office Use of Force Report on Frederick Jerome Williams; Mungin, 2005; Anglen, 2006).

Williams continued to struggle with jailers following the last application of the TASER pulses, indicating that the electric current did not immediately affect the rhythm of his heart. Because the TASER device did not have an immediate effect and the medical examiner attributed the cardiopulmonary arrest to causes unknown, but specifically discounted the TASER pulses as a cause, the device is excluded as the cause of death and as a contributing factor in the Williams case.

DARYL LAVON SMITH

AGE: 46
RACE/GENDER: AFRICAN-AMERICAN MALE
AGENCY: FULTON COUNTY (GEORGIA) SHERIFF'S OFFICE
ATLANTA POLICE (GEORGIA) DEPARTMENT
DATE OF INCIDENT: MAY 30, 2004
CAUSE OF DEATH: AGITATED DELIRIUM DUE TO ACUTE COCAINE POISONING
ROLE OF TASER DEVICE: EXCLUDED

Emergency Medical Services received a call of a man down, and paramedics found Smith lying in a street. They were able to get him up and into the ambulance, but Smith began kicking and swinging at them, so the paramedics called for police assistance. Atlanta police officers and Fulton County Sheriff's deputies responded. The officers struggled with Smith to pull him out of the ambulance, secure him, strap him onto a gurney, and put him back in the ambulance. A deputy from the Fulton County Sheriff's Office stunned Smith four times with a TASER device during the struggle. Because Smith continued to struggle in the ambulance, officers followed the ambulance to a local hospital. Smith died about six hours later.

The coroner's report listed the cause of Smith's death as agitated delirium associated with acute cocaine poisoning. The coroner noted that Smith arrived at the hospital with a narrow complex tachycardia and that his rectal temperature was 104.7 degrees Fahrenheit. He also reported that TASER-generated burns on Smith's body were superficial and not contributory to his death (Amnesty International, 2004; Fulton County (Georgia) Medical Examiner's Report 04–0974; Fulton County (Georgia) Sheriff's Office Report 04M059770; Mungin, 2004, June 3; Anglen, 2006).

Smith continued to struggle against officers and paramedics for several minutes following the last application of the TASER pulses, indicating that the electric current did not immediately affect the rhythm of his heart. Because the TASER device did not have an immediate effect, the medical examiner attributed the death to cocaine-induced excited delirium, and the coroner stated that the TASER device was not contributory, the device is excluded as the cause of death and as a contributing factor in the Smith case.

ANTHONY CARL OLIVER

AGE: 42
RACE/GENDER: AFRICAN-AMERICAN MALE
AGENCY: ORLANDO (FLORIDA) POLICE
    DEPARTMENT
DATE OF INCIDENT: MAY 31, 2004
DATE OF DEATH: JUNE 1, 2004
CAUSE OF DEATH: COCAINE EXCITED DELIRIUM;
    CARDIOMEGALY
CONTRIBUTING FACTORS: SUBDUED BY POLICE
    WITH STRUGGLE AND TASER DEVICE
ROLE OF TASER DEVICE: EXCLUDED/POSSIBLE

Oliver walked into traffic at a busy intersection, flagging down cars and screaming that someone was trying to kill him. He stopped a police officer by banging on the back window of her patrol car. Oliver was frothing at the mouth, and he told the officer that people were pursuing him and were going to shoot him. When the officer attempted to talk to Oliver and move him out of traffic, he began struggling and tried to pull her into traffic. The officer shocked Oliver eight times with a TASER device before she could put handcuffs on him. During the incident, Oliver suffered an apparent seizure. Paramedics took Oliver to a local hospital where he was placed in the intensive care unit. He died about 23 hours later.

A toxicology report detected marijuana and cocaine in Oliver's body. The coroner ruled that Oliver died as the result of complications of cocaine excited delirium. He was in a state of excited, delusional behavior from cocaine prior to being subdued by police with a struggle and TASER shocks. The medical examiner noted that Oliver was still combative after emergency medical services personnel arrived, but he had a witnessed arrest while on a stretcher. Upon Oliver's arrival at the hospital, he had a temperature of 108 degrees Fahrenheit. Finally, the medical examiner noted that Oliver suffered from another risk factor for sudden death, cardiomegaly (see Amnesty International, 2004; *Associated Press*, 2004, June 2; Office of the Medical Examiner, Ninth District (Florida) Report 04–2641; Orlando (Florida) Police Report 2004–193999; Barton, 2005; Anglen 2006).

Oliver suffered an apparent seizure during his struggle with police, and he suffered a cardiac arrest shortly after paramedics arrived. Because the medical examiner attributed the death to cocaine-induced excited delirium and cardiomegaly, the TASER device is excluded as the cause of death in the Oliver case. The medical examiner noted the use of the TASER device as a contributory factor in Oliver's death. The TASER shocks undoubtedly contributed to the physiological stress that Oliver experienced, and he did suffer a seizure during the time officers applied the device. That additional stress, considering Oliver's enlarged heart and hyperthermia, may have been more than Oliver could tolerate. Considering the extent of his medical problems, however, it is questionable that any other form of restraint would have yielded a different result. However, the role of the TASER device as a significant contributing factor in Oliver's death is possible.

JERRY W. PICKENS

AGE: 55
RACE/GENDER: WHITE MALE
AGENCY: JEFFERSON PARISH (LOUISIANA) SHERIFF'S OFFICE
DATE OF INCIDENT: JUNE 4, 2004
DATE OF DEATH: JUNE 7, 2004
CAUSE OF DEATH: BRAIN HEMORRHAGE
ROLE OF TASER DEVICE: EXCLUDED/ CONFIRMED

Deputies were called to the Pickens' residence following a disturbance between Pickens and his son. When the deputies arrived, Pickens confronted them in the front yard. Pickens refused to cooperate with the officers and refused to obey their orders not to go back in the house. When deputies used a TASER device to prevent him from entering the house, Pickens fell, striking his head on the driveway. Pickens was transported to a local hospital. Doctors declared him brain-dead that day. He died three days later, after doctors removed him from life support.

The autopsy report was not available, but media sources reported that the coroner ruled Pickens died of a brain hemorrhage because of the fall (see Amnesty International, 2004; Berenson, 2004; Jefferson Parish (Louisiana) Sheriff's Office Report F-03847-04; Anglen, 2006).

Clearly, Pickens died from injuries he sustained from the fall, which was caused by application of the TASER device. Because the coroner contributed the death to a brain hemorrhage, the device is excluded as the cause of death in the Pickens case; however, the role of the device as a significant contributing factor is confirmed.

JAMES ARTHUR COBB, JR.

AGE: 42
RACE/GENDER: AFRICAN-AMERICAN MALE
AGENCY: SAINT PAUL (MINNESOTA) POLICE DEPARTMENT
DATE OF INCIDENT: JUNE 9, 2004
CAUSE OF DEATH: EXCITED DELIRIUM ASSOCIATED WITH COCAINE USE, HEART DISEASE AND PULMONARY EMPHYSEMA
ROLE OF TASER DEVICE: EXCLUDED

Cobb was first spotted walking bare-chested down the middle of the street in the rain at about 1:42 A.M. He was waving his shirt and yelling as he tried to stop a female motorist and get into her car. She called police and said the man was displaying erratic behavior. Five officers reached the scene and found the man on a bridge. Cobb became combative as officers tried to escort him from the bridge. Officers first shocked Cobb him with probes from two TASER devices. When that did not work, officers got close enough to drive-stun Cobb. They also used chemical spray to try to subdue him, but those efforts were ineffective. Cobb continued to struggle, so officers struck him at least twice in the arms with batons, finally getting him to lower his arms so they could handcuff him. Shortly after they handcuffed him, Cobb became unresponsive. Officers noted that his pulse was weak, and they called for paramedics. Paramedics took Cobb to a local hospital, where he was pronounced dead at 3:14 A.M.

The autopsy report was not available, but media sources reported that the coroner listed Cobb's cause of death as excited delirium associated with cocaine toxicity. Cobb also had a preexisting heart condition, which the media sources did not define, and he suffered from pulmonary emphysema (see Amnesty International, 2004; Estrada & Gustafson, 2004; McKinney, 2004; Saint Paul (Minnesota) Police Report 04116367; Anglen, 2006).

Cobb continued to struggle with officers following the application of the TASER pulses, indicating that the electric current did not immediately affect the rhythm of his heart. Because the TASER device did not have an immediate effect and the medical examiner attributed the death to cocaine-induced excited delirium, heart disease, and emphysema, the device is excluded as the cause of death and as a contributing factor in the Cobb case.

JACOB JOHN LAIR

AGE: 26

RACE/GENDER: WHITE MALE
AGENCY: SPARKS (NEVADA) POLICE DEPARTMENT
  WASHOE COUNTY (NEVADA) SHERIFF'S OFFICE
DATE OF INCIDENT: JUNE 9, 2004
CAUSE OF DEATH: ACUTED METHAMPHETAMINE INTOXICATION WITH ASSOCIATED (PROBABLE) CARDIAC ARRHYTHMIA WHILE ENGAGED IN A PHYSICAL STRUGGLE WITH LAW ENFORCEMENT OFFICERS INVOLVING "TASER GUN," "PEPPER SPRAY," AND RESTRAINTS
ROLE OF TASER DEVICE: DOUBTFUL/POSSIBLE

The police reports on this incident were not available under Nevada law, but media sources reported that Sparks Police Department officers went to the home of Lair's girlfriend to question him about a crime. Lair became combative, and officers called for assistance. Deputies from the Washoe County Sheriff's Office arrived. Officers sprayed Lair with chemical spray and shocked him with a TASER device. During the struggle, Lair collapsed. Officers performed CPR and called for an ambulance. Paramedics transported Lair to a local hospital, where he died. The Reno Police Department conducted the investigation.

The medical examiner reported that Lair died from acute methamphetamine intoxication with associated cardiac arrhythmia while engaged in a physical struggle with law enforcement officers involving a TASER device, chemical spray, and restraints. The Washoe County Coroner later explained that the methamphetamine level, the police intervention, the restraint, and the use of the TASER device were all equivocal. He said the findings were a conglomerate of causes. He said he could neither include nor exclude the TASER device as a cause of death (see Amnesty International, 2004; *Associated Press*, 2004, June 12; Washoe County (Nevada) Medical Examiner's Report 59904; Gafni, 2005, February 2; Anglen, 2006).

By his own admission, the coroner could not give a physiological explanation of the TASER device's contribution to Lair's death. His conclusion was that the struggle with police, the TASER shocks, the restraints, and the methamphetamine were all part of the totality of the circumstances leading to Lair's cardiac arrhythmia. Lair did collapse immediately following the application of the TASER pulses, and the medical examiner and coroner both noted the TASER device as a significant contributing factor. However, neither the medical examiner nor the coroner could say that, but for the application of the TASER device, Lair would have survived. Consequently, the role of the device as a significant contributing factor in Lair's death is possible.

ABEL ORTEGA PEREZ

AGE: 36
RACE/GENDER: HISPANIC MALE
AGENCY: AUSTIN (TEXAS) POLICE DEPARTMENT
DATE OF INCIDENT: JUNE 16, 2004
CAUSE OF DEATH: COCAINE TOXICITY
ROLE OF TASER DEVICE: EXCLUDED

The residents of a home called 9-1-1 and reported that Perez, who they did not know, had forced his way into their home when they answered his knock at the door around 1:00 A.M. Perez began fighting two people who were trying to force him out of the house. When police arrived, Perez fought with the officers, who drive-stunned Perez one time with a TASER device to subdue him. The officers called for paramedics because Perez had a wound to his forehead, which witnesses said he already had when he knocked on the door. Before the ambulance could arrive, Perez went into cardiac arrest. The officers attempted CPR, but neither they nor the paramedics could revive Ortega. He died at the scene.

The medical examiner reported that Perez died of cocaine toxicity, and he ruled that the death was an accident (see Austin (Texas) Police Report 2004–1680065; Plohetski, 2004; Travis County (Texas) Medical Examiner's Report ME-04–1053; Humphrey & Osborn, 2005).

Perez did not collapse for several minutes following the application of the TASER pulse, indicating that the electric current did not immediately affect the rhythm of his heart. Because the TASER device did not have an immediate effect and the medical examiner attributed the death to cocaine toxicity, the device is excluded as the cause of death and as a contributing factor in the Perez case.

KRIS J. LIEBERMAN

AGE: 32
RACE/GENDER: WHITE MALE
AGENCY: BUSHKILL TOWNSHIP (PENNSYLVANIA) POLICE DEPARTMENT
DATE OF INCIDENT: JUNE 24, 2004
CAUSE OF DEATH: COCAINE-INDUCED EXCITED DELIRIUM
CONTRIBUTING FACTORS: EXERTION, INCLUDING SHOCKS FROM A TASER DEVICE AND RESTRAINT, AND HIGH LEVELS OF COCAINE
ROLE OF TASER DEVICE: EXCLUDED/ UNDETER- MINED

Neither the Bushkill Township Police Department nor the Northampton County Coroner's Office responded to requests for information regarding this incident, so the police report and autopsy report were not available. However, media accounts reported that, for about 45 minutes, Lieberman crawled around naked in a cornfield, talking to himself, moaning, and banging his head on the ground. When officers arrived and tried to speak with Lieberman, he lunged at them. They shocked him with a TASER device three times. He fought briefly and then collapsed. Officers tried to revive him, but he was pronounced dead a short time later.

Media sources reported that the medical examiner found that Lieberman had high levels of cocaine in his system. He also claimed that the exertion of Lieberman's fight with police, including shocks from a TASER device and physical restraint, contributed to Lieberman's death (see Amnesty International, 2004; *Associated Press*, 2004, June 25; *Associated Press*, 2004, December 13; Berenson, 2004; Anglen, 2006).

Media accounts were not specific in describing how much time passed between the application of the TASER pulses and Lieberman's collapse, but he did continue to struggle with officers for a while after the last application of the pulses, indicating that the electric current did not immediately affect the rhythm of his heart. Because the TASER device did not have an immediate effect and the medical examiner attributed the death to cocaine-induced excited delirium, the device is excluded as the cause of death in the Lieberman case. The medical examiner did list the TASER device as a contributing factor, but media sources did not explain how the device contributed to Lieberman's death. Therefore, without further information explaining its effects on Lieberman, the role of the TASER device as a significant contributing factor is undetermined.

ERIC BERNARD CHRISTMAS

AGE: 36
RACE/GENDER: AFRICAN-AMERICAN MALE
AGENCY: DAYTON (OHIO) POLICE DEPARTMENT
DATE OF INCIDENT: JUNE 30, 2004
CAUSE OF DEATH: COCAINE-INDUCED FATAL EXCITED DELIRIUM
CONTRIBUTING FACTORS: BODY TEMPERATURE IN EXCESS OF 108 DEGREES FAHRENHEIT, AND A HISTORY OF COCAINE ABUSE
ROLE OF TASER DEVICE: EXCLUDED

Witnesses called police after they saw Christmas running in circles in the middle of the street. One of the witnesses was able to track down a police officer in a cruiser nearby and alert him about Christmas' behavior. When the first officer arrived on the scene, Christmas ran over and jumped into the front passenger seat of the police car. Two other officers arrived and were able to remove Christmas from the cruiser. Once he was out of the car, Christmas struggled with all three officers. One officer placed the TASER device against Christmas' body and pulled the trigger several times. He then stepped back and fired the device at Christmas. However, the TASER's cartridge dislodged, and no current emitted. Once they secured him, officers called for medical assistance for Christmas, because he was sweating profusely. When paramedics arrived, Christmas quit breathing. Medical personnel took Christmas to a local hospital, where he died.

The coroner's report listed Christmas' cause of

death as cardiac arrest due to cocaine-induced excited delirium. The report also stated that Christmas had a core temperature near the time of cardiac arrest of 108.9 degrees Fahrenheit, and that he had a history of reported chronic cocaine abuse (see Amnesty International, 2004; Dayton (Ohio) Police Report 0406300555; Grieco, 2004; Montgomery County (Ohio) Coroner Report 04–2130; Wynn, 2004; Anglen, 2006).

Christmas continued to struggle with officers after the application of the TASER pulses, indicating that the electric current did not immediately affect the rhythm of his heart. Because the TASER device did not have an immediate effect and the medical examiner attributed the death to cocaine-induced excited delirium and hyperthermia, the device is excluded as the cause of death and as a contributing factor in the Christmas case.

DEMETRIUS TILLMAN NELSON

AGE: 45
RACE/GENDER: AFRICAN-AMERICAN MALE
AGENCY: OKALOOSA COUNTY (FLORIDA) SHERIFF'S OFFICE
DATE OF INCIDENT: JULY 3, 2004
CAUSE OF DEATH: COCAINE-ASSOCIATED EXCITED DELIRIUM
ROLE OF TASER DEVICE: EXCLUDED

Nelson was driving with a girlfriend and her three children when their car overheated. He pulled into the parking lot of a shopping center. An off-duty deputy sheriff, who was working security for the shopping center, was trying to help Nelson find a lost radiator cap when Nelson began arguing with his girlfriend, throwing things around and accusing her of stealing the radiator cap. Nelson became incoherent and aggressive, grabbing one of the children and struggling with the deputy and other security staff. A deputy shocked Nelson four times with the TASER device, secured him, placed him in the back seat of a patrol car, and called for medical assistance for Nelson, who had suffered a cut to his face during the struggle. Paramedics treated Nelson and released him to the deputy. When the deputy returned him to the back seat of the patrol car, Nelson quit breathing. Paramedics then transported Nelson to a local hospital, where he died the next day.

The autopsy report was not available, but media sources reported that deputies at the hospital noted Nelson had an extremely high body temperature. The coroner found no sign of blunt force trauma that could have caused Nelson's death. Instead, she noted that Nelson had a large quantity of cocaine in his system, and she ruled that he died from cocaine-associated excited delirium (see Okaloosa County (Oklahoma) Sheriff's Office Report OCSO04OFF 007901; Amnesty International, 2004; *Associated Press*, 2004, July 3; *Associated Press*, 2004, August 7; Brannon, 2004; Barton, 2005; Anglen, 2006).

Nelson did not collapse for several minutes following the application of the TASER pulse, indicating that the electric current did not immediately affect the rhythm of his heart. Because the TASER device did not have an immediate effect and the medical examiner attributed the death to cocaine-induced excited delirium, the device is excluded as the cause of death and as a contributing factor in the Nelson case.

WILLIE SMITH, III

AGE: 48
RACE/GENDER: AFRICAN-AMERICAN MALE
AGENCY: AUBURN (WASHINGTON) POLICE DEPARTMENT
DATE OF INCIDENT: JULY 11, 2004
DATE OF DEATH: JULY 13, 2004
CAUSE OF DEATH: ACUTE COCAINE INTOXICATION AND PHYSICAL RESTRAINT
CONTRIBUTING FACTORS: HEART DISEASE, PHYSICAL STRUGGLE
ROLE OF TASER DEVICE: EXCLUDED

Smith pinned his wife down in their apartment and told her he wanted to get the devil out of her as he tried to pry out her eyes with his fingers. She broke free, crawled out of their bedroom window, and called for help. When officers arrived, Smith charged them. Two officers shocked Smith with TASER devices six times, but the charges had no effect. Once officers secured Smith, they called for an ambulance to treat Smith for injuries he sustained in the struggle. Officers hogtied Smith and put him on a gurney. Smith suffered a heart attack in the ambulance. Paramedics stabilized him and transported him to a local hospital, where he died two days later.

The autopsy report was not available under Washington law, but media sources reported that the coroner concluded Smith died of a combination of acute cocaine intoxication, heart disease, and the stress brought on by the fight with police officers. An inquest jury found that acute cocaine intoxication was more to blame for Smith's death than the TASER device, the struggle with officers, or Smith's diseased heart. All six jury members said the TASER device was not a primary cause, and three of them said it was not even a contributing factor (see Auburn (Washington) Police Report 04–079192; *Seattle Times*, 2004; Le & Castro, 2004; Anglen, 2006; Castro, 2006).

Smith continued to struggle with officers for several minutes following the application of the TASER pulses, indicating that the electric current did not immediately affect the rhythm of his heart. Because the TASER device did not have an immediate effect and the medical examiner attributed the death to cocaine intoxication, the device is excluded as the cause of death and as a contributing factor in the Smith case.

MILTON FRANCISCO SALAZAR

AGE: 29
RACE/GENDER: HISPANIC MALE
AGENCY: MESA (ARIZONA) POLICE DEPARTMENT
DATE OF INCIDENT: JULY 21, 2004
DATE OF DEATH: JULY 23, 2004
CAUSE OF DEATH: COMPLICATIONS OF EXCITED DELIRIUM DUE TO COCAINE ADVERSE EFFECT
CONTRIBUTING FACTORS: STRESS FROM PHYSICAL STRUGGLE AND TASER GUN INJURIES
ROLE OF TASER DEVICE: EXCLUDED/DOUBTFUL

Just hours after Salazar was released from a state prison, he began throwing rocks at passing motorists. Salazar then went into a convenience store, where he threatened to shoot the attendant and began throwing candy bars at him. When an officer arrived and tried to arrest him, Salazar lay on the floor, folded his hands underneath his body, and refused to obey commands. The officer shocked Salazar between four and ten times with a TASER device, but after each shock Salazar yelled for the officers to give him more, and he continued to resist officers' attempts to handcuff him. Once they secured him, officers turned Salazar over and found he was pale. Salazar was breathing but unresponsive, so officers called for an ambulance. Paramedics took Salazar to a local hospital where he died two days later.

The autopsy report was not available, but media sources reported that the medical examiner ruled Salazar's death was the result of complications of excited delirium due to cocaine intoxication. He listed the stress from the physical struggle with police and injuries from the TASER shocks as contributing factors. The medical examiner ruled the death to be an accident (see Amnesty International, 2004; *Associated Press*, 2004, July 31; Mesa (Arizona) Police Report 2004–2030697; *Associated Press*, 2005, January 12; Anglen, 2006).

Salazar did collapse during the struggle with police and during the application of the TASER device. The medical examiner attributed the death to cocaine-induced excited delirium, but he listed the TASER device as a contributing factor. The TASER pulses would have contributed to the physiological stress that Salazar experienced. However, it is difficult to see how another form of restraint would have yielded a different result. Therefore, the TASER device is excluded as the cause of death, and, because the medical examiner could not say that, but for the use of the TASER device Salazar would have survived, the role of the device as a significant contributing factor is doubtful.

KEITH TUCKER

AGE: 47
RACE/GENDER: MALE (RACE NOT DETERMINABLE FROM THE AVAILABLE RECORDS)
AGENCY: LAS VEGAS METROPOLITAN (NEVADA) POLICE DEPARTMENT
DATE OF INCIDENT: AUGUST 2, 2004
CAUSE OF DEATH: CARDIAC ARREST DURING

RESTRAINT PROCEDURES
CONTRIBUTING FACTORS: USE OF TASER AND BATONS; COCAINE AND MUSCLE RELAXANTS
ROLE OF TASER DEVICE: DOUBTFUL

The police and autopsy reports in this incident were not available, but media sources reported that Tucker's roommate called the police, telling officers that Tucker was acting erratically, tearing the house apart, and talking to someone who was not there. The roommate had used a chair to block the bathroom door and keep Tucker detained, but Tucker broke through, went into the roommate's bedroom, and broke his desk. When officers arrived, they found Tucker sitting on his bed. As they approached, Tucker struck and kicked an officer. The officers used a baton and a TASER device to subdue Tucker. They struggled for a while, and one officer applied two drive-stuns to Tuckers chest. When those shocks had no effect, he applied the TASER prongs once to Tucker's thigh and once to his back. The struggle continued for a short time until the officers could get handcuffs on Tucker. Shortly thereafter, the officers noticed that Tucker was not breathing. The officers summoned medical assistance, and paramedics took Tucker to a local hospital, where he died.

Media sources reported that the coroner listed Tucker's cause of death as cardiac arrest during restraint procedures. He added that the use of the TASER device and baton were factors in the restraint tactics that were factors in Tucker's death. He found fibrous tissue in Tucker's heart, which can be the result of long-term cocaine use and can cause disturbances in the heartbeat. He also noted that Tucker had cocaine and prescription muscle relaxants in his body when he died, which he listed as additional contributing factors (see Amnesty International, 2004; Geary, 2004, August 4; Geary, 2004, October 23; *Las Vegas Review-Journal*, 2004; Anglen, 2006).

Tucker did collapse immediately following the use of the TASER device to his chest. The TASER pulses would have contributed to the physiological stress that Tucker experienced, but, considering the level of cocaine and pharmaceuticals in his system and his diseased heart, it is questionable whether another form of restraint would have yielded a different result. Because the medical examiner could not say that, but for the use of the TASER device Tucker would have survived, the role of the device as a cause of death and as a significant contributing factor is doubtful.

DAVID RILEY

AGE: 41
RACE/GENDER: WHITE MALE
AGENCY: JOPLIN (MISSOURI) POLICE DEPARTMENT
DATE OF INCIDENT: AUGUST 10, 2004
DATE OF DEATH: AUGUST 11, 2004
CAUSE OF DEATH: THERMAL BURNS
ROLE OF TASER DEVICE: EXCLUDED/ UNDETERMINED

Riley, who was apparently upset over having signed his final divorce papers the previous day, phoned his ex-wife's lawyer's office. The secretary there called police because she said that Riley sounded suicidal. Officials said Riley turned on the natural gas before officers arrived at his home. The house exploded and burned as an officer fired his TASER device at Riley while trying to subdue him inside the house. The next day, Riley died of injuries received in the explosion. About one month later, one of the officers died of his wounds. Fire marshals were unable to determine whether the TASER pulses, or some other ignition source, triggered the explosion.

The autopsy report was not available, but media sources reported that the coroner reported that Riley died of the thermal burns he received in the explosion and fire. He reported that the TASER probes were not in Riley's body, and it is unknown whether the probes ever contacted Riley (see Amnesty International, 2004; *Associated Press*, 2004, August 12; *St. Louis Post-Dispatch*, 2004, October 17; Joplin (Missouri) Police Report 1-04-038498; Anglen, 2006).

Because there is no evidence the TASER prongs ever contacted Riley and the medical examiner attributed his death to burns, the TASER device is excluded as the cause of death in the Riley case. Whether sparks from the device triggered the explosion and fire is unknown. Therefore, the role of the TASER device as a significant contributing factor is undetermined.

ERNEST J. BLACKWELL

AGE: 29
RACE/GENDER: AFRICAN-AMERICAN MALE
AGENCY: SAINT LOUIS COUNTY (MISSOURI) POLICE DEPARTMENT
DATE OF INCIDENT: AUGUST 11, 2004
CAUSE OF DEATH: AGITATED DELIRIUM
ROLE OF TASER DEVICE: EXCLUDED

Blackwell shot his 9-year-old stepdaughter in the chest with a 12-gauge shotgun, then ran outside and rushed three teenage girls standing in a driveway several houses down the street. The girls ran. Two jumped a fence and escaped, but Blackwell caught the third girl, who was 14 years old. Blackwell then began beating her in the face while yelling "Touchdown!" with each blow. He was also screaming that he hated women, and he kept repeating the number 6-6-6. When Blackwell walked away, the girl got up and tried to run, but he caught her again and beat her some more. Then Blackwell broke into the girl's home, knocking down the front door and beating the girl's stepmother. When an officer arrived, Blackwell, a 6-foot-3-inch, 235-pound former football player, tackled him, and grabbed his pistol. The officer managed to eject the ammunition magazine to keep Blackwell from firing it. Blackwell was choking the officer when other officers arrived and pushed him off. The officers used batons and fists to strike Blackwell about the shoulders, knees, legs and head. In all, ten officers fought with Blackwell, one using a TASER device to no effect. The officers managed to handcuff Blackwell, but he continued to struggle, broke free, and ran outside. Police used the TASER device again. When an officer tripped Blackwell, several others held him down while paramedics injected two shots of a sedative and loaded him into the ambulance. By the time they reached a local hospital, Blackwood was dead.

The autopsy report was not available, but media sources reported that the coroner determined Blackwell died of agitated delirium. The only drug he had in his system was marijuana (see Amnesty International, 2004; Ratcliffe, 2004; Saint Louis County (Missouri) Police Report 04–0073861; Ratcliffe, 2005; Anglen, 2006).

Blackwell struggled with officers for several minutes following the application of the TASER pulses, indicating that the electric current did not immediately affect the rhythm of his heart. Because the TASER device did not have an immediate effect and the coroner attributed the death to excited delirium, the device is excluded as a cause of death and as a significant contributing factor in the Blackwell case.

ANTHONY LEE MCDONALD

AGE: 46
RACE/GENDER: WHITE MALE
AGENCY: CABARRUS COUNTY (NORTH CAROLINA) SHERIFF'S OFFICE
DATE OF INCIDENT: AUGUST 13, 2004
CAUSE OF DEATH: CARDIAC ARRHYTHMIA DUE TO HYPERTENSIVE AND ATHEROSCLEROTIC CORONARY ARTERY DISEASE
ROLE OF TASER DEVICE: EXCLUDED

McDonald's mother called the police to report a disturbance with her son, who was damaging his home and cursing loudly in the yard. When the first deputy arrived, he observed McDonald throwing things through the windows and holding what appeared to be a rifle. Additional officers arrived, and when two officers entered the house, McDonald became aggressive. A deputy fired a TASER cartridge through a broken entry into the home, striking McDonald, but having no effect. Deputies then fired two beanbag rounds, striking McDonald, but, again, having no effect. The deputies then wrestled McDonald to the ground, using a carotid restraint, and handcuffed him. Shortly after they handcuffed McDonald, the deputies noticed that he was having difficulty breathing. Emergency medical services, which had staged a block away, responded and transported McDonald to the hospital, where he died.

The medical examiner concluded that McDonald had died of cardiac arrhythmia due to hypertensive and atherosclerotic coronary artery disease. He also noted McDonald's history of schizophrenia and depression, and he reported that toxicological reports showed Mirtazapine, an antidepressant used for the treatment of depression, in McDonald's body. He also reported that McDonald had mild emphysema. He concluded that the use of the carotid restraint may have contributed to hypoxia and precipitated a cardiac arrhythmia, but that there was no evidence of strangulation or compressional asphyxia. He also noted that Mirtazapine is among a class of antipsychotic medications known to increase the risk for ventricular arrhythmia. He said it was difficult to determine whether the TASER device may have contributed (see Amnesty International, 2004; Cabarrus County (North Carolina) Sheriff's Office Report 2004–006203; Mecklenburg County (North Carolina) Medical Examiner Report B2004–2092; Anglen, 2006).

McDonald continued to struggle with police for several minutes after the application of the TASER pulses, indicating that the electric current did not immediately affect the rhythm of his heart. Because the TASER device did not have an immediate effect and the medical examiner attributed the cardiac arrhythmia to hypertensive and atherosclerotic coronary artery disease, the device is excluded as a cause of death and as a significant contributing factor in the McDonald case.

WILLIAM MALCOLM TEASLEY

AGE: 31
RACE/GENDER: WHITE MALE
AGENCY: ANDERSON COUNTY (SOUTH CAROLINA) SHERIFF'S OFFICE
DATE OF INCIDENT: AUGUST 16, 2004
CAUSE OF DEATH: CARDIAC ARRHYTHMIA DUE TO THE COMBINATION OF PULMONARY, CARDIAC, AND VASCULAR DISEASE FOLLOWING TASER ELECTRICAL SHOCK
CONTRIBUTING FACTORS: POST-TRAUMATIC BRAIN INJURY AND EMOTIONAL AND PHYSICAL STRESS RESPONSE
ROLE OF TASER DEVICE: EXCLUDED/POSSIBLE

Officers had arrested Teasley for disorderly conduct. He initially had threatened officers, so they had handcuffed him, but at the jail he calmed down, so jailers released him from the cuffs. Jailers asked Teasley if he had any medical problems, and Teasley responded only that he had some back pain. When the jailers gave him a pen to sign a medical release, Teasley became violent, trying to stab the jailers with the pen. Jailers subdued Teasley and escorted him to a cell. When they removed the handcuffs, Teasley again began to resist jailers. One jailer used a TASER device in the drive-stun mode to subdue Teasley. The first attempt was a glancing blow that did not make good contact and was ineffective. The second attempt made contact on the right side of Teasley's body, but Teasley continued to resist. Several jailers swarmed Teasley and put the handcuffs back on him. A few minutes later, jailers discovered Teasley unresponsive, and they began resuscitation. Paramedics took Teasley to a local hospital, where he died a short time later. The South Carolina State Law Enforcement Division conducted the investigation.

The coroner noted a number of problems with Teasley's health. He noted that Teasley had cardiomegaly with atherosclerotic coronary vascular disease. He had a partial tracheal obstruction due to tracheal stenosis from a previous tracheotomy. He had edema and congestion of the lungs, liver, and spleen. He had steatosis of the liver and steatohepatitis. He also had remote blunt force trauma to the brain with brain injuries centered in the frontal lobes and deep brain structures from a previous automobile collision. The coroner concluded that Teasley died of cardiac arrhythmia due to the combination of pulmonary, cardiac, and vascular disease following TASER electrical shock. He also reported that contributing factors included the post-traumatic brain injury and emotional and physical stress response. He argued that, although Teasley had serious cardiac and vascular disease that would have placed him at risk for arrhythmia, the physical and emotional stress with the acidosis caused by partial airway obstruction would have lowered the threshold for cardiac dysfunction and electrical arrhythmia. He said that the added stress of the

TASER shock, with its electrical current, was proximal to the cardiac arrhythmia and must be considered contributory (see Amnesty International, 2004; Anderson County (South Carolina) Coroner Report OA-04-0000143; *Associated Press*, 2004, August 17; *Associated Press*, 2004, August 19; South Carolina Law Enforcement Division Report 34–04–167; Anglen, 2006).

Jailers stated that Teasley continued to struggle following the application of the TASER pulses, and the coroner noted that the only TASER injury site was on the right abdomen, away from the heart. Had the electrical impulses caused ventricular fibrillation or arrhythmia, they would have done so when jailers applied the charge, not after a continued struggle. Consequently, the TASER device is excluded as the cause of death in the Teasley case. Application of the TASER pulses could have added to the physiological stress that Teasley experienced, although, considering his extensive medical history, it is questionable whether another form of restraint would have yielded a different result. Nevertheless, the role of the TASER device as a significant contributing factor in the Teasley case is possible.

RICHARD "KEVIN" KARLO

AGE: 44
RACE/GENDER: WHITE MALE
AGENCY: DENVER (COLORADO) POLICE DEPARTMENT
DATE OF INCIDENT: AUGUST 19, 2004
CAUSE OF DEATH: ACUTE COCAINE AND NORTRIPTYLINE TOXICITY
ROLE OF TASER DEVICE: EXCLUDED

Police received a call about a man who, witnesses said, was frothing at the mouth, looking into cars and swinging a pole. Two police officers responded to the report that the man had overdosed on cocaine. When they arrived, they saw Karlo making threatening gestures to a woman on the other side of a fence. The officers ordered Karlo to step back, but he lunged at them instead. One of the officers used a TASER device on Karlo, shocking him three times. Karlo yanked the two probes off and grabbed a metal fence post to use as a weapon. An off-duty parole officer joined in, striking Karlo in the leg with a baton to force him to the ground. Meanwhile, the police officer reloaded his TASER device with another cartridge and shocked Karlo again. When the officers handcuffed Karlo, they saw that he was having trouble breathing. They summoned an ambulance, and paramedics transported Karlo to a local hospital, where he died about one hour later. Family members reported that Karlo had broken his back more than 20 years before, and that the only way he could control the pain was with cocaine. They also said that he had overdosed on at least three previous occasions.

The medical examiner ruled that Karlo died of acute cocaine and Nortriptyline intoxication. Nortriptyline is an antidepressant sometimes used for chronic pain modification. He reported that Karlo was in a state of agitated delirium, adding that the TASER device neither caused Karlo's death, nor served as a contributing factor (see Amnesty International, 2004; Crecente, 2004; Denver (Colorado) Coroner's Report 2004–3347; Denver (Colorado) Police Report 2004–38306; Kelly, 2004; Mitchell, 2004; Anglen, 2006).

Because the medical examiner stated that the TASER device was not a factor in the death, which he attributed to cocaine and Nortriptyline toxicity, the device is excluded as a cause of death and as a significant contributing factor in the Karlo case.

MICHAEL LEWIS SANDERS

AGE: 40
RACE/GENDER: AFRICAN-AMERICAN MALE
AGENCY: FRESNO (CALIFORNIA) POLICE DEPARTMENT
DATE OF INCIDENT: AUGUST 20, 2004
CAUSE OF DEATH: COMPLICATIONS OF COCAINE INTOXICATION
ROLE OF TASER DEVICE: EXCLUDED

The police report in this incident was not available, but media sources reported that police responded to a disturbance call at Sanders' home. When they arrived, officers heard yelling inside, entered, and found an unclothed Sanders struggling with his wife. Sanders reportedly made statements that the police were there to kill him, and officers described him as delusional. When his wife moved away, officers fired a TASER cartridge at him. Sanders appeared unaffected by the TASER pulses after at least seven shocks to the chest, abdomen, back, groin, arm, and leg. The officers wrestled Sanders to the ground, and he stabbed a sergeant on the arm several times with an unidentified object. The officers handcuffed Sanders, placed him on a gurney, and put him into an ambulance. Sanders stopped breathing in the ambulance. Efforts to resuscitate him were not effective, and Sanders died about two and a half hours later.

The coroner reported that Sanders died of complications of cocaine intoxication. The coroner noted that Sanders suffered from moderate cardiomegaly, right ventricular dilation, mild left ventricular hypertrophy, and focal severe atherosclerosis of the right and left anterior descending arteries. Toxicology tests showed the presence of cocaine and marijuana in Sanders' body (see Amnesty International, 2004; Davis, 2004; *Fresno Bee*, 2004; Fresno County (California) Coroner's Report 04–08.172; Anglen, 2006).

Sanders did not collapse until he got into the ambulance, long after the final application of the TASER pulses, which indicated that the electric current did not immediately affect the rhythm of his heart. Because the TASER device did not have an immediate effect and the medical examiner attributed the death to cocaine intoxication, the device is excluded as a cause of death and as a significant contributing factor in the Sanders case.

LAWRENCE SAMUAL DAVIS

AGE: 40
RACE/GENDER: AFRICAN-AMERICAN MALE
AGENCY: PHOENIX (ARIZONA) POLICE
   DEPARTMENT
DATE OF INCIDENT: AUGUST 24, 2004
CAUSE OF DEATH: COMPLICATIONS OF EXCITED
   DELIRIUM
ROLE OF TASER DEVICE: EXCLUDED

Police were conducting a surveillance of a suspected drug house. Officers saw a subject matching the description of the main suspect walking away from the house with another man. Undercover officers called for a marked police unit to stop the suspect. As the officers approached, the other man, later identified as Davis, came up and threw himself against the side of the police car. At first, officers ignored him while they tried to stop the other suspect, but Davis continued to throw himself against their car. He jumped on the windshield of a patrol car and began yelling incoherently. Officers followed Davis as he walked away from the car and rounded a corner. When Davis saw the police, he again ran toward the vehicle and jumped on the bumper before officers attempted to detain him. When Davis pushed the officers, an officer discharged a TASER cartridge at him, but the probe missed, and the charge had no effect. Officers forced Davis to the ground and shocked him twice in drive-stun, but the TASER device had no effect. A sergeant used a carotid restraint temporarily to render Davis unconscious. Once officers had secured Davis, they called for an ambulance. In the ambulance, Davis calmed down, but the paramedics called ahead to warn the hospital that they were bringing in a combative patient. At the hospital, however, as paramedics rolled Davis into the emergency room, a doctor noticed that Davis was not breathing. Davis died less than an hour later.

The medical examiner said Davis died of excited delirium. Toxicology tests revealed cocaine and methamphetamine in Davis' system (see Amnesty International, 2004; Maricopa County (Arizona) Medical Examiner's Report 04–02994; Phoenix (Arizona) Police Report 2004–41618520; Anglen, 2006).

Davis continued to struggle with officers and paramedics for several minutes following the application of the TASER pulses, which indicated that the electric current did not immediately affect the rhythm of his heart. Because the TASER device did not have an immediate effect, and because the medical examiner attributed the death to excited delirium, the device is excluded as a cause of death and as a significant contributing factor in the Davis case.

JASON DAVID YEAGLEY

AGE: 32

RACE/GENDER: WHITE MALE
AGENCY: POLK COUNTY (FLORIDA) SHERIFF'S OFFICE
DATE OF INCIDENT: AUGUST 27, 2004
CAUSE OF DEATH: EXCITED DELIRIUM DUE TO ALPRAZOLAM WITHDRAWL
CONTRIBUTING FACTORS: STRUGGLE DURING ARREST
ROLE OF TASER DEVICE: EXCLUDED

Several motorists called the 9-1-1 emergency center and reported a man wandering in an intersection, acting erratically. When a deputy got to the intersection, Yeagley was still wandering in and out of traffic, mumbling to himself and appearing distressed. The deputy tried several times to coax Yeagley out of the roadway, and when he refused, she tried to remove him from the intersection. A struggle ensued. Yeagley grabbed the deputy's handcuffs and threw them, prompting her to fire a TASER cartridge. When the TASER device showed no effect, the deputy fired again. When the deputy and other officers finally handcuffed Yeagley, they noticed he was in medical distress, so they called for an ambulance. Paramedics revived Yeagley at the scene, but he went into cardiac arrest again at the hospital, and was pronounced dead a little more than an hour later.

The medical examiner listed Yeagley's death as an accident, caused by excited delirium brought on by his withdrawal from the antianxiety drug Alprazolam, also known by the brand name Xanax. The medical examiner noted that a significant contributing cause was Yeagley's struggle during the arrest. The medical examiner had noted that Yeagley had stopped taking his medication, that he had become agitated and sleepless, and that he had begun having auditory and visual hallucinations (see Amnesty International, 2004; Greenwood & Bernard, 2004; Office of the Medical Examiner, Tenth District (Florida) Report 2004–10–FA–394; Polk County (Florida) Sheriff's Office Report 2004–145578; Schmidt, 2005; Anglen, 2006).

Yeagley continued to struggle with officers following the application of the TASER pulses, indicating that the electric current did not immediately affect the rhythm of his heart. Because the TASER device did not have an immediate effect and the medical examiner attributed the death to excited delirium, the device is excluded as the cause of death and as a significant contributing factor in the Yeagley case.

MICHAEL ROBERT ROSA

AGE: 38
RACE/GENDER: HISPANIC MALE
AGENCY: SEASIDE (CALIFORNIA) POLICE DEPARTMENT
DATE OF INCIDENT: AUGUST 29, 2004
CAUSE OF DEATH: VENTRICULAR ARRHYTHMIA WHILE IN AN AGITATED DELIRIOUS OR PSYCHOTIC STATE CAUSED BY ACUTE METHAMPHETAMINE INTOXICATION
CONTRIBUTING FACTORS: TASER DEVICE APPLICATION AND ARREST BY POLICE
ROLE OF TASER DEVICE: EXCLUDED/DOUBTFUL

The police incident report was not available in this incident, but media accounts reported that residents in Rosa's neighborhood called police because there was a disturbance. They claimed that Rosa was wandering through yards and screaming for no apparent reason. An officer tried to talk to him, but Rosa avoided him, running through several backyards. Officers from other agencies arrived and caught up with Rosa, who began swinging a two-by-four board that he had picked up. Police told him to drop it, but he continued swinging. Officers shocked him twice with a TASER device. After handcuffing Rosa, officers noticed he was having trouble breathing. Rosa was taken to a local hospital where he later died.

A coroner's investigation discovered that Rosa had significant amounts of methamphetamine in his system. The medical examiner stated that Rosa died of a ventricular arrhythmia while in an agitated delirious or psychotic state caused by acute methamphetamine intoxication. He added that the added stress and/or psychologic effects of the TASER application and the arrest very likely contributed to the death (see Amnesty International, 2004; Monterey County Coroner's Report 04–256; Ravn, 2004; Aljentera, 2005; Anglen, 2006).

Because the medical examiner attributed the death to the effects of methamphetamine intoxication, the TASER device is excluded as the cause of death in the Rosa case. The application of the TASER pulses could have contributed to the physiological and psychological stress that Rosa experienced. However, the medical examiner provided no pathological evidence to support that conclusion, and he could not say that, but for the application of the TASER device, Rosa would have survived. Whether another form of restraint would have produced a different result is questionable. Consequently, the role of the TASER device as a significant contributing factor is doubtful.

SAMUEL RAMON WAKEFIELD

AGE: 22
RACE/GENDER: HISPANIC MALE
AGENCY: JOHNSON COUNTY (TEXAS) SHERIFF'S OFFICE
DATE OF INCIDENT: SEPTEMBER 12, 2004
CAUSE OF DEATH: COCAINE OVERDOSE
ROLE OF TASER DEVICE: EXCLUDED

Wakefield was a passenger in a car stopped by police for suspicion of drunken driving. The officer called for backup because Wakefield was behaving furtively and trying to leave the car. When other officers arrived, Wakefield tried to run, but he fell into a ditch. An officer drive stunned Wakefield three times with a TASER device. After the officers got handcuffs on Wakefield, they sat him up and began asking him questions. A few minutes later, Wakefield vomited and appeared to have a seizure. Officers called for paramedics, who transported Wakefield to a local hospital, where he was pronounced dead. Other passengers in the car later told investigators that they had been to a party in a nearby town. Wakefield had been drinking and smoking marijuana. They said he also swallowed a large quantity of cocaine about one hour before the initial traffic stop, and he was acting strangely when they left the party.

The medical examiner ruled that Wakefield died from a cocaine overdose. He found cocaine in Wakefield's stomach, and toxicology tests revealed a high concentration of cocaine in his blood (see Amnesty International, 2004; Johnson County (Texas) Sheriff's Office Report S0432593; Johnson County (Texas) Medical Examiner Report WA04–27; Anglen, 2006).

Wakefield continued to talk to officers for several minutes following the application of the TASER pulses, indicating that the electric current did not immediately affect the rhythm of his heart. Because the TASER device had no immediate effect and the medical examiner attributed the death to a cocaine overdose, the device is excluded as a cause of death and as a significant contributing factor in the Wakefield case.

ANDREW LAMAR WASHINGTON

AGE: 31
RACE/GENDER: AFRICAN-AMERICAN MALE
AGENCY: VALLEJO (CALIFORNIA) POLICE DEPARTMENT
DATE OF INCIDENT: SEPTEMBER 16, 2004
CAUSE OF DEATH: CARDIAC ARREST ASSOCIATED WITH EXCITEMENT DURING A POLICE CHASE AND COCAINE AND ALCOHOL INTOXICATION, OCCURING SHORTLY AFTER TASER SHOCK
ROLE OF TASER DEVICE: EXCLUDED

The police and autopsy reports for this incident were not available, but media sources reported that police were called after Washington's car struck a parked car and left the area. Later, police located Washington, but he fled on foot, initiating a chase that lasted more than 30 minutes. Officers caught up to him as he climbed a chain link fence in a canal area. An officer shocked him with a TASER device, and he fell into the canal. In an attempt to secure him, the officer shocked Washington 17 times in three minutes. Officers finally secured Washington, and, as they led him away, they noticed he was not breathing normally. Officers called for an ambulance, which transported Washington to a local hospital, where he died.

According to media reports, the coroner listed Washington's cause of death as cardiac arrest associated with excitement during the police chase and cocaine and alcohol intoxication, occurring shortly after a TASER shock, but he did not know whether the TASER device caused Washington's death. Washington also had a blood alcohol content of 0.19 (see Amnesty International, 2004; Gafni, 2005, January 6; Ginsburg, 2005; Anglen, 2006).

Following the last application of the TASER pulses, Washington was still able to walk from the scene with officers, indicating that the electric current did not immediately affect the rhythm of his heart. The association that the coroner listed in the official cause of death is a statement of temporal correlation, not a statement of causation, and he later admitted that he could not make that causal connection. Because the TASER device did not have an immediate effect and the medical examiner attributed the cause of the cardiac arrest to cocaine and alcohol, the device is excluded as the cause of death and as a significant contributing factor.

JOHN ALEX MERKLE

AGE: 40
RACE/GENDER: WHITE MALE
AGENCY: MIAMI-DADE (FLORIDA) POLICE DEPARTMENT
DATE OF INCIDENT: SEPTEMBER 20, 2004
CAUSE OF DEATH: EXCITED DELIRIUM ASSOCIATED WITH COCAINE INTOXICATION
ROLE OF TASER DEVICE: EXCLUDED

Merkle, who had a history of cocaine use and drug arrests, was running through backyards in his neighborhood and acting erratically. Police found him inside an abandoned house, where he was beating the walls and windows with a large stick and screaming that someone was trying to kill him. Officers were able to get Merkle to drop the stick, but he started swinging his fists at them when they attempted to arrest him. They shocked him with a TASER device. The officers reported that he was feverish and excited and that he repeatedly attempted to lie down. Soon after officers handcuffed him, he became unresponsive. He died at a local hospital.

Toxicology tests showed that Merkle had significant levels of cocaine and cocaine metabolites in his system at the time of his death. A medical examiner found that Merkle suffered from cardiomegaly and that he had a partially blocked coronary artery. He ruled that Merkle's death was an accident caused by excited delirium associated with cocaine intoxication (see Miami-Dade County (Florida) Medical Examiner's Report 04–2086; Miami (Florida) Police Report 26415714; Barton, 2005; Anglen, 2006).

Merkle continued to struggle with officers following the application of the TASER pulses, indicating that the electric current did not significantly affect the rhythm of his heart. Because the TASER device did not have an immediate effect and the coroner attributed the death to cocaine-induced excited delirium, the device is excluded as the cause of death and as a significant contributing factor in the Merkle case.

DWAYNE ANTHONY DUNN

AGE: 33
RACE/GENDER: AFRICAN-AMERICAN MALE
AGENCY: LAFAYETTE (LOUISIANA) POLICE DEPARTMENT
DATE OF INCIDENT: OCTOBER 4, 2004
CAUSE OF DEATH: ACUTE CARDIAC ARRHYTHMIA DUE TO COCAETHYLENE CARDIOTOXICITY DUE TO ACUTE AND CHRONIC SELF-USE OF COCAINE AND ETHANOL
CONTRIBUTING FACTORS: COCAINE-RELATED ENTEROISCHEMIA
ROLE OF TASER DEVICE: EXCLUDED

Police received a complaint that Dunn was intoxicated at a local supermarket. When officers arrived, they found the glass door of the supermarket broken and Dunn sleeping on the ground in the parking lot. When officers ordered him to get up, he refused. The officers arrested Dunn and tried to place him in a police car, but he resisted. An officer drive-stunned Dunn one time with the TASER device to get him into the car. The officer booked Dunn into the Lafayette Parish jail, where he signed a refusal for medical treatment. Approximately four hours later, jailer staff put Dunn on medical watch after they saw that he was acting strangely. Later, jailers called for an ambulance to transport Dunn to a local hospital when his condition deteriorated. He died later that day.

A coroner's report noted that Dunn died from acute cardiac arrhythmia due to cocaethylene cardiotoxicity due to acute and chronic self-use of cocaine and ethanol. The coroner listed a contributing factor of cocaine-related enteroischemia. The coroner noted that the TASER device had nothing to do with Dunn's death (see Amnesty International, 2004; *Associated Press*, 2004, October 6; *Associated Press*, 2004, November 19; Lafayette (Louisiana) Police Report 04-00231203; Lafayette Parish (Louisiana) Coroner Report 230-04AP; Burgess, 2005; Anglen, 2006).

Dunn continued to walk and talk for several hours following the application of the TASER pulses, indicating that the electric current did not immediately affect the rhythm of his heart. Because the TASER device did not have an immediate effect and the coroner attributed the cardiac arrhythmia to the effects of cocaine and alcohol, the device is excluded as the cause of death and as a significant contributing factor in the Dunn case.

CHRISTI MICHELE BALL

AGE: 35
RACE/GENDER: WHITE FEMALE
AGENCY: FORT WORTH (TEXAS) POLICE
  DEPARTMENT
DATE OF INCIDENT: OCTOBER 14, 2004
DATE OF DEATH: OCTOBER 21, 2004
CAUSE OF DEATH: HYPERTROPHIC CARDIOMYOPATHY
ROLE OF TASER DEVICE: EXCLUDED

Neither the police report nor the autopsy report in this incident was available, but a newspaper account reported that Ball, who had a long history of mental illness, had quit taking her medications. On October 14, Ball presented herself to a local hospital claiming to be pregnant and seeking help for lightheadedness, irritability, and mood swings. When a nurse asked Ball about her suffering from mania, Ball refused treatment, and she refused to leave the emergency room. An officer arrived to escort her from the property. When she resisted, he shocked her once in the leg with a TASER device to get her into his car. He then took Ball to a local psychiatric care facility. For the next several days, Ball was in and out of many different hospitals. Finally, when she refused to leave one hospital after refusing treatment, an officer arrested her for trespassing. At the jail, she refused to take prescribed medications, and the jail staff had her on enhanced supervision as a suicide risk. Despite several trips to the infirmary and visits with doctors, on October 21, jail staff found her slumped against the doorway, not breathing. Paramedics took Ball to a local hospital, where she died.

Media sources reported that the coroner's report found Ball died of a genetic condition that had gone undetected–hypertrophic cardiomyopathy. There was no reported mention of the TASER device in the autopsy report (Autrey, 2005).

For seven days following application of the TASER pulses, Ball was walking and talking, proof that the electric current did not immediately affect the rhythm of her heart. Because the TASER device did not have an immediate effect and the coroner attributed the death to hypertrophic cardiomyopathy, the device is excluded as the cause of death and as a significant contributing factor in the Ball case.

GRESHMOND GRAY

AGE: 25
RACE/GENDER: AFRICAN-AMERICAN MALE
AGENCY: LAGRANGE (GEORGIA) POLICE
  DEPARTMENT
DATE OF INCIDENT: NOVEMBER 2, 2004
CAUSE OF DEATH: COMBINED EFFECTS OF THE
  PHYSIOLOGIC STRESS OF A PHYSICAL ALTERCATION (INCLUDING HAVING BEEN SHOT BY A
  TASER DEVICE) AND ENLARGEMENT AND
  FIBROSIS OF THE HEART
CONTRIBUTING FACTORS: HISTORY OF NONRECENT COCAINE USE
ROLE OF TASER DEVICE: EXCLUDED/POSSIBLE

Police received a call from the female residents of an apartment because Gray was intoxicated, cursing, threatening people, and refusing to leave.

Officers tried to convince Gray to leave, but he would not go. When the officers decided to arrest Gray, he would not follow orders to place his hands behind his back. When Gray bent down to pick up a hibachi loaded with hot coals, an officer shocked him with a TASER device. Officers shocked him twice more as he was trying to run away. After the third TASER stun, officers realized Gray was unresponsive. They started CPR, applied an automated defibrillator, which applied two discharges without effect, and called for an ambulance. Gray died shortly after arriving at a local hospital. The Georgia Bureau of Investigation investigated the incident.

The coroner reported that Gray died of the combined effects of the physiologic stress of a physical altercation, including having been shocked with a TASER device, and enlargement and fibrosis of the heart. He found that Gray had an abnormal heart, with significant abnormalities of the heart muscle that he probably had since birth. Toxicology tests were positive for metabolites of marijuana, and Gray's blood alcohol level was 0.14. The coroner explained that the emotional and physical stress that Gray underwent during the struggle with police led to the abnormal and lethal heart rhythm disturbance. He explained that, in individuals with significant underlying heart disease, physiologic stress, such as being shocked with a TASER device, can trigger the heart disease and result in a heart attack. He concluded that the electrical current from the TASER pulses did not directly cause the electrical abnormality in Gray's heart rhythm (see *Associated Press*, 2004, November 3; Georgia Bureau of Investigation Record of Medical Examiner Report 2004–1028709; LaGrange (Georgia) Police Report 041131154; Amnesty International, 2006; Anglen, 2006).

Gray became unresponsive during the application of the TASER pulses during the struggle. The medical examiner explained how the added stress caused by the TASER device, coupled with the additional stresses of the struggle and his underlying heart disease, contributed to Gray's arrhythmia. However, he could not say that, but for the application of the TASER pulses, Gray would have survived. Considering Gray's medical condition, it is questionable whether some other form of restraint would have yielded a different result. Consequently, the TASER device is excluded as the cause of death, but role of the device as a significant contributing factor is possible.

ROBERT GUERRERO

AGE: 21
RACE/GENDER: HISPANIC MALE
AGENCY: FORT WORTH (TEXAS) POLICE DEPARTMENT
DATE OF INCIDENT: NOVEMBER 2, 2004
CAUSE OF DEATH: COCAINE INTOXICATION
CONTRIBUTING FACTORS: MYOCARDITIS AND HEPATIC STEATOSIS
ROLE OF TASER DEVICE: EXCLUDED

Guerrero fled from police after an apartment manager reported that he was stealing electricity for a friend's apartment. Officers discovered Guerrero hiding in a bedroom closet. When he refused to come out, officers shocked him four times with a TASER device, striking him in the chest. The officers handcuffed Guerrero and carried him down the stairs. Witnesses said Guerrero was not breathing as officers carried him out of the apartment. Officers administered CPR and called for medical assistance. Paramedics transported Guerrero to a local hospital where he died.

The medical examiner reported that Guerrero died from heart failure due to cocaine intoxication. He listed myocarditis and hepatic steatosis as contributing factors. He also noted that, although there was a temporal relationship between application of the TASER pulses and unconsciousness, he could establish no causal affect of the TASER device to the death. He ruled the death an accident (see Fort Worth (Texas) Police Report 04132367; Olsen & Davis, 2004; Tarrant County (Texas) Medical Examiner's Report 0409265; *Associated Press*, 2005, April 27; Amnesty International, 2006; Anglen, 2006).

Guerrero collapsed very shortly following the application of the TASER pulses, but the medical examiner reported he was unable to find any evidence that the TASER device contributed to Guerrero's heart failure. Consequently, the device is excluded as the cause of death and as a significant contributing factor in the Guerrero case.

KEITH RAYMOND DRUM

AGE: 41
RACE/GENDER: WHITE MALE
AGENCY: CLEARLAKE (CALIFORNIA) POLICE DEPARTMENT
DATE OF INCIDENT: NOVEMBER 7, 2004
CAUSE OF DEATH: CARDIORESPIRATORY ARREST ASSOCIATED WITH BLUNT FORCE INJURIES, METHAMPHETAMINE INTOXICATION, AND THE PHYSIOLOGICAL STRESS OCCURRING DURING A STRUGGLE WITH AND RESTRAINT BY POLICE
CONTRIBUTING FACTORS: HYPERTENSIVE AND ARTERIOSCLEROTIC CARDIOVASCULAR DISEASE AND CIRRHOSIS OF THE LIVER
ROLE OF TASER DEVICE: EXCLUDED/DOUBTFUL

The police report on this incident was exempt from disclosure under California law, but media sources reported that Drum called 9-1-1 from his residence. He was difficult to understand, but he was trying to report something about a body or bodies in the trunk of a vehicle. As officers were responding, they discovered that there was a warrant outstanding for Drum's arrest. Two officers from the Clearlake Police Department responded and spoke with Drum. Once the officers were inside his house, Drum charged the officers. One of the officers used a TASER device on Drum, but it had no effect. The officer also sprayed chemical spray in Drum's face, but it also had no effect at first. After the officer sprayed Drum a second time with chemical spray, he calmed down, and the officers were able to get one handcuff on him. Drum started to fight again, and, as the fight continued, Clearlake officers called for backup from the Lake County Sheriff's Department. Two deputies responded. Once the deputies arrived, Drum continued to fight with all four of the officers. The struggle continued for more than nine minutes. Only after the officers handcuffed Drum and restrained his legs, did he calm down. Drum stopped breathing, and the officers called for rescue personnel. Paramedics responded and transported Drum to a local hospital, where he died.

The medical examiner found that the cause of Drum's death was complicated. He ruled that Drum died of cardiorespiratory arrest associated with blunt force injuries, methamphetamine intoxication, and the physiological stress occurring during a struggle with and restraint by police officers. He listed significant contributing conditions as hypertensive and arteriosclerotic cardiovascular disease and cirrhosis of the liver. According to the medical examiner, each condition in and of itself would not have caused death. However, the struggle with the officers combined with all the other conditions caused Drum's death (see Lake County (California) Coroner's Report 10621; Stoneberg, 2005; Anglen, 2006).

Drum continued to struggle with officers for nine minutes following the application of the TASER pulses, indicating that the electric current did not immediately affect the rhythm of his heart. Because the TASER device did not have an immediate effect, the device is excluded as the cause of death. The medical examiner noted that the TASER shocks created some of the physiological stress that led, ultimately, to Drum's death. He also admitted that the TASER pulses, alone, would not have killed Drum. Moreover, Drum continued to fight following the last application of the TASER device. It is questionable that some other form of restraint would have produced a different result. Consequently, the role of the TASER device as a significant contributing factor in the Drum case is doubtful.

RICARDO REYES ZARAGOZA

AGE: 40
RACE/GENDER: HISPANIC MALE
AGENCY: SACRAMENTO COUNTY (CALIFORNIA) SHERIFF'S OFFICE
DATE OF INCIDENT: NOVEMBER 8, 2004
CAUSE OF DEATH: SUDDEN CARDIAC ARREST IMMEDIATELY FOLLOWING A PROLONGED STRUGGLE WITH ATTEMPTS AT FORCEFUL RESTRAINT DUE TO EXCITED DELIRIUM DUE TO SCHIZOPHRENIA TREATED WITH OLANZAPINE
CONTRIBUTING FACTORS: FOCAL SEVERE CORONARY ARTERY ATHEROSCLEROSIS; OBESITY
ROLE OF TASER DEVICE: EXCLUDED/DOUBTFUL

The police report on this incident was exempt from disclosure under California law, but media sources reported that Zaragoza's parents called sheriff's deputies to their home to report his erratic behavior, which included tearing boards off the fence. Zaragoza, who had a history of paranoid schizophrenia, had been taking his medicine, but he had not eaten for five days. His incoherent statements led the deputies to decide that Zaragoza needed a mental health evaluation. While the deputies spoke with his parents, Zaragoza left the room to enter his bedroom. Deputies followed, one using his foot to block Zaragoza from closing the bedroom door. When Zaragoza resisted their efforts to remove him from the room, the deputies sprayed Zaragoza with chemical spray, shocked him twice with a TASER device, and handcuffed him. After being handcuffed, Zaragoza collapsed forward and struck his head on concrete as officers were taking him out of the house, but he continued to struggle before becoming motionless. Zaragoza's father noticed that Zaragoza was no longer breathing. Paramedics arrived and transported Zaragoza to a local hospital where he died.

The medical examiner reported that Zaragoza was taking a relatively new antipsychotic medication, Olanzapine. He noted that normal therapeutic ranges of Olanzapine were 20 to 40 ng/mL of blood. Concentrations of 80 ng/mL were considered the threshold for developing adverse effects. Although the toxicology lab had difficulty quantifying the level of Olanzapine in Zaragoza's system, they were able to give an approximate level of 170 ng/mL. The medical examiner noted that Zaragoza's mild hepatic steatosis might have had some affect on Zaragoza's ability to metabolize the drug. The medical examiner ruled that Zaragoza died from a heart attack brought on by excited delirium and schizophrenia. He noted contributing factors of focal severe coronary artery atherosclerosis and obesity (see Fletcher & Jewett, 2004; Sacramento County (California) Coroner's Report 04–05571; Hume, 2005; Amnesty International, 2006; Anglen, 2006).

Zaragoza continued to struggle with officers following the application of the TASER pulses, and he was walking after they handcuffed him and started to walk him out of the house, indicating that the electric current did not immediately affect the rhythm of his heart. Because the TASER device did not have an immediate effect and the coroner attributed the death to excited delirium, the device is excluded as the cause of death and as a significant contributing factor in the Zaragoza case.

JESSIE ROBERT TAPIA

AGE: 37
RACE/GENDER: HISPANIC MALE
AGENCY: POMONA (CALIFORNIA) POLICE
    DEPARTMENT
DATE OF INCIDENT: NOVEMBER 15, 2004
CAUSE OF DEATH: EXCITED DELIRIUM WITH
    HYPERTHERMIA DUE TO THE COMBINED EFFECTS
    OF COCAINE, METHAMPHETAMINE, AND
    HYDROCODONE
CONTRIBUTING FACTORS: ATHEROSCLEROTIC
    HEART DISEASE, HEPATIC CIRRHOSIS, AND
    HEPATITIS C VIRUS
ROLE OF TASER DEVICE: EXCLUDED

The police report in this incident was exempt from disclosure under California law, but media sources reported witnesses called 9-1-1 reporting that Tapia was walking into traffic and screaming that his life was in danger. When Tapia noticed a woman, who was a civilian employee of the Pomona police department, in her car looking at him, he began pounding on her vehicle and screaming that someone was trying to kill him. She radioed the police. The first officer on the scene tried to calm Tapia, but he would not listen to the officer. He kept claiming that people in passing cars were pointing rifles at him. The officer used a TASER device in an attempt to secure Tapia, but it had little effect. Tapia remained very aggressive, and, when he saw the female employee of the department get out of her car, he started toward her. She used a TASER device on Tapia, and he dropped to the ground. Officers handcuffed him, bound his feet together, and called for medical assistance. When paramedics arrived, Tapia was breathing and had a pulse. While the paramedics were checking him, Tapia complained that he could not breathe. When he suddenly stopped breathing, paramedics transported him to a local hospital, where he died.

The coroner reported that, upon arrival at the hospital, Tapia had a rectal temperature of 104.6 degrees Fahrenheit. The autopsy revealed that Tapia suffered from occlusive atherosclerotic coronary artery disease, cardiomegaly, left ventricular hypertrophy, hepatic cirrhosis, and hepatitis C virus. Toxicology tests also revealed significant amounts of cocaine, methamphetamine, and hydrocodone in Tapia's system. The coroner ruled that Tapia died of excited delirium with hyperthermia due to the combined effects of cocaine, methamphetamine, and hydrocodone. He also concluded that atherosclerotic heart disease, hepatic cirrhosis, and hepatitis C virus were contributing factors. He concluded that the restraint maneuvers showed no evidence of asphyxia, and he made a specific notice that there was no temporal relationship with the use of the TASER devices and the death (see County of Los Angeles (California) Coroner's Report 2004–08657; Sholley, 2004).

For several minutes following the last application of the TASER pulses, Tapia showed no signs of medical distress, indicating that the electrical current did not immediately affect his heart. Because the TASER device did not have an immediate effect and the coroner attributed the death to excited delirium and hyperthermia due to a combination of cocaine, methamphetamine, and hydrocodone, the device is excluded as the cause of death and as a significant contributing factor in the Tapia case.

CHARLES CHRISTOPHER KEISER

AGE: 47
RACE/GENDER: WHITE MALE
AGENCY: LIVINGSTON COUNTY (MICHIGAN) SHERIFF'S OFFICE
DATE OF INCIDENT: NOVEMBER 25, 2004
CAUSE OF DEATH: DROWNING
ROLE OF TASER DEVICE: EXCLUDED

Keiser drove a bulldozer from a nearby road construction site onto a highway just before morning rush hour. When police arrived, he was on a backhoe apparently trying to start it so he could move it onto the freeway also. Two Michigan State Police officers chased Keiser into the woods nearby and scuffled with him. Keiser was attempting to strangle one of the officers when four sheriff's deputies from the Livingston County Sheriff's Office arrived to assist. They wrestled with Keiser in a swampy area. One deputy used a TASER device to control Keiser, delivering a drive stun three times before subduing him. The officers finally subdued Keiser, and he ended up face down and handcuffed in one foot of water. About one minute after officers removed him from the water, he stopped breathing.

The medical examiner ruled that Keiser died of an accidental drowning. The autopsy findings showed Keiser had ingested a considerable amount of muddy water. The exam revealed Keiser's trachea and bronchial area were filled with a thick, black, muddy material. The toxicology report came out negative for alcohol, but a urine sample tested positive for marijuana. Keiser's family requested a second autopsy. That pathologist concluded that use of a TASER device could have contributed to the drowning. The report stated that Keiser died of drowning in muddy water, adding that the use of a TASER device while immersed would have enhanced the drowning process. However, a Michigan State Police investigation showed that Keiser was standing when the deputies used the TASER device. After using the device, and while officers were trying to handcuff him, Keiser kept deliberately placing his face into the water. The independent autopsy and the official report both indicated Keiser had extensive amounts of mud in his airway, nostrils and mouth (see Livingston County (Michigan) Medical Examiner's Report 04–209; Tolen, 2004; Tolen, 2005, March 18; Tolen, 2005, April 15; Amnesty International, 2006; Anglen, 2006).

Keiser continued to struggle with officers following the application of the TASER pulses, indicating that the electric current did not immediately affect the rhythm of his heart. Although the use of a TASER device while immersed might contribute to the drowning process, the evidence shows that Keiser was not immersed when the officers used the device on him. Consequently, because the TASER device did not have an immediate effect and the coroner attributed the death to drowning, the device is excluded as the cause of death and as a significant contributing factor in the Keiser case.

BYRON W. BLACK

AGE: 39
RACE/GENDER: WHITE MALE
AGENCY: LEE COUNTY (FLORIDA) SHERIFF'S OFFICE
DATE OF FIRST INCIDENT: NOVEMBER 23, 2004
DATE OF SECOND INCIDENT: NOVEMBER 27, 2004
DATE OF DEATH: NOVEMBER 27, 2004
CAUSE OF DEATH: CARDIAC ARRHYTHMIA CAUSED BY A SEIZURE AND ALCOHOL WITHDRAWL
CONTRIBUTING FACTORS: TASER AND STRUGGLE WITH DEPUTIES
ROLE OF TASER DEVICE: EXCLUDED/UNDETERMINED

The Lee County Sheriff's Office report was not available, but media sources reported that on November 23, 2004, state fire marshals responded to Black's residence to investigate a suspected arson. Black had apparently set fire to his van inside his garage. When the fire marshal attempted to arrest Black, he resisted. A deputy with the Lee County Sheriff's Office touch-stunned Black one time with a TASER device. The deputy secured Black and took him to jail. On November 27, 2004, jailers thought Black was having a seizure and tried to remove him from his cell. When Black attacked a nurse and began to struggle with jailers, they sprayed him with chemical spray and shocked him with a TASER device before he collapsed.

The autopsy report was not available, but media sources reported that the medical examiner listed the cause of Black's death as cardiac arrhythmia caused by a seizure and alcohol withdrawal. He added that the struggle with jailers and the TASER shocks were contributing factors (see Florida Division of State Fire Marshal's Report 04-3821; Barton, 2005; Amnesty International, 2006; Anglen, 2006; *Associated Press*, 2006, January 11).

The first application of the TASER device, four days before his death clearly did not affect Black. The media reports do not specify how much time passed between the application of the TASER pulses and Black's collapse on the day he died. Because the medical examiner attributed the cardiac arrhythmia to a seizure caused by alcohol withdrawal, the TASER device is excluded as the cause of death. However, without additional information such as the police report or the autopsy, the role of the device as a significant contributing factor is undetermined.

PATRICK FLEMING

AGE: 35
RACE/GENDER: MALE (RACE NOT DETERMINABLE FROM THE RECORDS AVAILABLE)
AGENCY: JEFFERSON PARISH (LOUISIANA) SHERIFF'S OFFICE
DATE OF INCIDENT: DECEMBER 4, 2004
DATE OF DEATH: DECEMBER 5, 2004
CAUSE OF DEATH: Unknown
ROLE OF TASER DEVICE: UNDETERMINED

The police and autopsy reports for this incident were not available, but media sources reported that deputies stopped Fleming for driving erratically. When the deputies ordered him out of the vehicle, Fleming refused, so the deputies dragged him out of the car. Fleming, who was wanted on an outstanding warrant, became combative after being removed from the vehicle. One of the deputies used a TASER device to drive-stun Fleming. Once they had secured him, the deputies took Fleming to jail. While going through the booking procedure, Fleming became enraged and started fighting with the booking officer, hitting him several times. That deputy shocked Fleming with a TASER device, but he continued to resist until deputies eventually subdued him. At that point, Fleming began having breathing problems. Deputies called for medical assistance, and paramedics transported Fleming to a local hospital, where he died the next day. Media sources did not report a cause of death (see Perlstein, 2004; Philbin, 2004; Amnesty International, 2006; Anglen, 2006).

Without additional information as to the cause of death, the role of the TASER device in the Fleming case is undetermined.

KEVIN DOWNING

AGE: 36
RACE/GENDER: WHITE MALE
AGENCY: HOLLYWOOD (FLORIDA) POLICE DEPARTMENT
DATE OF INCIDENT: DECEMBER 15, 2004
CAUSE OF DEATH: COCAINE PSYCHOSIS AND EXCITED DELIRIUM DUE TO A LETHAL LEVEL OF COCAINE DUE TO A COCAINE OVERDOSE
CONTRIBUTING FACTORS: RARE MYOCARDIAL FIBROSIS, PULMONARY EDEMA, ALVEOLAR HEMORRHAGES, CEREBRAL EDEMA, CARDIOMEGALY, AND MYOCARDIAL CONTRACTION BAND NECROSIS
ROLE OF TASER DEVICE: EXCLUDED

Downing was grunting incoherently in the middle of a busy street, his van blocking two lanes of traffic. Paramedics, who happened by, stopped to help Downing. He became agitated and began screaming at them, so they called the police. When officers arrived, Downing was sitting in the van, clutching the steering wheel, making grunting noises, and sweating profusely. He refused to communicate with rescue workers or the police. When officers tried to pull Downing from the van, he punched and kicked them. An officer shocked Downing with a TASER device, but it had no effect. It took three officers to control Downing in handcuffs and leg restraints. Paramedics loaded Downing into an ambulance, although he continued to struggle, and took him to a local hospital. Upon arrival, he had a body temperature of 105.5 degrees Fahrenheit, and a urine toxicology screen was positive for cocaine and amphetamines.

An autopsy revealed that Downing died of cocaine psychosis and excited delirium due to a lethal level of cocaine due to a cocaine overdose. He had approximately five times the toxic level of cocaine in his system. The medical examiner also noted in the autopsy report several significant contributing factors: pulmonary edema, alveolar hemorrhages, cerebral edema, cardiomegaly, and myocardial contraction band necrosis. He noted that it was unlikely that the TASER device contributed to Downing's death (see Hollywood (Florida) Police Report 04–117086; Broward County (Florida) Medical Examiner Report 04–1798; Barton, 2005; *Associated Press*, 2005, January 6; Amnesty International, 2006; Anglen, 2006).

Downing continued to struggle with officers following application of the TASER pulses, indicating that the electric current did not immediately affect the rhythm of his heart. Because the TASER device did not have an immediate effect and the medical examiner attributed the death to cocaine-induced excited delirium, the device is excluded as the cause of death and as a significant contributing factor in the Downing case.

LYLE LEE NELSON

AGE: 35
RACE/GENDER: WHITE MALE
AGENCY: COLUMBIA (ILLINOIS) POLICE DEPARTMENT
DATE OF INCIDENT: DECEMBER 16, 2004
DATE OF DEATH: DECEMBER 17, 2004
CAUSE OF DEATH: ACUTE COCAINE TOXICITY
CONTRIBUTING FACTORS: HISTORY OF COCAINE ABUSE
ROLE OF TASER DEVICE: EXCLUDED

Police received a 9-1-1 hang-up call of a disturbance at Nelson's home. When dispatchers called back, Nelson's 8-year-old daughter answered the phone and told them that her mother and father were in the bathroom fighting. When officers arrived, they found that Nelson had been arguing with his wife over his smoking crack cocaine in the house. An officer drive-stunned Nelson eight times with a TASER device as Nelson fought to prevent officers from removing him from the bathroom. Once Nelson was subdued, officers said he appeared normal, and he answered all the questions and seemed fine during the booking process. He admitted that he had been smoking crack cocaine. About five hours later, however, Nelson began ripping at papers and the mattress in his cell and splashing himself with water from the toilet. Officers called for an ambulance when Nelson collapsed against a wall. Before the ambulance arrived, jail staff attached an automatic defibrillator, but it never authorized a shock. Paramedics took Nelson to a local hospital, where he died. The Illinois State Police Public Integrity Unit conducted the investigation.

The autopsy report was not available, but media sources reported that the medical examiner ruled Nelson died of acute cocaine toxicity, and a coroner's jury found that Nelson's death was an accident (see Columbia (Illinois) Police Report 04–16602; Goodrich, 2004; Goodrich, 2005; Amnesty International, 2006; Anglen, 2006).

Nelson continued to struggle with officers following the application of the TASER pulses, and he showed no signs of medical distress for more than five hours, indicating that the electric current did not immediately affect the rhythm of his heart. Because the TASER device did not have an immediate effect and the medical examiner attributed the death to cocaine toxicity, the device is excluded as the cause of death and as a significant contributing factor in the Nelson case.

DOUGLAS G. MELDRUM

AGE: 37
RACE/GENDER: MALE (RACE NOT DETERMINABLE FROM THE AVAILABLE RECORDS)
AGENCY: HEBER (UTAH) POLICE DEPARTMENT
DATE OF INCIDENT: DECEMBER 17, 2004
CAUSE OF DEATH: EXCITED DELIRIUM DUE TO ACUTE EPHEDRINE INTOXICATION
CONTRIBUTING FACTORS: LACERATED LIVER
ROLE OF TASER DEVICE: EXCLUDED

The police report and the autopsy report were exempt from disclosure under Utah law. However, media sources reported that Meldrum was driving erratically and that he ignored efforts by police to pull him over. A short chase ensued in which Meldrum ran four traffic lights, finally leaving the road and stopping his pickup in a snow-covered field. When officers attempted to remove Meldrum from his vehicle, he resisted. One officer shocked Meldrum with a TASER device, but he continued to fight. Officers then used chemical spray to subdue Meldrum and take him into custody. After officers handcuffed Meldrum, they noticed he was not breathing. An ambulance took Meldrum to a local hospital, where he died.

According to media reports, the medical examiner ruled that Meldrum died of cardiopulmonary arrest due to excited delirium due to acute ephedrine intoxication. He also noted that Meldrum had a lacerated liver due to blunt force trauma, but he did not mention the TASER shocks in the report (see Canham, 2004; Dobner & Choate, 2004; Hyde, 2005; Rosetta, 2005; Amnesty International, 2006; Anglen, 2006).

Meldrum continued to struggle for a short time following application of the TASER pulses, indicating that the electric current did not immediately affect the rhythm of his heart. Because the TASER device did not have an immediate effect and the medical examiner attributed the cardiopulmonary arrest to excited delirium, the device is excluded as the cause of death and as a significant contributing factor in the Meldrum case.

DAVID J. COOPER

AGE: 40
RACE/GENDER: MALE (RACE NOT DETERMINABLE FROM THE AVAILABLE RECORDS)
AGENCY: WHITELAND (INDIANA) POLICE DEPARTMENT
DATE OF FIRST INCIDENT: DECEMBER 19, 2004
AGENCY: JOHNSON COUNTY (INDIANA) SHERIFF'S OFFICE
DATE OF SECOND INCIDENT: DECEMBER 21, 2004
DATE OF DEATH: DECEMBER 30, 2004
CAUSE OF DEATH: CARDIAC ARREST DUE TO PROBABLE CARDIAC DYSRHYTHMIA
CONTRIBUTING FACTORS: SCHIZOPHRENIA

The police and autopsy reports in this incident were not available, but media sources reported that on December 19, Cooper asked a neighbor for a ride to church. However, he got out of the car en route, dropped to his knees, and prayed in the middle of the street. Eventually, the neighbor got Cooper to the church, but, before the celebration could begin, Cooper attacked and strangled the pastor, whom he had known for more than 20 years.

When police arrived, Cooper charged them as they tried to resuscitate the pastor. The officers shocked Cooper seven times with a TASER device to subdue him. After they handcuffed him, it took six officers to take Cooper to the transport van. In a later interrogation, Cooper claimed he was killing the devil.

On December 21, when his attorney came to the jail to speak with him, Cooper again became violent, and he began banging his head against the wall of his cell. Jailers again had to use a TASER device to subdue Cooper to get him medical attention. Because of his mental condition, on December 27 officials transferred Cooper to the Indiana Department of Correction diagnostic center at Wishard Memorial Hospital. Shortly after his arrival at Wishard, Cooper stopped breathing, six days after the last application of the TASER device. Medical staff placed Cooper on a respirator, but he died on December 30.

Media sources reported that the coroner ruled Cooper died of cardiac dysrhythmia, a condition he had long before encountering the officers. He added that no injury that Cooper suffered, either from the TASER device or from banging his head on the wall was sufficient to cause his death (see *Associated Press*, 2004, December 21; *Associated Press*, 2004, December 22; *Associated Press*, 2004, December 30; *Associated Press*, 2005, January 5; Bird, 2005; Amnesty International, 2006; Anglen, 2006).

Cooper showed no signs of medical distress for six days following the last application of the TASER pulses, indicating that the electric current had no immediate affect on his heart. Because the TASER device had no immediate effect and the medical examiner attributed the death to cardiac dysrhythmia, the device is excluded as the cause of death and as a significant contributing factor in the Cooper case.

JEANNE MARIE HAMILTON

AGE: 46
RACE/GENDER: WHITE FEMALE
AGENCY: INYO COUNTY (CALIFORNIA) SHERIFF'S OFFICE
DATE OF INCIDENT: DECEMBER 22, 2004

CAUSE OF DEATH: UNKNOWN
ROLE OF TASER DEVICE: UNDETERMINED

California Highway Patrol officers spotted a vehicle coming up behind them at 91 miles per hour. The vehicle slowed briefly and then accelerated reaching speeds nearing 100 miles per hour as the officers pursued it. Eventually, the vehicle pulled over, but the driver, Hamilton, refused to get out. The officers sprayed Hamilton with chemical spray, but it had no effect. With the help of an Inyo County Sheriff's Office deputy, the troopers pulled Hamilton out of the vehicle. Hamilton struggled with the officers, and they forced her to the ground, but she refused to put her hands behind her back. The deputy used a TASER device to shock Hamilton twice. She finally put her hands behind her back, and the officers handcuffed her. The deputy took her to a local hospital to obtain a blood sample for alcohol testing. After a doctor had medically cleared Hamilton, the troopers took her to jail. At the jail, Hamilton struggled throughout the booking process. Jailers found a small amount of methamphetamine and drug paraphernalia on Hamilton during booking. The jailers placed her in a cell, but they soon noticed that she had stopped breathing. Jail staff performed CPR on Hamilton, and paramedics took her back to the local hospital, where she died (see Department of California Highway Patrol Report 05553 UN; Botonis, 2005, January 6; Amnesty International, 2006; Anglen, 2006).

The Inyo County Coroner did not respond to a public records request for a copy of the autopsy, and media sources do not report a cause of death. Without additional information, the role of the TASER device in the Hamilton case is undetermined.

RONNIE JAMES PINO

AGE: 31
RACE/GENDER: HISPANIC MALE
AGENCY: SACRAMENTO COUNTY (CALIFORNIA) SHERIFF'S OFFICE
DATE OF INCIDENT: DECEMBER 22, 2004
DATE OF DEATH: DECEMBER 23, 2004
CAUSE OF DEATH: SUDDEN UNEXPECTED DEATH

SYNDROME DUE TO PROBABLE SEIZURE DUE TO CHRONIC SEIZURE DISORDER DUE TO POST-SURGICAL REMOVAL OF GANGLIOGLIOMA
CONTRIBUTING FACTORS: CARDIOMEGALY, ATRIAL FLUTTER, CHRONIC HYPERAMMONEMIA AND HYPONATREMIA, OBESITY, DEVELOPMENTAL DELAY, NONSPECIFIC PSYCHIATRIC DISORDER WITH HALLUCINATIONS
ROLE OF TASER DEVICE: EXCLUDED

The police report regarding this incident was exempt from disclosure under California law, but media sources reported that Pino, a patient at a mental hospital, was taking psychotropic drugs three times a day for a variety of medical conditions, including antipsychotic, heart, and epilepsy medications. He also had a vagus nerve stimulator implanted near his heart to control grand mal seizures, a result of abnormal electrical discharges in his brain. Pino wanted to go outside to smoke a cigarette, but staff members tried to stop him. As staff members started to close in on him, he smashed a glass pane of a door with his foot. The staff called for police. When officers arrived, they found Pino with cuts on his arms and legs. He flailed his arms when officers tried to arrest him, striking one of them in the face. Officers shocked Pino twice with a TASER device to subdue him. Paramedics arrived and checked Pino, giving medical clearance to transport him to jail. Pino underwent a medical screening during booking and was placed on the jail's medical floor, where his condition was checked every half hour. Jail staff found him dead about 17 hours after his struggle with police.

The coroner's report lists Pino's cause of death as sudden unexpected death syndrome due to probable seizure due to chronic seizure disorder due to post-surgical removal of ganglioglioma. The coroner also noted a number of other significant conditions: cardiomegaly, atrial flutter, chronic hyperammonemia, hyponatremia, obesity, developmental delay, and a nonspecific psychiatric disorder with hallucinations (see Hume, 2004; Sacramento County (California) Coroner Report 04–06432; Jewett, 2005, April 14; Amnesty International, 2006; Anglen, 2006).

Pino displayed no signs of medical distress for more than 17 hours following the application of the TASER pulses, indicating that the electric current had no immediate affect on his heart. Because the TASER had no immediate effect and the medical examiner attributed the death to a seizure disorder resulting from surgery to remove a brain tumor, the TASER is excluded as the cause of death and as a significant contributing factor in the Pino case.

TIMOTHY BOLANDER

AGE: 31
RACE/GENDER: WHITE MALE
AGENCY: DELRAY BEACH (FLORIDA) POLICE DEPARTMENT
DATE OF INCIDENT: DECEMBER 23, 2004
CAUSE OF DEATH: COCAINE TOXICITY
CONTRIBUTING FACTORS: CARDIAC HYPERTROPHY
ROLE OF TASER DEVICE: EXCLUDED

Bolander's estranged wife called 9-1-1 to report that he was outside her home in the backyard banging his head against a metal fence post. She had filed court papers two days prior seeking a restraining order to keep him away from her. Before the officers arrived, Bolander ran to the front yard and began banging his head on his car. When two officers arrived, Bolander was agitated and combative. He ran across the street and struggled with officers when they caught him. Two officers shocked Bolander four times with TASER devices to subdue him. As the officers were walking Bolander to a patrol car, he collapsed. Paramedics, who had been waiting down the street, treated him within a few minutes. They transported him to a local hospital, but he died shortly after arrival. Half an hour after his death, Bolander's body temperature was 102 degrees Fahrenheit.

A medical examiner listed Bolander's cause of death as accidental cocaine toxicity. Bolander had four ruptured bags of cocaine in his stomach. The medical examiner also reported that Bolander had the opiates hydrocodone and morphine in his system. Bolander also suffered from cardiac hypertrophy (see Davies, 2004; Davies, 2005; Delray Beach (Florida) Police Report 04–35821; Office of the

Medical Examiner, Fifteenth District (Florida) Report ME 04–1325; Barton, 2005; Amnesty International, 2006; Anglen, 2006).

Bolander continued to resist officers following the application of the TASER pulses, and he was able to walk part of the way to the police car, indicating that the electric current did not immediately affect his heart. Because the TASER device did not have an immediate effect and the medical examiner attributed the death to the effects of cocaine, the device is excluded as the cause of death and as a significant contributing factor in the Bolander case.

CHRISTOPHER DEARLO HERNANDEZ

AGE: 19
RACE/GENDER: AFRICAN-AMERICAN MALE
AGENCY: COLLIER COUNTY (FLORIDA) SHERIFF'S OFFICE
DATE OF INCIDENT: DECEMBER 28, 2004
CAUSE OF DEATH: METHYLENEDIOXYMETHAMPHETAMINE AND COCAINE
CONTRIBUTING FACTORS: SICKLING OF RED BLOOD CELLS
ROLE OF TASER DEVICE: EXCLUDED

The driver of a car in which Hernandez was a passenger was driving recklessly through the parking lot of a nightclub after closing. Deputies, who had responded to a call of a fight at the club, ordered the driver to stop, but he left the parking lot. Other deputies stopped the car in the parking lot of a nearby convenience store. While one deputy spoke with the driver, another deputy, who had seen Hernandez acting strangely in the back seat, ordered him out of the car. A fight ensued between Hernandez and three deputies. The deputies hit and kicked Hernandez while trying to subdue him. One deputy sprayed Hernandez with chemical spray, and two other deputies used TASER devices. Later tests would show that deputies activated the devices 17 times, although the number of stuns Hernandez received is unknown. It took the deputies several minutes to get the handcuffs on Hernandez. Once they had him secured, the deputies put Hernandez in the back of a police car while they waited for emergency medical assistance. Because Hernandez had received a cut on his head, paramedics transported him to a local hospital.

While at the hospital, Hernandez continued to be uncooperative, cursing, refusing to give information to the medical staff, and trying to leave the hospital. Doctors noted that he had hyperkalemia, and a toxicology screen was positive for cocaine and methylenedioxymethamphetamine (Ecstasy). A CT scan revealed that Hernandez had a subdural hematoma. About six hours after his confrontation with the deputies, the medical staff transferred Hernandez to the intensive care unit. Shortly after his arrival in ICU, Hernandez's heart stopped beating, and doctors could not revive him.

A coroner ruled that Hernandez died from a cardiac arrest caused by drugs. Toxicology results showed that Hernandez had Ecstasy, cocaine, and Dextromethorphan, an over-the-counter antitussive commonly used as a recreational drug, in his system. The autopsy also revealed that Hernandez suffered from sickle cell anemia (see *Associated Press*, 2004, December 30; Collier County (Florida) Sheriff's Office Report 0400040700; Office of the Medical Examiner, Twentieth District (Florida) Report ME 2004–385; *Associated Press*, 2005, June 9; Barton, 2005; Amnesty International, 2006; Anglen, 2006).

Hernandez continued to struggle with officers and medical staff for several hours following the application of the TASER pulses, indicating that the electric current did not immediately affect his heart. Because the TASER device did not have an immediate effect and the medical examiner attributed the death to the combined effects of Ecstasy, cocaine, and Dextromethorphan, the device is excluded as the cause of death and as a significant contributing factor in the Hernandez case.

GREGORY SAULSBURY

AGE: 30
RACE/GENDER: AFRICAN-AMERICAN MALE
AGENCY: PACIFICA (CALIFORNIA) POLICE DEPARTMENT

Date of Incident: January 2, 2005
Cause of Death: Cardiopulmonary arrest due to agitated state due to cocaine intoxication, with sequelae of struggles, forcible restraints, and TASER stun gun applications
Contributing Factors: Fatty infiltration of atrioventricular bundle
Role of TASER Device: Doubtful

The police report in this case was exempt from disclosure under California law. Media sources reported that family members called 9-1-1 for paramedics, saying Saulsbury was acting paranoid and occasionally kicking and swinging at family members who were trying to calm him. They said that Saulsbury was claiming someone had slipped him drugs and that people were plotting to kill him. Police arrived first to secure the scene for paramedics. Saulsbury lunged at an officer, and both men fell to the floor. Two other officers then used their TASER devices on Saulsbury in drive-stun mode at least eleven times, with little initial effect. Once officers had Saulsbury secured and handcuffed, he quit breathing. Officers and paramedics were unable to revive him. Paramedics took him to a local hospital, where he died about one hour later.

A coroner concluded that Saulsbury had died of cardiopulmonary arrest due to an agitated state due to cocaine intoxication, with sequelae of struggles, forcible restraints, and TASER stun gun applications. The coroner explained that the struggles with his family and with police created for Saulsbury an increased oxygen demand and adrenal output. Forcible restraints by his family and the police, including bear hugs and handcuffing in the prone position, restricted Saulsbury's respiration. Muscle spasms and the pain concurrent with the application of the TASER shocks contributed to the physiological stresses that led to cardiopulmonary arrest, a combination of cardiac and respiratory failure. However, the coroner also reported that the history of deaths due to cocaine-induced psychosis without TASER application suggested that the contribution of the device in Saulsbury's death was probably minor. The coroner also noted that Saulsbury had fatty infiltration of the atrioventricular bundle, a group of specialized muscle cells that assist in conducting electrical impulses in the heart, which raised his heart attack risk (see Gathright, 2005, January 4; Gathright, 2005, March 18; Gathright, 2005, June 25; San Mateo County (California) Coroner Report 05–0022-A; Amnesty International, 2006; Anglen, 2006).

Saulsbury continued to struggle with officers following the application of the TASER pulses, indicating that the electric current did not immediately affect his heart. Although the autopsy report does mention the possible contribution of the TASER device, the medical examiner noted that the contribution would be minor. Considering Saulsbury's medical condition, it is questionable whether another form of restraint would have yielded a different result. The coroner could not say that, but for the application of the TASER device, Saulsbury would have survived. Consequently, the role of the device as a cause of death or as a contributing factor in the Saulsbury case is doubtful.

DENNIS S. HYDE

Age: 30
Race/Gender: White Male
Agency: Akron (Ohio) Police Department
Date of Incident: January 5, 2005
Cause of Death: Probable cardiac arrhythmia due to acute methamphetamine intoxication and electrical pulse incapacitation
Contributing Factors: Psychiatric disorder with agitated behavior; blood loss by arterial injury
Role of TASER Device: Excluded

Hyde broke into a home by smashing a window. He called out to the resident that he was the devil and that he was coming upstairs. The resident had been sleeping upstairs, but when she heard Hyde breaking in she called 9-1-1. When police arrived, they found the house dark because Hyde had disabled the electric service. They followed a blood trail that led to the basement. There, they found

Hyde hiding behind the furnace. He refused to comply with their commands to come out. Instead, he held his hand behind his back and told officers that he had a gun. An officer tried to subdue Hyde with a TASER device, but it had no effect. Officer wrestled with Hyde and finally handcuffed him. Because of a severely lacerated wrist, which he suffered when he broke into the house, officers transported Hyde to a local hospital. There, his condition deteriorated, and he died about one and a half hours later.

The medical examiner noted that Hyde was recovering from burn wounds, and that he had skin grafts that were healing, but that Hyde was taking oxycodone, and a tablet of oxycodone was found in his gastric system. The toxicology report indicated that Hyde also had methamphetamine and amphetamine in his system. The medical examiner concluded that Hyde died due to a probable cardiac arrhythmia caused by the combined effects of methamphetamine intoxication and electrical pulse incapacitation. Blood loss from injuries sustained prior to the altercation with police and an apparent underlying, though unspecified, psychiatric illness contributed to his death (see Akron (Ohio) Police Report 05-000475; County of Summit (Ohio) Medical Examiner's Report N–008–05; Farkas, 2005; Amnesty International, 2006; Anglen, 2006).

Hyde continued to struggle with officers following the application of the TASER pulses, indicating that the electric current did not immediately affect his heart. Hyde's condition did not begin to deteriorate until he was in the hospital. The medical examiner provided no forensic evidence to support the conclusion that the TASER shock affected the rhythm of Hyde's heart, nor did he explain how the electric current would have had a delayed effect. However, the potential cardiac effects of sympathomimetic drugs, such as methamphetamine, are well known. Consequently, the TASER device is excluded as the cause of death and as a significant contributing factor in the Hyde case.

CARL NATHANIEL TROTTER

AGE: 33

RACE/GENDER: AFRICAN-AMERICAN MALE
AGENCY: ESCAMBIA COUNTY (FLORIDA) SHERIFF'S OFFICE
DATE OF INCIDENT: JANUARY 8, 2005
CAUSE OF DEATH: UNDETERMINED
ROLE OF TASER DEVICE: DOUBTFUL

Trotter broke into several homes and attacked at least two residents in a neighborhood. After the residents of one home fought with him and chased him away, he picked up an elderly woman walking on the sidewalk and carried her, kicking and screaming, to a church parking lot. When a neighbor forced Trotter to let her go, he crashed through the glass door of another house and attacked another woman there. When deputies arrived, Trotter was fighting with the woman's family members. Sheriff's deputies and residents struggled with Trotter. More deputies arrived, and one of them shocked Trotter with a TASER device. The first shock was a drive-stun to the neck. The second charge was a probe charge to the back. Neither of those charges was effective, and Trotter continued to fight. A third shock, a drive-stun applied to Trotter's stomach was effective, and Trotter quit actively resisting, although he would try to get up when officers were not holding him down. About two minutes later, paramedics arrived and found that Trotter was not breathing. They were unable to resuscitate Trotter, and he died at the scene.

Toxicology reports disclosed the presence of alcohol and cocaine in Trotter's system, but the medical examiner determined that the cause and manner of Trotter's death were unknown. He reported that excited delirium alone may have caused Trotter's death, but the potential effects of the TASER shocks delivered just prior to Trotter's collapse could not be ignored (see *Associated Press*, 2005, January 8; Barton, 2005; Escambia County (Florida) Sheriff's Office Report 05–000663; Office of the Medical Examiner, First District (Florida) Report MLA05–023; Amnesty International, 2006; Anglen, 2006).

Trotter continued to resist officers for only a very short time following the last application of the TASER pulses. The medical examiner did mention the temporal relationship between the last charge and Trotter's cessation of breathing, however, he provided no forensic evidence of how the TASER device was the cause of death, and he stated that he could say that but for the application of the device Trotter would have survived. Therefore, the role of the TASER device as the cause of death and as a significant factor in Trotter's death is doubtful.

JERRY JOHN MORENO

AGE: 33
RACE/GENDER: HISPANIC MALE
AGENCY: LOS ANGELES COUNTY (CALIFORNIA) SHERIFF'S OFFICE
DATE OF INCIDENT: JANUARY 10, 2005
DATE OF DEATH: JANUARY 14, 2005
CAUSE OF DEATH: SEQUELAE OF HYPOXIC/ISCHEMIC ENCEPHALOPATHY DUE TO THE EFFECTS OF METHAMPHETAMINE INTAKE
ROLE OF TASER DEVICE: EXCLUDED

The Sheriff's Office report was exempt from disclosure under California law, but media sources reported that on January 10, while at a detention center under the influence of drugs, Moreno began climbing atop some bunk beds, yelling incoherently. He armed himself with a piece of aluminum torn from a light fixture, and began swinging it around wildly, injuring another inmate. Deputies fired pepper balls at Moreno and shocked him three times with a TASER device, but it did not subdue him. Deputies then entered the cell and attempted to handcuff him, but he struggled and his head struck either the bottom rail of the bed or the concrete floor. Deputies finally handcuffed and hobbled him, and they covered his head with a towel to prevent him from spitting on or biting anyone. Moreno remained on his stomach during transport to an area hospital. At the hospital, medical personnel found Moreno in full cardiac arrest. The medical staff revived him, but four days later, he was declared brain-dead and taken off life support.

A coroner ruled that Moreno had suffered a cardiac arrest, and that he died of the sequelae of hypoxic/ischemic encephalopathy, a lack of blood flow and oxygen in the brain, due to the effects of methamphetamine. The toxicology reports disclosed that Moreno was under the influence of marijuana and methamphetamine (see Anderson, 2005; Baeder, 2005; County of Los Angeles (California) Coroner Report 2005–00453; Aidem, 2006; Amnesty International, 2006).

Moreno continued to struggle with jailers for several minutes following the application of the TASER pulses, indicating that the electric current did not immediately affect his heart. Because the TASER device did not have an immediate effect and the medical examiner attributed the cardiac arrest to the effects of methamphetamine, the device is excluded as the cause of death and as a significant contributing factor in the Moreno case.

JAMES EDWARD HUDSON

AGE: 33
RACE/GENDER: AFRICAN-AMERICAN MALE
AGENCY: CHICKASHA (OKLAHOMA) POLICE DEPARTMENT
DATE OF INCIDENT: JANUARY 28, 2005
CAUSE OF DEATH: COCAINE OVERDOSE
ROLE OF TASER DEVICE: EXCLUDED

Hudson tried to grab a shotgun from a police officer during a narcotics raid, which included agents from the Federal Bureau of Investigation, the Grady County Sheriff's Office, and the Chickasha Police Department. As Hudson tried to flee, a Chickasha police officer used a TASER device to prevent his escape. For a while, Hudson was walking and talking to officers, but he suddenly started throwing up. Hudson threw up a rock of crack cocaine that he had apparently swallowed during the raid. Officers tried to recover the rock, but Hudson chewed on it until he could swallow it again. Officers called for an ambulance for Hudson who became unresponsive. Paramedics took Hudson to a local hospital, where he died.

The autopsy report was not available, but media sources reported that the coroner concluded reported Hudson died of a cocaine overdose (see Chickasha (Oklahoma) Police Report 05003259; Amnesty International, 2006; Anglen, 2006; Epperson, 2006).

Hudson did not exhibit any signs of medical distress for several minutes following the application of the TASER pulses, indicating that the electric current did not immediately affect his heart. Because the TASER device did not have an immediate effect and the medical examiner attributed the death to an overdose of cocaine, the device is excluded as the cause of death and as a significant contributing factor in the Hudson case.

JEFFREY A. TURNER

AGE: 41
RACE/GENDER: AFRICAN-AMERICAN MALE
AGENCY: TOLEDO (OHIO) POLICE DEPARTMENT
   LUCAS COUNTY (OHIO) SHERIFF'S OFFICE
DATE OF INCIDENT: JANUARY 31, 2005
CAUSE OF DEATH: HYPERTENSIVE HEART DISEASE
CONTRIBUTING FACTORS: ALTERCATION WITH JAIL PERSONNEL, INCLUDING MULTIPLE SHOCKS WITH A TASER WEAPON
ROLE OF TASER DEVICE: DOUBTFUL

Toledo police approached Turner outside the Toledo Museum of Art after receiving a call about a suspicious person. Officers attempted to search Turner, but he resisted and started swinging his elbows at officers. The officers warned Turner that they were going to shock him, but he continued struggling with them. They then shocked Turner three times with a TASER device until they could get him handcuffed. Once cuffed, Turner began to kick at the officers. The officers used the TASER device two more times in drive-stun to get Turner's legs hobbled. A transport van arrived and took Turner to the Lucas County Correction Center.

At the corrections center, Turner settled down, but he later became agitated and banged on a security window. Jail officers tried to restrain him and used a TASER device. Turner grabbed at the device's wires and was shocked four more times. The jail officers put handcuffs on Turner and called a nurse. The nurse found Turner unresponsive, and he was taken to a local hospital, where he died.

The coroner reported that Turner had heart conditions, cardiomegaly, left ventricular hypertrophy, and right ventricular dilation, which were the primary causes of death. However, he added that the TASER shocks and the struggle with police and sheriff's deputies contributed to the death because they took place just before Turner died. The coroner said that it was not possible to determine whether Turner would have survived if he had not been shocked, but the investigation found that Turner appeared not to suffer ill effects for several minutes after police and the deputies shocked him (see *Associated Press*, 2005, February 2; Lucas County (Ohio) Coroner's Report 69–05; Seewer, 2005; Toledo (Ohio) Police Report 006063–05; Amnesty International, 2006; Anglen, 2006).

Turner continued to struggle with jailers for a short time following the last application of the TASER pulses, indicating that the electric current did not immediately affect his heart. The medical examiner attributed the death to Turner's heart disease. No doubt, the TASER shocks contributed to the physiological stress that Turner experienced, but the medical examiner could not say that, but for the use of the TASER device, Turner would have survived. Considering Turner's heart condition, it is questionable whether any other form of restraint would have produced a different result. Consequently, the role of the TASER device in Turner's death is doubtful.

RONALD ALAN HASSE

AGE: 54
RACE/GENDER: WHITE MALE
AGENCY: CHICAGO (ILLINOIS) POLICE
   DEPARTMENT
DATE OF INCIDENT: FEBRUARY 10, 2005
DATE OF INCIDENT: FEBRUARY 10, 2005
CAUSE OF DEATH: ELECTROCUTION FROM THE TASER DEVICE
CONTRIBUTING FACTORS: METHAMPHETAMINE INTOXICATION
ROLE OF TASER DEVICE: EXCLUDED

The police report in this incident was not available, but media sources reported that Hasse had been visiting friends at a Chicago apartment. He started acting strangely, so the friends called for medical assistance. When paramedics and firefighters arrived, Hasse threatened to infect them with human immunodeficiency virus (HIV), so they called for police assistance. When the police arrived, Haas tried to kick and bite the officers. The officers gave Hasse three warnings that they were going to use a TASER device, but he continued to resist. When he charged a sergeant, the sergeant shocked him, once for five seconds, then another time for 57 seconds. Paramedics saw that Hasse was in distress after the second charge. They took him to a local hospital where he died.

In the first ruling of its kind in the nation, a medical examiner ruled that the primary cause of Hasse's death was electrocution from the TASER device, but a contributing cause was methamphetamine intoxication. An autopsy found Hasse had .55 micrograms per milliliter of methamphetamine in his blood, more than the recognized lethal level. Nevertheless, the medical examiner claimed that the drug probably would not have killed Hasse without the TASER jolts. The medical examiner reviewed information provided by TASER International, but he said that he was persuaded by the arguments of an engineer who had published a report claiming that the electric current from a TASER electronic control device is enough to cause cardiac arrhythmia despite its low energy output.

However, the engineer's qualifications to make that determination were suspect. As an expert witness in a lawsuit, TASER International's attorney questioned that engineer about his credentials. In that deposition, he said he did not complete high school and instead had obtained a GED. He claimed to have earned his undergraduate and graduate degrees in computer science through an online correspondence course, but he did not have degrees or any formal training in medicine or in engineering. He also said that, at the time he wrote the report, he had never tested a TASER device (see Main, 2005, July 29; Main, 2005, July 30; Tarm, 2005; Amnesty International, 2006; Anglen, 2006).

In his report, the engineer made simple mathematical mistakes that overestimated the power of the TASER electric output more than 69,000 times. Additionally, his calculations showed that the TASER pulses could be expected to cause ventricular fibrillation about 50 percent of the time, which the empirical evidence clearly disproves. Researchers in the United States, Canada, Great Britain, and Australia have contradicted his claims. Because the engineer's report upon which the medical examiner based his conclusion has been discredited and the medical examiner noted a lethal level of methamphetamine in Hasse's system, the TASER device is excluded as the cause of death and as a significant contributing factor in the Hasse case.

ROBERT FIDALGO CAMBA

AGE: 45
RACE/GENDER: WHITE MALE
AGENCY: SAN DIEGO (CALIFORNIA) POLICE DEPARTMENT
DATE OF INCIDENT: FEBRUARY 10, 2005
DATE OF DEATH: FEBRUARY 12, 2005
CAUSE OF DEATH: HYPOXIC/ISCHEMIC ENCEPHALOPATHY DUO TO CARDIOPULMONARY ARREST DURING LAW ENFORCEMENT RESTRAINT FOLLOWING APPLICATION OF TASER DEVICE DUE TO EXCITED DELIRIUM DUE TO COCAINE AND METHAMPHETAMINE INTOXICATION
CONTRIBUTING FACTORS: HYPERTENSIVE ATHEROSCLEROTIC HEART DISEASE
ROLE OF TASER DEVICE: DOUBTFUL

The police report in this incident was exempt from disclosure under California law. However, media sources reported that police received a call to investigate a disturbance at a hotel apartment. Camba was nude, yelling and threatening guests at the hotel where he lived. On arrival, officers found Camba thrashing around on the floor in his room, sweating profusely and foaming at the mouth. Camba threw things and kicked at the officers, who tried to subdue him with a baton. When that failed, an officer shocked him with a TASER device. Camba temporarily stopped thrashing about, but soon began rolling around, yelling and kicking, and, in so doing, he dislodged one of the TASER probes. One officer placed the dislodged probe on his abdomen and applied another charge. Officers were able to place handcuffs on him, but, as they tried to apply a maximum restraint, the use of a cord cuff that restricts the movement of a person's hands and feet, he began to struggle again. Officers dragged him into the hall of the hotel. One officer drive-stunned Camba as he was thrashing about hitting his head on a pillar. About 30 seconds later, just as paramedics arrived, officers noticed that Camba was pale, unresponsive, and not breathing. The paramedics stabilized Camba and took him to a local hospital. Medical staff diagnosed Camba with fever, rhabdomyolysis, hyperkalemia, acute renal failure, and respiratory distress that required ventilation support. Toxicology screens were positive for cocaine and methamphetamine. According to his mother, Camba had also been diagnosed as schizophrenic. Camba's condition deteriorated, and he died two days later.

The autopsy revealed the presence of acute hypoxic/ischemic encephalopathy with cerebral edema and cerebellar tonsillar herniation resulting in brain death following resuscitation form cardiopulmonary arrest. The heart showed left ventricular hypertrophy and severe multifocal coronary artery atherosclerosis consistent with hypertensive atherosclerotic heart disease. The medical examiner ruled that Camba died of hypoxic/ischemic encephalopathy due to cardiopulmonary arrest during law enforcement restraint following application of a TASER device due to excited delirium due to cocaine and methamphetamine intoxication. The medical examiner also noted that a significant contributing factor to Camba's death was longstanding hypertensive atherosclerotic heart disease (see *City News Service*, 2005, February 15; County of San Diego (California) Medical Examiner Report 05–00304; Hughes, 2005; Amnesty International, 2006; Anglen, 2006).

The TASER shocks certainly contributed to the physiological stress that Camba experienced during his restraint. That stress likely resulted in his rhabdomyolysis and subsequent hyperkalemia. How much more the TASER device contributed than the blow with the baton, the physical force used by officers, and Camba's struggle against his restraints the medical examiner does not report. However, considering Camba's mental state, physical condition, and state of cocaine and methamphetamine intoxication, it is questionable whether another form of restraint would have caused a different result. Consequently, because the medical examiner could not say that but for the use of the TASER device Camba would have survived, the role of the device as the cause of death and as a significant contributing factor is doubtful.

JOEL DON CASEY

AGE: 52
RACE/GENDER: WHITE MALE
AGENCY: HARRIS COUNTY (TEXAS) CONSTABLE'S OFFICE, PRECINCT 1
DATE OF INCIDENT: FEBRUARY 18, 2005
CAUSE OF DEATH: PSYCHOTIC DELIRIUM WITH PHYSICAL RESTRAINT ASSOCIATED WITH HYPERTENSIVE AND ATHEROSCLEROTIC CARDIOVASCULAR DISEASE
ROLE OF TASER DEVICE: DOUBTFUL

The police report on this incident, which the Houston Police Department investigated, was exempt from disclosure under Texas law. However, news sources reported that constables from the Harris County Constable's Office mental health unit shocked Casey, a psychiatric patient, during a confrontation at his mother's house as they attempted to serve a mental health commitment warrant. Casey was scheduled to enter a drug and alcohol abuse treatment facility. He resisted when the constables attempted to restrain him, and he continued to struggle even after the constables shocked him three times with a TASER device. After they placed him in handcuffs and put him in a patrol car, the constables noticed that Casey was not breathing. The constables initiated CPR and called for medical assistance. Paramedics transported Casey to a local hospital, where he died.

A preliminary report found some evidence to suggest that Casey might have strangled when he suffered a thyroid cartilage fracture. The autopsy report, however, listed Casey's cause of death as psychotic delirium with physical restraint associated with hypertensive and atherosclerotic cardiovascular disease (see Harris County Medical Examiner's Office Report ML2005–0552; *Houston Chronicle*, 2005; Khanna, 2005; Amnesty International, 2006; Anglen, 2006).

Casey continued to struggle with constables following the application of the TASER pulses, indicating that the electric current did not immediately affect his heart. Because the TASER device did not have an immediate effect and the medical examiner attributed the death to psychotic delirium complicated by Casey's preexisting heart condition, the device is excluded as the cause of death and as a significant contributing factor in the Casey case.

ROBERT CLARK HESTON, JR.

AGE: 40
RACE/GENDER: WHITE MALE
AGENCY: SALINAS (CALIFORNIA) POLICE
 DEPARTMENT
DATE OF INCIDENT: FEBRUARY 19, 2005
DATE OF DEATH: FEBRUARY 20, 2005
CAUSE OF DEATH: MULTIPLE ORGAN FAILURE DUE TO CARDIAC ARREST DUE TO AGITATED STATE DUE TO METHAMPHETAMINE INTOXICATION
CONTRIBUTING FACTORS: LEFT VENTRICULAR HYPERTROPHY AND DILATION: TASER APPLICATION AND STRUGGLE WITH POLICE
ROLE OF TASER DEVICE: EXCLUDED

The police report in this incident was exempt from disclosure under California law. Media sources, however, reported that Heston's father called police because of a dispute with Heston. The elder Heston had ordered his son to move out of his house because the younger Heston was using drugs. The younger Heston tried to tip over a grandfather clock. When the elder Heston tried to stop him, the younger Heston knocked him down. When police arrived to a second call within 50 minutes, they found objects strewn about the front yard, the front door open, and the younger Heston dragging his 66-year-old father along the floor. Heston began throwing things at the officers, who fired two TASER devices at him, but failed to subdue him. Assistance arrived, and the officers finally subdued Heston, after firing their devices ten times. After they had subdued Heston, they realized that he had stopped breathing. Officers applied CPR and revived Heston. Paramedics took Heston to a local hospital, where he died two days later.

An autopsy was performed, and two other pathologists reviewed the report, each of whom wrote a report. All three medical examiners agreed that methamphetamine was the cause of Heston's death, but one medical examiner also cited TASER shocks as a cause of death while the other two cited TASER shocks as a contributing factor in Heston's death (see Burress, 2005; Gathright, 2005, February 24; Monterey County Coroner's Report 20050159; Amnesty International, 2006; Anglen, 2006).

All three medical examiners, to some degree, attributed the use of the TASER devices with Heston's death. However, Heston fought violently with officers for several minutes during the application of the TASER devices. The final applications were drive-stun applications to Heston's back. No study has shown that TASER pulses, when applied to the back, can affect the heart. Additionally, two of the medical examiners either could not or did not try to explain how the electric current could have a delayed effect on Heston's heart. One medical examiner did try to explain that the cause might have been rhabdomyolysis, but he admitted that his conclusion was speculative and without support from peer-reviewed medical literature. Moreover, electric current can cause ventricular fibrillation, but paramedics found Heston in asystole, a condition not generally caused by electricity but common to drug-induced excited delirium cases. Consequently, the available evidence supports that the TASER device was not the cause of Heston's death, nor did it significantly contribute to his death.

SHIRLEY ANDREWS

AGE: 38
RACE/GENDER: AFRICAN-AMERICAN FEMALE
AGENCY: CINCINNATI (OHIO) POLICE
 DEPARTMENT
DATE OF INCIDENT: FEBRUARY 24, 2005

DATE OF DEATH: MARCH 4, 2005
CAUSE OF DEATH: PULMONARY THROMBOEMBOLISM
CONTRIBUTING FACTORS: OBESITY
ROLE OF TASER DEVICE: EXCLUDED

Police responded to a group home for the mentally ill where Andrews was out of control and assaulting staff members. On arrival, officer found Andrews destroying property in the house. Officers tried to talk with Andrews to get her to calm down, but she threw a plastic case at them. After warning her, officers shocked her with a TASER device. Andrews pulled out the barbs and continued fighting, so the officers shocked her again. Again, Andrews pulled out the barbs. The officers cycled their TASER devices several times, one officer for a total of 98 seconds, and one officer for 32 seconds. None of the shocks had any effect on Andrews. Following several verbal commands, Andrews finally allowed the officers to place her in handcuffs, and they took her to jail. She died eight days later.

The autopsy was not available, but media sources reported that the coroner found Andrews died of a pulmonary thromboembolism. The coroner reported that Andrew's inactivity and obesity were contributing factors, but the TASER shocks were not (see Cincinnati (Ohio) Police Report 50501162; Cincinnati (Ohio) Police Use of Force Form 2005–62523.1; Flannery, 2005; Anglen, 2006).

Following the application of the TASER pulses, Andrews survived eight days without signs of medical distress. Because the TASER device did not have an immediate effect and the coroner attributed the death to a pulmonary thromboembolism, the device is excluded as the cause of death and as a significant contributing factor in the Andrews case.

DAVID LEVI EVANS, JR.

AGE: 45
RACE/GENDER: MALE (RACE NOT DETERMINABLE FROM THE AVAILABLE RECORDS)
AGENCY: LOS ANGELES COUNTY (CALIFORNIA) SHERIFF'S OFFICE
DATE OF INCIDENT: MARCH 4, 2005
DATE OF DEATH: MARCH 6, 2005
CAUSE OF DEATH: UNKNOWN
ROLE OF TASER DEVICE: UNDETERMINED

The Sheriff's Office report was exempt from disclosure under California law, but media sources reported that deputies, working on a tip that Evans was selling drugs from a hotel room, watched as several people entered and left the room, and they decided to arrest him. When deputies entered the room, a woman, who had registered for the room, complied with their commands, but Evans became combative. Deputies shocked him with a TASER device before subduing him. Because department policy required that any time force was used in an arrest the suspect must be examined before being booked into jail, the deputies took Evans to a local hospital. There his condition deteriorated, and he died two days later. Deputies believed that Evans ingested cocaine before the deputies entered the room (see Botonis, 2005, March 9).

The Los Angeles County Medical Examiner's Office was not able to produce an autopsy report, and news accounts do not report a cause of death in this incident. Consequently, without further information, the role of the TASER device in the Evans case is undetermined.

WILLIE MICHAEL TOWNS

AGE: 30
RACE/GENDER: AFRICAN-AMERICAN MALE
AGENCY: DELAND (FLORIDA) POLICE DEPARTMENT
DATE OF INCIDENT: MARCH 6, 2005
CAUSE OF DEATH: ACUTE COCAINE INTOXICATION

Officers were searching for a vehicle burglary suspect, and thought Towns was suspicious because he ran away when they shined a light at him. They pursued Towns to the roof on a nearby building. When he saw the officers, Towns broke through a window into the building. Officers pursued him, and a struggle began. Officers shocked Towns three times with a TASER device, they sprayed him with chemical spray, and they delivered several body blows to Towns, who continued to fight with the officers. During the fight, Towns told the officers he had been using cocaine. Additional officers arrived, and they finally handcuffed Towns. Once officers subdued Towns, they called for an ambulance because Towns was bleeding from several cuts. Towns resisted being strapped to a backboard, and he was yelling and screaming as paramedics and firefighters carried him from the building. He continued to fight with the paramedics in the ambulance. The paramedics administered a sedative to calm him. About ten minutes later, Towns stopped breathing. Paramedics transported Towns to a local hospital, where he died.

A medical examiner ruled that Towns died of acute cocaine intoxication. Although the medical examiner noted several injuries on Towns, he said that they were superficial and did not contribute to his death (see *Associated Press*, 2005, March 8; *Associated Press*, 2005, July 8; Barton, 2005; Deland (Florida) Police Report 0DL050001569; Office of the Medical Examiner, Seventh and Twenty-Fourth Districts (Florida) Report ME 2005–0185V; Amnesty International, 2006; Anglen, 2006).

Towns continued to struggle with officers and paramedics for several minutes following the application of the TASER pulses, indicating that the electric current did not immediately affect his heart. Because the TASER device did not have an immediate effect and the medical examiner attributed the death to the effects of cocaine, the device is excluded as the cause of death and as a significant contributing factor in the Towns case.

MARK YOUNG

AGE: 25
RACE/GENDER: AFRICAN-AMERICAN MALE
AGENCY: INDIANAPOLIS (INDIANA) POLICE DEPARTMENT
DATE OF INCIDENT: MARCH 10, 2005
CAUSE OF DEATH: COCAINE OVERDOSE
ROLE OF TASER DEVICE: EXCLUDED

Young was arrested at a suspected drug house. When officers entered the house, Young resisted officers, and they shocked him with a TASER device to get him into handcuffs. Young continued talking with officers after they had shocked him. During the search, police discovered and confiscated cocaine in the house. About 45 minutes after the arrest, Young had a seizure. Paramedics took Young to a local hospital, where he died about one hour later. Police reported that Young had swallowed drugs during the search of the house and that preliminary tests at the hospital were positive for cocaine and marijuana.

The autopsy report was not available, but the coroner's verdict report showed that Young died of an overdose of cocaine. He said that Young had swallowed twice the amount of cocaine it would take to kill a person of his size, and an hour after his death, Young still had large amounts of cocaine in his stomach (see Indianapolis (Indiana) Police Report 05–0030179; Horne & Spalding, 2005; Marion County (Indiana) Coroner's Verdict Report 05–0248; Tuohy, 2005; Amnesty International, 2006; Anglen, 2006).

Young continued to talk to officers for about 45 minutes following the application of the TASER pulses, and he showed no signs of medical distress, indicating that the electric current did not immediately affect his heart. Because the TASER device did not have an immediate effect and the coroner attributed the death to an overdose of cocaine, the device is excluded as the cause of death and as a significant contributing factor in the Young case.

MILTON WOOLFOLK, JR.

AGE: 39
RACE/GENDER: AFRICAN-AMERICAN MALE
AGENCY: COLUMBIA COUNTY (FLORIDA) SHERIFF'S OFFICE
DATE OF INCIDENT: MARCH 11, 2005
CAUSE OF DEATH: ISCHEMIC AND HYPERTROPHIC CARDIOMYOPATHY
CONTRIBUTING FACTORS: RESTRAINT STRESS
ROLE OF TASER DEVICE: EXCLUDED

Deputies were attempting to take Woolfolk, who had a history of mental problems, into custody for a court-ordered psychiatric evaluation. Woolfolk resisted and refused to follow commands. The deputies shocked him with a TASER device four times, but each time Woolfolk pulled out the prongs, and the device had no effect. Several deputies took him to the floor and placed him in handcuffs. Shortly thereafter, Woolfolk stopped moving and went unconscious. The deputies called for medical assistance and started CPR. Paramedics transported Woolfolk to a local hospital, where he died.

The medical examiner noted that Woolfolk was obese and suffered from hypertrophic cardiomyopathy. He concluded that Woolfolk died from ischemic and hypertrophic cardiomyopathy, and that the stress of restraint was a contributing factor (see Barton, 2005; Columbia County (Florida) Sheriff's Office Report 2005–009094; Office of the Medical Examiner, Third District (Florida) Report ME 05-0377; Pinkham, 2005; Amnesty International, 2006; Anglen, 2006).

Woolfolk continued to struggle with officers for several minutes following the application of the TASER pulses, indicating that the electric current did not immediately affect his heart. Although the application of the TASER shocks certainly contributed to the physiological stress that Woolfolk experienced, in his physical and mental state, it is questionable whether another form of restraint would have produced a different result. Consequently, the TASER device is excluded as the cause of death and as a significant contributing factor in the Woolfolk case.

ERIC J. HAMMOCK

AGE: 43
RACE/GENDER: WHITE MALE
AGENCY: FORT WORTH (TEXAS) POLICE DEPARTMENT
CAUSE OF DEATH: COCAINE INTOXICATION; CORONARY ATHEROSCLEROSIS WITH CARDIOMEGALY AND HEPATIC STEATOSIS
ROLE OF TASER DEVICE: EXCLUDED

Hammock, an architect, drove past a no-trespassing sign into a waste management facility. An off-duty police officer working at the facility as a security guard tried to stop him, but he continued to drive past the officer. The officer radioed for on-duty help, then chased Hammock in his personal vehicle. The officer followed Hammock after Hammock drove out of the plant. Hammock abandoned his vehicle and ran into the backyard of a house with the officer chasing him on foot. When the officer caught Hammock, Hammock tried to hit the officer on the head. The officer shocked Hammock with a TASER device, but Hammock pulled out the probes and continued to resist. On-duty officers arrived to assist, and one of them shocked Hammock several times, but he continued to resist. After the officers handcuffed Hammock, they noticed that he was having trouble breathing. They called an ambulance, and paramedics transported Hammock to a local hospital, where he died about one hour later.

A medical examiner listed Hammock's cause of death as cocaine intoxication and coronary atherosclerosis with cardiomegaly and hepatic steatosis. Toxicology tests indicated the presence in Hammock's body of ethanol, cocaine, cannabinoids, and cocaethylene. The medical examiner noted that, although a temporal relationship between the application of the TASER device and uncounsciousness existed, he could not establish a causal relationship (see *Associated Press*, 2005, April 4; *Associated Press*, 2005, May 4; Fort Worth (Texas) Police Report 05038655; Tarrant County (Texas) Medical Examiner's Report 0503238; Amnesty International, 2006; Anglen, 2006).

Hammock continued to resist officers following the application of the TASER pulses, indicating that the electric current did not immediately affect his heart. Because the TASER device did not have an immediate effect and the medical examiner attributed the death to the effects of cocaine and Hammock's preexisting heart and liver conditions, the device is excluded as the cause of death and as a significant contributing factor in the Hammock case.

JAMES FLOYD WATHAN, JR.

AGE: 32
RACE/GENDER: WHITE MALE
AGENCY: LIVINGSTON (CALIFORNIA) POLICE

DEPARTMENT,
MERCED COUNTY (CALIFORNIA) SHERIFF'S
OFFICE
DATE OF INCIDENT: APRIL 3, 2005
CAUSE OF DEATH: PROBABLE CARDIAC DYSRHYTHMIA DUE TO PHYSIOLOGIC STRESS DURING RESTRAINT AND ATTEMPTED ARREST, MILD MYOCARDIAL HYPERTRPHY AND ACUTE METHAMPHETAMINE INTOXICATION
ROLE OF TASER DEVICE: EXCLUDED

Wathan's father summoned deputies to his home to help in controlling him. He reported that the younger Wathan was on narcotics, had forced his way into the house, and had assaulted his mother. Deputies discovered that Wathan was the subject of a felony warrant for giving false identification and four misdemeanor warrants on drug and alcohol charges, and they tried to arrest him. Three deputies wrestled with Wathan through several rooms, struck him with batons and applied chemical spray, but those methods were ineffective in subduing Wathan. A Livingston police officer arrived as backup and twice shocked Wathan with a TASER device, once on the shoulder and once in the small of the back. After the shock in the back, the deputies handcuffed Wathan, but they noticed that he was not breathing. The deputies performed CPR on Wathan, but they were unable to revive him. The deputies called for medical assistance, but paramedics could not revive Wathan, and he died at the scene.

The pathologists conducting the autopsy concluded that Wathan died of probable cardiac dysrhythmia due to physiologic stress during restraint and attempted arrest, mild myocardial hypertrophy, and acute methamphetamine intoxication. He added that Wathan's enlarged heart and liver were consistent with long-term drug and alcohol abuse (see Conway, 2005; Stern, 2005; Livingston (California) Police Report 050700; Merced County (California) Coroner's Report 20050011633; Merced County (California) Sheriff's Office Report 20050011640; Pathology Associates (Fresno, California) Autopsy Report A05–018; Amnesty International, 2006; Anglen, 2006).

Wathan continued to resist officers following the application of the TASER pulse to his shoulder, but he stopped resisting immediately following the shock to the small of his back. Current studies indicate that a TASER charge applied to the back and away from the heart does not affect the rhythm of the heart. Consequently, because the application of the TASER shock to the shoulder did not have an immediate effect and the application of the shock to the small of the back was too far from the heart to have an effect, the TASER device is excluded as the cause of death and as a significant contributing factor in the Wathan case.

RICKY PAUL BARBER

AGE: 46
RACE/GENDER: WHITE MALE
AGENCY: CARTER COUNTY (OKLAHOMA) SHERIFF'S OFFICE
DATE OF INCIDENT: APRIL 5, 2005
DATE OF DEATH: APRIL 8, 2005
CAUSE OF DEATH: PROBABLE FATAL CARDIAC DYSRHYTHMIA
CONTRIBUTING FACTORS: INTERVENTRICULAR HYPERTROPHY WITH ABERRANT CORONARY ARTERY, HISTORY OF PSYCHOSIS
ROLE OF TASER DEVICE: EXCLUDED

On April 5, Barber assaulted several campers at a campsite, and when deputies arrived, he attacked them. He struck one deputy, and another used a TASER device to subdue him. The deputies took Barber to jail. Barber was combative at that time. He would throw his food around the cell, and he would not keep his clothing on. His mother stated that he had a history of psychosis. Three days later, April 8, jail staff found him dead in his jail cell with no apparent cause of death.

The medical examiner found no illicit drugs or alcohol in Barber's system. He ruled that Barber died of probable cardiac dysrhythmia caused by an aberrant coronary artery with occlusive sclerosis located in the centrum of the interventricular septum, or a blocked artery in the walls between the ventricles of the heart (see Office of the Chief Medical Examiner (Oklahoma) Report 0500692; Anglen, 2006).

For three days following the application of the TASER pulses, Barber showed no signs of medical distress, indicating that the electric current did not immediately affect his heart. Because the TASER device did not have an immediate effect and the medical examiner determined that the cause of Barber's dysrhythmia was a deformed coronary artery and a blocked artery in the walls of his heart, the device is excluded as the cause of death and as a significant contributing factor in the Barber case.

JOHN COX

AGE: 39
RACE/GENDER: MALE (RACE NOT DETERMINABLE FROM THE AVAILABLE RECORDS)
AGENCY: SUFFOLK (NEW YORK) POLICE DEPARTMENT
DATE OF INCIDENT: APRIL 22, 2005
CAUSE OF DEATH: UNKNOWN
ROLE OF TASER DEVICE: UNDETERMINED

Neither the police report nor the autopsy report were available under New York law, but media sources reported that Cox, who had a history of erratic behavior and who was taking antipsychotic medication, became upset while visiting his girlfriend. Police responded to a 9-1-1 call about an intoxicated man who was tearing up a house. When officers arrived, they found Cox bare-chested and yelling. They learned he had been fighting with four friends. When police ordered him to get on his knees, he charged them. Officers shocked Cox with a TASER device five times, but it had no effect. Eventually nine officers suffered injuries trying to subdue Cox. The officers loaded him, face down, onto a stretcher for paramedics to take him to the hospital. At the hospital, officers noticed that Cox was not breathing.

Media sources reported that the preliminary autopsy results indicated that Cox's blood-alcohol content was between 0.07 and 0.09 and that he had cocaine in his system, although the amount of cocaine was unclear. Cox had no broken bones or skull fractures and only a few abrasions. He also had an enlarged heart and significant blockage in his coronary arteries. The final report was awaiting toxicology results, but the media accounts did not report the final cause of death (see Kelleher, 2005; *New York Times*, 2005; *Newsday*, 2005; Amnesty International, 2006; Anglen, 2006).

Although the length of Cox's struggle with officers following the application of the TASER pulses indicates that the electric current could not have affected his heart, without additional information regarding the autopsy report, the role of the TASER device in the Cox case is undetermined.

JESSE CLEON COLTER, III

AGE: 31
RACE/GENDER: AFRICAN-AMERICAN MALE
AGENCY: PHOENIX (ARIZONA) POLICE DEPARTMENT
DATE OF INCIDENT: APRIL 24, 2005
CAUSE OF DEATH: ARRHYTHMIA DUE TO EXCITED DELIRIUM, PHENCYCLIDINE TOXICITY, AND CARDIOMEGALY
ROLE OF TASER DEVICE: EXCLUDED

A woman called 9-1-1 about a disturbance that possibly involved a shooting at an apartment building. When officers arrived, they saw a nude man, later identified as Colter, covered in blood, standing in the parking lot, and screaming that someone was shooting at him. Colter either fell or jumped from a window and landed on the hood of a vehicle below. As the officers approached, Colter jumped up and ran into a nearby parking lot, where he found himself cornered next to an eight-foot wall. Colter was bleeding badly and shrieking, ignoring officers' efforts to calm him. Two officers tried to shoot him with TASER devices, but they each missed. A third officer fired his TASER device, striking Colter in the chest. Colter dropped to the ground, but continued to resist officers' efforts to handcuff him. It took officers more than a minute to subdue him.

Phoenix firefighters arrived about two minutes after police arrested Colter. While paramedics were examining Coulter, he slipped into unconsciousness. The paramedics took Colter to a local hospital, where he died approximately 47 minutes after the 9-1-1 call from the apartment complex. Police investigators later learned that Colter was a chronic drug user, and that just before the incident he had been smoking a cigarette laced with phencyclidine (PCP), which can induce delusions of invulnerability and superhuman strength.

The autopsy report was not available, but according to media and police accounts, the medical examiner reported that in addition to ingesting a large amount of PCP and methylenedioxymethamphetamine (Ecstasy) Colter had a badly damaged and enlarged heart. An assistant medical examiner ruled that Colter's death was due to excited delirium, PCP toxicity, and cardiomegaly (see Phoenix (Arizona) Police Report 2005–50769124; Rubin, 2007).

Colter continued to struggle with officers for at least one minute following the final application of the TASER pulses, and he showed no signs of medical distress for several more minutes until paramedics began examining him, indicating that the electric current did not immediately affect his heart. Because the TASER device did not have an immediate effect and the medical examiner attributed the cause of death to excited delirium, PCP toxicity, and a diseased heart, the device is excluded as the cause of death and as a significant contributing factor in the Colter case.

KEITH EDWARD GRAFF

AGE: 24
RACE/GENDER: WHITE MALE
AGENCY: PHOENIX (ARIZONA) POLICE
    DEPARTMENT
DATE OF INCIDENT: MAY 2, 2005
DATE OF DEATH: MAY 3, 2005
CAUSE OF DEATH: EXCITED DELIRIUM DUE TO
    METHAMPHETAMINE INTOXICATION
ROLE OF TASER DEVICE: EXCLUDED

Just before midnight, police approached Graff in his apartment. They wanted to question him regarding an assault on a police officer three weeks earlier during which he fled. He also had a warrant for failing to report to probation officers after serving nine months in jail on a car theft conviction. Graff provided the officers with a false identification card, and when they questioned him about it, he tried to run out the patio door. When the officers caught him, he began fighting. One officer fired a TASER cartridge, but missed with the probes. She then applied a drive-stun on Graff. Another officer fired his TASER device into Graff's back. Police continued to shock Graff until they could place him in handcuffs. The officers shocked Graff five times until he stopped fighting, four times with the first TASER device in short durations of five to eight seconds, and once with the second device continuously for 84 seconds. Graff stopped breathing a short time later. He was transported to a local hospital where he was pronounced dead within an hour.

An autopsy report listed excited delirium due to methamphetamine intoxication as the cause of death and made no mention of the TASER shocks, except to note the positions of probe marks on Graff's body (see *Associated Press*, 2005, August 13; DeFalco, 2005; Maricopa County (Arizona) Medical Examiner's Report 05–01639; Phoenix (Arizona) Police Report 2005–50832168; Amnesty International, 2006; Anglen, 2006).

Graff continued to struggle with officers following the several applications of the TASER device. He did not show signs of medical distress for several minutes following the final application of the TASER pulses, which lasted 84 seconds, indicating that the electric current did not immediately affect his heart. Because the TASER device did not have an immediate effect and the medical examiner attributed the cause of death to methamphetamine-induced excited delirium, the device is excluded as the cause of death and as a significant contributing factor in the Graff case.

LAWRENCE BERRY

AGE: 33
RACE/GENDER: AFRICAN-AMERICAN MALE
AGENCY: JEFFERSON PARISH (LOUSIANA)
    SHERIFF'S OFFICE
DATE OF INCIDENT: MAY 6, 2005
CAUSE OF DEATH: UNKNOWN
ROLE OF TASER DEVICE: UNDETERMINED

Officers had arrested Berry on a parole violation, drug distribution and battery on a police officer. He began having seizures, but refused treatment and was under observation in a medical area of the jail. A week later, deputies said he began screaming that he was on fire. He became violent and combative, and he broke a shatterproof glass window. Deputies used a TASER device on Berry and placed him in an emergency restraint chair for his own protection. Jailers removed him from the chair an hour later, and then put him back when he continued to be disruptive. When Berry went limp, deputies performed CPR and had Berry taken to a local medical center, where he died (see *Associated Press*, 2005a, May 6; Jefferson Parish (Louisiana) Sheriff's Office Report E–06413–05; Amnesty International, 2006; Anglen, 2006).

The autopsy report in this case was not available, and media sources did not report the cause of death. Consequently, the role of the TASER device in the Berry case is undetermined.

STANLEY WILSON

AGE: 44
RACE/GENDER: AFRICAN-AMERICAN MALE
AGENCY: MIAMI-DADE (FLORIDA) POLICE DEPARTMENT
DATE OF INCIDENT: MAY 6, 2005
CAUSE OF DEATH: ACUTE COCAINE PSYCHOSIS
ROLE OF TASER DEVICE: EXCLUDED

Neither the police report nor the autopsy report was available in this incident, but media sources reported that officers discovered Wilson on a busy street, dodging traffic and acting irrationally, claiming that the devil was after him. When officers approached him, Wilson tried to hit them. Officers shocked him with a TASER device and arrested him. The officers took him to a local hospital, where medical staff examined and released him. The officers then took Wilson to jail, where he became violent and combative. They placed him in a cell, where jailers later found him unconscious. Medical personnel took Wilson to another local hospital, where he died.

Media sources reported that the medical examiner's report listed Wilson's cause of death as acute cocaine psychosis (see *Associated Press*, 2005b, May 6; Barton, 2005; Ovalle, 2005; Amnesty International, 2006; Anglen, 2006).

For about two and a half hours following application of the TASER pulses, Wilson showed no signs of medical distress. In fact, doctors at a hospital checked him and released him. Because the TASER device did not have an immediate effect and the medical examiner attributed the cause of death to the effects of cocaine, the device is excluded as the cause of death and as a significant contributing factor in the Wilson case.

VERNON ANTHONY YOUNG

AGE: 31
RACE/GENDER: AFRICAN-AMERICAN MALE
AGENCY: UNION TOWNSHIP (OHIO) POLICE DEPARTMENT
DATE OF INCIDENT: MAY 13, 2005
CAUSE OF DEATH: ACUTE COCAINE INTOXICATION
ROLE OF TASER DEVICE: EXCLUDED

Young went on a rampage in an apartment complex, claiming that he was hearing voices. He fired a gun into his own closet, forced his way into the manager's unit, and threatened her with a knife. The manager locked herself in her bathroom and called police. When officers arrived, witnesses told them that Young was agitated, as if he were on drugs or having an emotional breakdown. Officers ordered Young to the floor, and he complied, but when he started to rise, an officer shocked him with a TASER device. Because of the way Young was acting, as a precaution, officers took Young to a local hospital. At the hospital, his condition began to deteriorate, and he died about an hour later.

The coroner said preliminary tests showed Young had ingested enough cocaine that he would have died with or without the use of a TASER device. The final report listed the cause of death as acute cocaine intoxication, with cocaine and benzoylecgonine in his blood. He also noted that the internal organs were abnormally warm. Young's liver temperature was 101.5 degrees Fahrenheit almost three hours postmortem. He ruled the death was a result of acute cocaine intoxication (see *Associated Press*, 2005, May 14; Hamilton County Coroner's Office Autopsy Case Number CC05–01487; Union Township Police Department Report 05–4552; Amnesty International, 2006; Anglen, 2006).

Young continued to struggle with officers for a while following the application of the TASER pulses, and until he got to the hospital several minutes later he displayed no signs of medical distress. Because the TASER device did not have an immediate effect and the medical examiner attributed the death to the effects of cocaine, the device is excluded as the cause of death and as a significant contributing factor in the Young case.

LEROY PIERSON

AGE: 55
RACE/GENDER: MALE (RACE NOT DISCERNABLE FROM THE AVAILABLE RECORDS)
AGENCY: SAN BERNARDINO COUNTY (CALIFORNIA) SHERIFF'S OFFICE
DATE OF INCIDENT: MAY 17, 2005
DATE OF DEATH: MAY 18, 2005
CAUSE OF DEATH: LACK OF BLOOD AND OXYGEN TO THE BRAIN
CONTRIBUTING FACTORS: CAUSED BY STRUGGLE AND APPLICATION OF THE TASER DEVICE, METHAMPHETAMINE
ROLE OF TASER DEVICE: EXCLUDED/POSSIBLE

The Sheriff's Office report in this incident was exempt from disclosure under California law, but media sources reported that a witness called deputies and reported that a man was walking along a road, behaving oddly, waving a belt and striking the road with a piece of wood. A deputy who stopped and questioned Pierson determined that he was under the influence of a controlled substance. He also learned that two warrants existed for Pierson's arrest. Pierson battled the arresting deputies and tried to kick out the window of a patrol car after they handcuffed him. The deputies placed a nylon mesh sack over Pierson's head to stop him from biting or spitting. Three deputies put Pierson into a patrol car and took him to the local detention center. When deputies ordered Pierson out of the car, he refused and again became combative. Officers gave Pierson two short bursts from a TASER device to prod him out of the car.

Inside the jail Pierson resisted again, and officers shocked him twice with another TASER device. The final shock lasted 47 seconds. After that shock, Pierson lost consciousness. Deputies and jail medical staff initiated first aid and called for medical assistance. Paramedics transported Pierson to a local hospital. Hospital staff placed Pierson on life support, but he died about 24 hours later.

The autopsy report was not available, but media sources reported that the Orange County coroner concluded Pierson died from lack of oxygen and blood to his brain caused by a combination of the struggle and the continuous application of the TASER device. Toxicology results also showed that Pierson had methamphetamine in his system, which the coroner said could have factored into his death (see Tenorio, 2005; Amnesty International, 2006; Anglen, 2006; Leveque, 2006).

Pierson did collapse immediately following the application of the TASER pulses, but, without information on the evidence the medical examiner found to make his decision, it was not possible fully to evaluate his claims that the continuous application of the TASER shock contributed to whatever condition created the lack of oxygen and blood in Pierson's brain. Consequently, because the medical examiner contributed the cause of death to lack of oxygen to the brain, the TASER device is excluded as the cause of death. However, because there is insufficient evidence to confirm the medical examiner's conclusion that the device was a significant contributing factor, its role is classified as possible.

RANDY MARTINEZ

AGE: 40
RACE/GENDER: HISPANIC MALE
AGENCY: ALBUQUERQUE (NEW MEXICO) POLICE DEPARTMENT

DATE OF INCIDENT: MAY 19, 2005
DATE OF DEATH: MAY 20, 2005
CAUSE OF DEATH: COMPLICATIONS OF COCAINE INTOXICATION
ROLE OF TASER DEVICE: EXCLUDED

The police report in this incident was not available, but media sources reported that Martinez's mother called 9-1-1 for help, saying her son was naked, sweating, and out of control. Responding officers reported that Martinez was out of control, combative, and belligerent, and that he refused to obey commands. An officer shocked Martinez once with a TASER device, but Martinez pulled out the darts and continued to fight. The officer stunned Martinez two more times. After the third stun, Martinez fell and hit his head. Officers placed Martinez on his stomach and handcuffed him. After they rolled him back over, he was breathing and responsive. Several minutes later, Martinez became unresponsive and stopped breathing. Paramedics transported Martinez in cardiac arrest to a local hospital, where he died the next day.

The medical examiner noted that Martinez's hospital course was complicated by multisystem organ failure, rhabdomyolysis, and coma. He reported that an ante mortem urine drug screen was positive for cocaine, and toxicology tests on ante mortem blood were positive for cocaine metabolites. The medical examiner concluded that Martinez died of complications of cocaine intoxication. He also reported that the fact that Martinez did not collapse immediately following the use of the TASER device made it more likely that cocaine, not the device, led to Martinez's cardiac arrest (see Office of the Medical Examiner (New Mexico) Report 2005–02996; Wilhelm, 2005; Amnesty International, 2006; Anglen, 2006).

For several minutes following the application of the TASER pulses, Martinez showed no signs of medical distress, indicating that the electric current did not affect his heart. Because the TASER device did not have an immediate effect and the medical examiner attributed the death to the effects of cocaine, the device is excluded as the cause of death and as a significant contributing factor.

LEE MARVIN KIMMEL

AGE: 38
RACE/GENDER: WHITE MALE
AGENCY: TOWNSHIP OF SPRING (PENNSYLVANIA) POLICE DEPARTMENT
DATE OF INCIDENT: MAY 22, 2005
DATE OF DEATH: MAY 23, 2005
CAUSE OF DEATH: ANOXIC ENCEPHALOPATHY DUE TO CARDIAC ARREST DURING POLICE RESTRAINT DUE TO AGITATED DELIRIUM WITH COCAINE TOXICITY
ROLE OF TASER DEVICE: EXCLUDED

The police reports in this incident were exempt from disclosure under Pennsylvania law, but media sources reported that Kimmel, wearing only a black T-shirt and white socks, punched through a window of a municipal building and was trying to climb into the building when police approached. Kimmel struggled with the officers who used a TASER device and chemical spray on him. Kimmel crawled over the broken glass, and officers shocked him again inside the building. Once restrained, Kimmel became unresponsive. The officers immediately removed the restraints and began CPR. Emergency medical personnel transported Kimmel to a local hospital. Hospital staff was able to reestablish Kimmel's heartbeat, but he died the next day. The Pennsylvania State Police investigated this incident.

The autopsy report was not available, but the coroner's investigation report noted that the coroner concluded Kimmel died of anoxic encephalopathy due to cardiac arrest during police restraint due to agitated delirium with cocaine toxicity. The TASER device was not mentioned in the investigation report (see *Associated Press*, 2005, May 23; *Associated Press*, 2005, August 24; Berks County (Pennsylvania) Coroner's Investigation Report 05–0336; Amnesty International, 2006; Anglen, 2006).

The media reports were not specific in saying how much time passed between the last application of the TASER pulses and Kimmel's collapse. However, since the coroner attributed the cardiac arrest to cocaine-induced agitated delirium, the TASER device is excluded as a cause of death or as a contributing factor in the Kimmel case.

RICHARD JAMES ALVARADO

AGE: 38
RACE/GENDER: HISPANIC MALE
AGENCY: TUSTIN (CALIFORNIA) POLICE DEPARTMENT
DATE OF INCIDENT: MAY 23, 2005
DATE OF DEATH: MAY 24, 2005
CAUSE OF DEATH: COMBINED EFFECTS OF DRUGS AND ETHANOL
ROLE OF TASER DEVICE: EXCLUDED

The police report in this incident was exempt from disclosure under California law, but media sources reported that Alvarado, who appeared to be under the influence of drugs, broke into an unoccupied apartment and struggled with the police when they arrived following a call for assistance. Officers shocked Alvarado once with a TASER device and placed him under arrest. Shortly after the officers called for paramedics to treat Alvarado for wounds he received while breaking through the apartment window, he stopped breathing. Paramedics took him to a local medical center, where he died the next day without regaining consciousness.

The medical examiner concluded that Alvarado died from the combined effects of drugs and alcohol. Toxicology tests revealed the presence of cocaine and cocaine metabolites, Diazepam, and morphine in Alvarado's system (see *City News Service*, 2005, May 23; Orange County (California) Coroner's Report 05–03930–SR); Amnesty International, 2006; Anglen, 2006).

Alvarado was able to speak with officers for a short time following the application of the TASER pulses, indicating that the electric current did not have an immediate affect on his heart. Because the TASER device did not have an immediate effect and the coroner attributed the death to the combined effects of alcohol and the drugs in Alvarado's system, the device is excluded as the cause of death and as a significant contributing factor in the Alvarado case.

WALTER LAMONT SEATS

AGE: 23
RACE/GENDER: AFRICAN-AMERICAN MALE
AGENCY: NASHVILLE (TENNESSEE) POLICE DEPARTMENT
DATE OF INCIDENT: MAY 26, 2005
CAUSE OF DEATH: ASPHYXIA DUE TO FOREIGN BODY ASPIRATION (PLASTIC BAG WITH COCAINE "ROCKS")
ROLE OF TASER DEVICE: EXCLUDED

An undercover officer bought a rock of crack cocaine from Seats. As other officers approached to arrest Seats, they saw him put what they thought was cocaine in his mouth, and they ordered him to spit it out. They were not sure if he swallowed it then or hid it under his tongue or in his jaw. The officers wrestled Seats to the ground. A sergeant shocked Seats on the arm with a TASER device when Seats refused to spit out what he had put in his mouth. Soon after Seats was subdued and handcuffed, officers noticed that he was acting strangely. An ambulance took him to a local hospital where doctors initially said his heartbeat was normal, but fast. His condition deteriorated, and he died about an hour later.

At the autopsy, the medical examiner discovered a plastic baggie of cocaine lodged in Seats' throat, and he found cocaine in Seats' stomach. The medical examiner ruled that Seats died of asphyxiation, choking on the plastic bag containing cocaine (see *Associated Press*, 2005, May 28; Nashville (Tennessee) Metropolitan Police Report 05–279630; Office of the Medical Examiner (Tennessee) Report MEC 05-1665; Amnesty International, 2006; Anglen, 2006).

For about an hour following the application of the TASER pulses, Seats had a heartbeat that was fast but regular, indicating that the electric current did not immediately affect his heart. Because the TASER device did not have an immediate effect and the medical examiner attributed the death to asphyxiation, the device is excluded as the cause of death and as a significant contributing factor in the Seats case.

RICHARD THOMAS HOLCOMB

AGE: 18
RACE/GENDER: WHITE MALE

AGENCY: SPRINGFIELD TOWNSHIP (OHIO) POLICE DEPARTMENT
DATE OF INCIDENT: MAY 27, 2005
CAUSE OF DEATH: CARDIAC ARRHYTHMIA DUE TO DRUG INDUCED PSYCHOSIS DUE TO DRUG (METHAMPHETAMINE AND MDMA/MDA) INTOXICATION, ACUTE
CONTRIBUTING FACTORS: ELECTRICAL PULSE INCAPACITATION
ROLE OF TASER DEVICE: EXCLUDED/DOUBTFUL

The police report in this incident was not available, but media sources reported that a homeowner called police to report someone trespassing on her property. The responding officer found Holcomb in her yard. Holcomb was walking toward the officer, shirtless, singing, and talking nonsense, and not answering basic questions, such as his name or address. When the officer asked his age, Holcomb incorrectly answered that he was 17. He walked away from the officer muttering phrases about killing someone. Holcomb sat down in a field, still incoherent and not answering questions, and then suddenly he charged the officer. She ordered him to stop, but he did not. She shocked him with a TASER device and called for an ambulance. When he tried to get up, she shocked him three more times. When backup officers arrived, they handcuffed Holcomb. The officers later noticed Holcomb's chest and chin twitching, so they took off the handcuffs. A little later, they noticed he was not breathing. When paramedics arrived, they discovered Holcomb was in ventricular fibrillation. The paramedics shocked Holcomb several times with a defibrillator without success. They also could not insert an airway tube because of his clenched jaw. Paramedics took Holcomb to a local hospital. He arrived in full cardiac arrest, and advanced cardiac life support was not successful in resuscitating him.

A medical examiner reported that Holcomb had a history of methamphetamine and cocaine abuse. Toxicology tests revealed the presence of methamphetamine, methylenedioxymethamphetamine (MDMA), and methylenedioxyamphetamine (MDA) in Holcomb's system. She attributed his death to a cardiac arrhythmia induced by the physiologic stress caused by the use of methamphetamine and MDMA/MDA. She added that, to some extent, the electrical impulses delivered by the TASER device while police attempted to arrest Holcomb contributed to his death (see County of Summit (Ohio) Medical Examiner's Report N–220-05; Iacoboni, 2005; Meyer, 2005; Amnesty International, 2006; Anglen, 2006).

The media reports did not specify how much time passed between the final application of the TASER pulses and when Holcomb stopped breathing. However, it was apparently at least a few minutes, indicating that the electric current did immediately affect Holcomb's heart. In her report, the medical examiner did not explain how, "to some extent," the TASER shocks contributed to Holcomb's death, and she did not document how the electrical discharges disrupted the rhythm of Holcomb's heart. She did comment on the added physiological stress of the electrical discharges, but she did not explain how that stress would have had a delayed effect. The roles of sympathomimetic drugs, such as methamphetamine and Ecstasy, in sudden death are well known. Considering Holcomb's state of methamphetamine and Ecstasy intoxication, it is questionable whether other forms of restraint would have produced less physiologic stress than did the TASER shocks.

Consequently, because the TASER device did not have an immediate effect and the medical examiner attributed the cause of the cardiac arrhythmia to the effects of methamphetamine and Ecstasy, the device is excluded as the cause of death. Additionally, a heart in fibrillation due to electrical stimulation is generally responsive to cardioversion with a defibrillator, but Holcomb's heart was not. Because the TASER pulses did not have an immediate effect on Holcomb's heart, because he clearly exhibited symptoms of Ecstasy intoxication, i.e., the clenched jaws, and because repeated attempts at defibrillation were ineffective, the role of the TASER device as a significant contributing factor is doubtful.

NAZARIO JAVIER SOLORIO

AGE: 38
RACE/GENDER: HISPANIC MALE

AGENCY: ESCONDIDO (CALIFORNIA) POLICE DEPARTMENT
DATE OF INCIDENT: MAY 28, 2005
DATE OF DEATH: MAY 2, 2005
CAUSE OF DEATH: ANOXIC ENCEPHALOPATHY, DUE TO PROLONGED HYPOXIA, DUE TO POSITIONAL ASPHYXIA
CONTRIBUTING FACTORS: SCHIZOPHRENIA, METHAMPHETAMINE USE
ROLE OF TASER DEVICE: EXCLUDED

The police report in this incident was exempt from disclosure under California law, but media sources reported that Solorio's mother made an emergency call to report her son's bizarre and threatening behavior. Solorio locked himself in a converted garage where he was living, and would not open the door when police asked. A relative used a garage door opener to let in officers. Solorio fought with police as they tried to take him into custody. The officers subdued Solorio by drive stunning him five times with a TASER device, four times between the shoulder blades and once on the left thigh. While officers were handcuffing Solorio and putting on a spit mask, he complained that he could not breathe, and he lost consciousness. Paramedics took Solorio to a local hospital. Solorio slipped into a coma and died five days later.

The medical examiner ruled Solorio's death was due to anoxic encephalopathy due to prolonged hypoxia due to positional asphyxia. The medical examiner also concluded that Solorio's methamphetamine use and schizophrenia were significant contributory factors in his death. He also noted the presence of alcohol in Solorio's system when he arrived at the hospital (see *City News Service*, 2005, August 26; Shearer, 2005; County of San Diego (California) Medical Examiner's Report 05–01049; Amnesty International, 2006; Anglen, 2006).

Solorio continued to speak with officers for a short time following the last application of the TASER pulses, indicating the electric current did not immediately affect his heart. Because the TASER device did not have an immediate affect and the medical examiner attributed the death to the effects of restraint asphyxia, the device is excluded as the cause of death and as a significant contributing factor in the Solorio case.

RAVAN JERMONT CONSTON

AGE: 33
RACE/GENDER: AFRICAN-AMERICAN MALE
AGENCY: SACRAMENTO (CALIFORNIA) POLICE DEPARTMENT
DATE OF INCIDENT: JUNE 4, 2005
CAUSE OF DEATH: SUDDEN CARDIAC DEATH FOLLOWING PHYSICAL ALTERCATION WITH POLICE OFFICERS DURING WHICH HE WAS TASERED, SUBDUED AND HANDCUFFED, DUE TO EXCITED DELIRIUM, DUE TO ACUTE COCAINE INTOXICATION
CONTRIBUTING FACTORS: CHRONIC ABUSE OF ILLICIT DRUGS
ROLE OF TASER DEVICE: EXCLUDED

The police report in this incident was exempt from disclosure under California law, but media sources reported that officers encountered Conston after a 9-1-1 caller said a man near downtown was assaulting people. On their arrival, Conston immediately drew the officers' attention. As the officers started to put him in handcuffs, Conston threw off one officer and turned to the others in a fighting stance. One of the officers then shot a TASER cartridge at Conston, but it had no effect. Conston began to run, and another officer fired an ineffective TASER shot. Officers caught Conston, and another TASER shot brought him to the ground. The officer deployed a second charge when Conston lay on his arms and refused to allow officers to handcuff him. Officers were transporting Conston to a hospital when he collapsed.

Toxicology results revealed the presence of a significant amount of cocaine in Conston's system. Additionally, the tests revealed the presence of Mirtazapine, an antidepressant used for the treatment of mild to severe depression. The medical examiner concluded that Conston died of sudden cardiac death following a physical altercation with police officers during which he was shocked with a TASER device, subdued and handcuffed, due to excited delirium, due to acute cocaine intoxication (see Jewett, 2005, June 7; Sacramento County (California) Coroner Report 05–02915; Amnesty International, 2006; Anglen, 2006).

Conston continued to struggle with officers following the last application of the TASER pulses, indicating that the electric current did not immediately affect his heart. Although the medical examiner mentioned the TASER device in the cause of death, that mention is a temporal statement regarding the altercation with the police. In his report, the medical examiner did not mention TASER shocks as a causal factor in Conston's death. Consequently, because the TASER device did not have an immediate effect and the medical examiner attributed the cardiac death to cocaine-induced excited delirium, the device is excluded as the cause of death and as a significant contributing factor.

RUSSELL WALKER

AGE: 47
RACE/GENDER: MALE (RACE NOT DISCERNABLE FROM THE AVAILABLE RECORDS)
AGENCY: LAS VEGAS METROPOLITAN (NEVADA) POLICE DEPARTMENT
DATE OF INCIDENT: JUNE 6, 2005
CAUSE OF DEATH: CARDIAC ARRHYTHMIA DURING RESTRAINT AND EXCITED DELIRIUM
CONTRIBUTING FACTORS: COCAINE ABUSE
ROLE OF TASER DEVICE: EXCLUDED

The police and autopsy reports in this incident were exempt from disclosure under Nevada law, but media sources reported that Walker was acting erratically outside a hotel in a section of downtown Las Vegas. Agitated and unresponsive to security guards at the casino, Walker was ripping up and throwing money on the ground as he walked in circles and stared into space. When police arrived, Walker exhibited incredible strength while wrestling with security guards, police officers and paramedics who tried to handcuff him, restrain him on a gurney, and transport him for medical attention. At one point, two security guards and three police officers could not pull Walker's arms behind him to handcuff him, and Walker was able to bend a pair of handcuffs during the struggle. An officer then shocked Walker once with a TASER device, and the officers were able to handcuff him. When paramedics arrived to take Walker to the hospital, the officers wrestled with him again as they tried to sit him up to remove the handcuffs. An officer shocked Walker again with a TASER device. Officers and paramedics eventually got Walker restrained on a gurney for transport to the hospital. However, Walker's heart stopped while in the ambulance, and he died at a local hospital.

Media sources reported that the medical examiner determined Walker died of a heart arrhythmia during restraint procedures, and that he was suffering from acute alcohol and cocaine intoxication with excited delirium syndrome (see Geary, 2005; Amnesty International, 2006; Anglen, 2006). Because the TASER device did not have an immediate effect and the medical examiner attributed the arrhythmia to cocaine-induced excited delirium, the device is excluded as the cause of death and as a significant contributing factor in the Walker case.

HORACE OWENS

AGE: 48
RACE/GENDER: AFRICAN-AMERICAN MALE
AGENCY: BROWARD COUNTY (FLORIDA) SHERIFF'S OFFICE
DATE OF INCIDENT: JUNE 11, 2005
CAUSE OF DEATH: ACUTE MYOCARDIAL INFARCT DUE TO THROMBOSED RIGHT CORONARY ARTERY
CONTRIBUTING FACTORS: COCAINE USE
ROLE OF TASER DEVICE: EXCLUDED

Owens broke into a house and was screaming that somebody was trying to kill him. The owner of the house fled and called the Sheriff's Office. Deputies entered the home and coaxed Owens outside, where he began fighting with them. The deputies shocked Owens six times with a TASER device, but he continued struggling with officers. During the struggle, Owens collapsed and stopped breathing. Deputies tried to revive him, but he was pronounced dead at a local hospital about one hour later.

The medical examiner determined that Owens suffered from cardiomegaly and that he had critical coronary atherosclerosis. Owens also had an acute coronary thrombosis. The medical examiner also noted that Owens had a body temperature of 102.6 degrees Fahrenheit more than one hour postmortem. The medical examiner reported that the cocaine in Owens' system caused his cardiac arrest. He also noted that only one TASER probe penetrated Owens' body, and there were no burns on Owens' body from the shocks (see Broward County (Florida) Medical Examiner Report 05-0924; *Associated Press*, 2005, June 13; Barton, 2005; Broward County (Florida) Sheriff's Office Report WP05-06-00467; Amnesty International, 2006; Anglen, 2006).

Owens continued to struggle with deputies following the application of the TASER pulses, indicating that the electric current did not immediately affect his heart. Because the TASER device did not have an immediate effect and the medical examiner attributed the death to an acute myocardial infarct, the device is excluded as the cause of death and as a significant contributing factor in the Owens case.

MICHAEL ANTHONY EDWARDS

AGE: 48
RACE/GENDER: AFRICAN-AMERICAN MALE
AGENCY: BROWARD COUNTY (FLORIDA) SHERIFF'S OFFICE
DATE OF INCIDENT: JUNE 13, 2005
CAUSE OF DEATH: EXCITED DELIRIUM DUE TO COCAINE INTOXICATION
ROLE OF TASER DEVICE: EXCLUDED

Deputies were called to Edwards' home twice on noise complaints. The second time they arrived, a deputy observed Edwards run through two windows, shattering the glass. Deputies found him naked, bleeding, and trying to run through a wooden privacy fence in the backyard. Deputies could not get Edwards to calm down, and when talking to him failed, deputies attempted to secure him by shocking him three times with a TASER device. A few minutes later, after deputies had handcuffed Edwards, they realized he was in distress. Paramedics called to the scene found that Edwards was not breathing and did not have a pulse. He was dead on arrival at a local hospital. Edwards' girlfriend later told investigating officers that he had been taking drugs and drinking.

The medical examiner concluded that Edwards was suffering from excited delirium due to cocaine intoxication. He noted high levels of cocaine and cocaine metabolites in Edward's system. Additionally, the toxicology tests revealed traces of marijuana in Edward's blood. His blood alcohol was negative (see Putnam County Sheriff's Office Report 2005-6485; Florida Medical Examiner's Report 05-23-322; Barton, 2005; *Florida Times-Union*, 2005; Amnesty International, 2006; Anglen, 2006).

Edwards continued to struggle with officers for a few minutes following application of the TASER pulses, indicating that the electric current did not immediately affect his heart. Because the TASER device did not have an immediate effect and the medical examiner attributed the death to cocaine-induced excited delirium, the device is excluded as the cause of death and as a significant contributing factor in the Edwards case.

SHAWN CHRISTOPHER PIROLOZZI

AGE: 30
RACE/GENDER: AFRICAN-AMERICAN MALE
AGENCY: CANTON (OHIO) POLICE DEPARTMENT
DATE OF INCIDENT: JUNE 13, 2005
CAUSE OF DEATH: EXCITED DELIRIUM, SELF-INFLICED INJURIES
CONTRIBUTING FACTORS: TRACE OF MARIJUANA, MULTIPLE BLUNT AND SHARP FORCE INJURIES TO HEAD, TRUNK, EXTREMITIES, LOSS OF BLOOD AND OXYGEN, HISTORY OF MENTAL ILLNESS
ROLE OF TASER DEVICE: EXCLUDED

The police and autopsy reports were exempt from disclosure under Ohio law, but media sources reported that police received several 9-1-1 calls reporting that a naked man was jumping on top of cars as they passed. Officers later discovered that Pirolozzi had gone berserk inside his apartment, breaking windows and splashing the inside of the residence with water to purify it from demons. He broke out a second story window, jumped out, and ran out into traffic. He was naked, bleeding, and jumping on cars when officers arrived. Pirolozzi jumped on a patrol car and began attacking the officer inside, trying to take the officer's gun. Officers shocked Pirolozzi once with a TASER device, but it had no effect. Officers and a firefighter finally pulled Pirolozzi from the police vehicle and subdued him. Medical personnel took Pirolozzi to a local hospital for treatment of his wounds, but he died shortly after arrival.

Media sources reported that the coroner listed Pirolozzi's immediate cause of death as excited delirium. The death certificate also listed multiple blunt and sharp force injuries to head, trunk and extremities, loss of blood and loss of oxygen as contributing factors. The injuries were the result of self-inflicted actions and asphyxia during legal intervention (see *Associated Press*, 2005, June 15a; *Associated Press*, 2005, July 23; *Associated Press*, 2005, August 28; Amnesty International, 2006; Anglen, 2006).

Pirolozzi continued to resist officers for several minutes following the application of the TASER pulses, indicating that the electric current did not immediately affect his heart. Because the TASER device did not have an immediate effect and the medical examiner attributed the death to excited delirium and self-inflicted injuries, the device is excluded as the cause of death and as a significant contributing factor in the Pirolozzi case.

ROBERT EARL WILLIAMS

AGE: 62
RACE/GENDER: AFRICAN-AMERICAN MALE
AGENCY: WACO (TEXAS) POLICE DEPARTMENT
DATE OF INCIDENT: JUNE 14, 2005
CAUSE OF DEATH: ACUTE PHYSIOLOGIC STRESS ASSOCIATED WITH MULTIPLE ELECTRICAL SHOCKS DURING ATTEMPTED RESTRAINT BY POLICE FOR SCHIZOPHRENIA AND EXCITED DELIRIUM
CONTRIBUTING FACTORS: HYPERTENSIVE AND ARTERIOSCLEROTIC CARDIOVASCULAR DISEASE, DIABETES MELLITUS AND OBESITY

ROLE OF TASER DEVICE: EXCLUDED

Police responded to a 9-1-1 call to the home of Williams' sister regarding a disturbance. When officers arrived, they found Williams had an outstanding warrant on a previous resisting arrest charge. When they tried to arrest him, Williams picked up a piece of rebar, threatening officers. Officers shocked him with a TASER device, but it had no effect. Eventually, after eight shocks, it took five officers physically to subdue and handcuff Williams. After the officers had handcuffed him, he complained that he could not breathe. Paramedics arrived on the scene and began treating Williams when he stopped breathing. The paramedics transported Williams to a local hospital, where he died.

The medical examiner reported that Williams died from acute physiological stress associated with multiple electrical shocks during attempted restraint by police. He explained that Williams was schizophrenic and in the throes of excited delirium. He added that Williams' hypertensive and arteriosclerotic cardiovascular disease, diabetes mellitus and obesity were significant contributing factors (see *Associated Press*, 2005, June 15b; Dallas County (Texas) Medical Examiner Report JP1922–05; Waco (Texas) Police Report 05–041101; Amnesty International, 2006; Anglen, 2006).

Williams continued to resist officers following the last application of the TASER pulses, indicating that the electric shocks did not immediately affect his heart. The medical examiner attributed the physiological stress to the electric shocks, but not to the physical struggle against five officers. However, because the TASER device did not have an immediate effect on Williams, it is clear that the physiological stress that he experienced because of the shocks were not sufficient to make him stop fighting. Additionally, Williams was suffering from a number of sudden death factors, including hypertensive and arteriosclerotic cardiovascular disease, diabetes mellitus and obesity. It is questionable whether any other form of restraint would have produced a different result. Consequently, because the TASER device did not have an immediate effect and there is no evidence to attribute the electrical charges to Williams' collapse, the device is excluded as the cause of death and as a significant contributing factor in the Williams case.

MELINDA KAYE (FAIRBANKS) NEAL

AGE: 33
RACE/GENDER: WHITE FEMALE
AGENCY: WHITFIELD COUNTY (GEORGIA) SHERIFF'S OFFICE
DATE OF INCIDENT: JUNE 22, 2005
DATE OF DEATH: JUNE 23, 2005
CAUSE OF DEATH: SYMPATHOMIMETIC POISONING SYNDROME (MALIGNANT HYPERTHERMIA) DUE TO METHAMPHETAMINE TOXICITY
ROLE OF TASER DEVICE: EXCLUDED

The Sheriff's Office report in this incident was not available, but media sources reported that Neal suffered from manic depression and paranoid schizophrenia, but she had been off her medications for about one year. On June 22, 2005, Neal smoked a quantity of methamphetamine and began to exhibit bizarre and delusional behavior. She walked into an elderly couple's home and began removing things from the cabinets. When deputies arrived, she became combative. The deputies handcuffed her and placed her in ankle restraints. When the deputies placed Neal into the back of a patrol car, she became combative and then reached into her pants. When the deputies took her out to search her for weapons, Neal began hitting her head on the side of the patrol car. After the deputies had put her back into the car, she got out again, fell to the ground, and hit her head. After the deputies put her back into the car, she kicked out the back window. The deputies reported shocking Neal with a TASER device, but they did not disclose the number of times. In a lawsuit filed later, the family claimed that deputies shocked Neal 13 times. The deputies called for an ambulance, but Neal would not let the paramedics treat her. At the jail, Neal passed out, and paramedics transported her to a local hospital. Neal's body temperature on arrival was 107 degrees Fahrenheit. She died the next morning, about 18 hours after her original contact with the deputies.

A medical examiner concluded that Neal died of sympathomimetic poisoning syndrome (malignant hyperthermia) due to methamphetamine toxicity. Essentially, Neal's body overheated because of the methamphetamine she had taken. The medical examiner explained that Neal's case was a classic example of amphetamine-related hyperthermia syndrome, and he noted that the level of methamphetamine in Neal's system was lethal. He also reported that the application of the TASER device did not contribute to Neal's death (see Anderson, M., 2005; Georgia Bureau of Investigation Record of Medical Examiner Report 2005-7002735; Swiney, 2005; Anderson, M., 2006; Anglen, 2006).

Because the TASER pulses did not have an immediate effect and the medical examiner attributed the death to the effects of methamphetamine, the TASER device is excluded as the cause of death and as a significant contributing factor.

CAROLYN J. DANIELS

AGE: 35
RACE/GENDER: AFRICAN-AMERICAN FEMALE
AGENCY: FORT WORTH (TEXAS) POLICE DEPARTMENT
DATE OF INCIDENT: JUNE 24, 2005
CAUSE OF DEATH: ACUTE COCAINE INTOXICATION
ROLE OF TASER DEVICE: EXCLUDED

Daniels approached a police officer and said someone was following her. She was acting strangely and not making sense, and the officer observed a crack pipe in her purse. When the officer tried to arrest Daniels, she resisted. He shocked her twice with a TASER device to subdue her. After the second shock, she calmed down. The officer arrested Daniels and took her to jail. While jailers were searching Daniels, they noticed that Daniels was having trouble breathing, and they called for paramedics. Paramedics took Daniels to a local hospital, where she died about 90 minutes after the initial confrontation with police.

The medical examiner ruled that Daniels died a sudden death due to cocaine intoxication. Toxicology results indicated high levels of cocaine and cocaine metabolites in Daniel's blood. The test also showed the presence of the antidepressant drugs

Amitriptyline and Nortriptyline (see Boyd, 2005; Fort Worth (Texas) Police Report 05075522; McDonald, 2005; Tarrant County (Texas) Medical Examiner Report 0505897; Amnesty International, 2006; Anglen, 2006).

For several minutes following the application of the TASER pulses, Daniels showed no signs of medical distress, indicating that the electric current did not immediately affect her heart. Because the TASER device did not have an immediate effect and the medical examiner attributed the death to the effects of cocaine, the device is excluded as the cause of death and as a significant contributing factor in the Daniels case.

PHOARAH KAREEM KNIGHT

AGE: 33
RACE/GENDER: AFRICAN-AMERICAN MALE
AGENCY: MIAMI-DADE (FLORIDA) POLICE DEPARTMENT
DATE OF INCIDENT: JUNE 24, 2005
CAUSE OF DEATH: COCAINE PSYCHOSIS
ROLE OF TASER DEVICE: EXCLUDED

The police report on this incident was not available, but media sources reported that neighbors called 9-1-1 to report Knight was acting irrationally and trying to break into their homes. When police responded, they encountered an extremely agitated Knight, who was banging on doors, climbing on roofs, and screaming, "Kill them." When officers confronted him, Knight ran and attempted to break into a house. Officers shocked him at least twice with a TASER device. Medical personnel took him to a local hospital, where he died.

The medical examiner noted the presence of cocaine and cocaine metabolites in Knight's body. He concluded that Knight died of cocaine psychosis (see Barton, 2005; Miami-Dade County (Florida) Medical Examiner's Report 2005–01598; *Miami Herald*, 2005; Yanez, 2005; Amnesty International, 2006; Anglen, 2006). Because the medical examiner attributed the death to the effects of cocaine, the TASER device is excluded as the cause of death and as a significant contributing factor in the Knight case.

TOMMY VALENTINE GUTIERREZ

AGE: 38
RACE/GENDER: HISPANIC MALE
AGENCY: SACRAMENTO COUNTY (CALIFORNIA) SHERIFF'S OFFICE
DATE OF INCIDENT: JULY 2, 2005
CAUSE OF DEATH: INCISED WOUND OF LEFT WRIST
CONTRIBUTING FACTORS: EXCITED DELIRIUM DUE TO ACUTE METHAMPHETAMINE INTOXICATION
ROLE OF TASER DEVICE: EXCLUDED

The Sheriff's Office report on this incident was exempt from disclosure under California law, but media sources reported that Gutierrez entered a gas station convenience store and told the clerk that someone was following him. He was clearly agitated, and he locked himself in the restroom. When the clerk saw blood leaking under the door, he called 9-1-1. Sheriff's deputies arrived and kicked open the restroom door. Gutierrez, who had cut both his wrists, was slumped in a pool of blood. As a deputy approached him, Gutierrez crawled towards him and tried to bite through the officer's boot. The deputy shocked him with a TASER device, and, when only one prong stuck, he shot again. Gutierrez collapsed and died at a local hospital about 50 minutes later.

The coroner observed several minor blunt impact injuries and several superficial incised wounds on Gutierrez, but none was sufficient to cause death. However, the incised wound on Gutierrez's left wrist severed his ulnar artery and resulted in significant blood loss. Toxicology tests were positive for alcohol and methamphetamine. The coroner ruled that Gutierrez died from extensive blood loss from the severed ulnar artery, and he ruled the death a suicide. He noted a significant contributing factor of excited delirium due to methamphetamine intoxication, and he added that the TASER device did not contribute to Gutierrez's death (see ACLU of Northern California, 2005; Ginsburg, 2005; Sacramento County (California) Coroner Report 05–03421). The TASER device is excluded as the cause of death and as a contributing factor in the Gutierrez case.

ROCKEY BRYSON

AGE: 41
RACE/GENDER: WHITE MALE
AGENCY: BIRMINGHAM (ALABAMA) POLICE DEPARTMENT
DATE OF INCIDENT: JULY 6, 2005
DATE OF DEATH: JULY 7, 2005
CAUSE OF DEATH: PROBABLE ARRHYTHMIA SECONDARY TO FOCAL SEVERE CORONARY ATHEROSCLEROSIS
CONTRIBUTING FACTORS: CHRONIC ETHANOL ABUSE
ROLE OF TASER DEVICE: EXCLUDED

The police report in this incident was not available, but media sources reported that police officers had arrested Bryson for driving under the influence of alcohol and booked him in the city jail. Three days later, jail officers observed Bryson in his cell hallucinating. When a nurse came to examine him, Bryson tried to leave the cellblock and was involved in a confrontation with jailers. The jailers shocked Bryson twice with a TASER device and sprayed him with chemical spray. Medical personnel took Bryson to a local hospital, where he was treated and released. He was up and responsive when returned to the jail. Later that night, jailers found him unresponsive. Paramedics arrived, but Bryson was dead at the scene.

The coroner ruled that Bryson died of a heart attack caused by coronary heart disease and delirium tremens, and he noted significant narrowing of Bryson's coronary arteries, which meant he was not getting sufficient blood flow to his heart. Bryson's heart disease was complicated by his chronic alcohol abuse. The coroner reported that the TASER shocks did not contribute to Bryson's death (see *Birmingham News*, 2005; Jefferson County (Alabama) Coroner's Report 05-0746; Robinson, 2005; Anglen, 2006).

For several hours following the application of the TASER pulses, Bryson displayed no signs of medical distress, even when he was being treated at a hospital emergency room. Because the TASER device did not have an immediate effect and the coroner attributed the cause of Bryson's heart attack to preexisting heart disease and delirium tremens, the device is excluded as the cause of death and as a significant contributing factor in the Bryson case.

KEVIN RAY OMAS

AGE: 17
RACE/GENDER: WHITE MALE
AGENCY: EULESS (TEXAS) POLICE DEPARTMENT
DATE OF INCIDENT: JULY 12, 2005
DATE OF DEATH: JULY 14, 2005
CAUSE OF DEATH: NEUROLEPTIC MALIGNANT SYNDROME DUE TO THERAPEUTIC ADMINISTRATION OF HALDOL DUE TO ECSTACY INTOXICATION
ROLE OF TASER DEVICE: EXCLUDED

Responding to a call of a disturbance, officers found Omas in the playground of a school, wearing only boxer shorts and socks, and screaming that he was God. Friends told officers that Omas had ingested four hits of lysergic acid diethylamide (LSD), two tablets of methylenedioxymethamphetamine (Ecstasy), and some alcohol at a party. When officers approached Omas, he charged them. One officer deployed his TASER device, shocking Omas three times before subduing him. Paramedics sedated Omas with Haloperidol (Haldol) and transported him to a local hospital, where, after about 40 minutes, he went into a coma and died two days later.

A medical examiner ruled that Omas died from neuroleptic malignant syndrome due to the therapeutic administration of Haldol. The autopsy revealed no LSD or alcohol in Omas' system, but it did reveal the presence of Ecstasy (see *Associated Press*, 2005, July 13; Euless (Texas) Police Report 0500043238; Sanchez, 2005; Tarrant County (Texas) Medical Examiner's Report 0506455; Amnesty International, 2006; Anglen, 2006).

Omas continued to struggle with officers and paramedics following the application of the TASER device, thereby necessitating the administration of Haldol. The electric current of the TASER pulses obviously did not immediately affect Omas' heart. Because the TASER device did not have an immediate effect and the medical examiner attributed the death to neuroleptic malignant syndrome, the device is excluded as the cause of death and as a significant contributing factor in the Omas case.

OTIS GENE THRASHER

AGE: 42
RACE/GENDER: WHITE MALE
AGENCY: BUTTE-SILVER BOW COUNTY (MONTANA) SHERIFF'S OFFICE
DATE OF INCIDENT: JULY 12, 2005
DATE OF DEATH: JULY 15, 2005
CAUSE OF DEATH: HEART ATTACK DUE TO HEART DISEASE
CONTRIBUTING FACTORS: TASER SHOCK, METHAMPHETAMINES, DIAZEPAM AND MARIJUANA MIGHT HAVE CONTRIBUTED
ROLE OF TASER DEVICE: EXCLUDED/ UNDETERMINED

Thrasher's mother called 9-1-1 to report that Thrasher was threatening his wife and daughter with a butcher knife. When police arrived, Thrasher began to wave the knife at them and put it to his own throat. Thrasher's wife and daughter left the house, and deputies called for an officer with a TASER device to respond to the scene. The initial shot missed. Thrasher threw the knife and a chair at the deputies and tried to run away. One deputy sprayed him with chemical spray, which had no effect. The other deputy reloaded the TASER device and fired it at Thrasher again, this time hitting him and knocking him to the ground. One probe from the stun gun hit Thrasher in the chest and the other hit him in the forehead. Deputies called for paramedics, but, about 30 minutes later, Thrasher went into cardiac arrest. Doctors pronounced him brain-dead three days later and disconnected him from life support.

The autopsy report was exempt from disclosure under Montana law, but media sources reported that the medical examiner concluded, although the TASER shocks and the methamphetamine probably contributed to his death, the primary cause was Thrasher's weak heart. The autopsy report showed that Thrasher's coronary arteries were both clogged, and that one artery was clogged more than 90 percent (see *Associated Press*, 2005, July 15; Butte-Silver Bow County (Montana) Sheriff's Office Report CR05–03479; Amnesty International, 2006; Anglen, 2006; *Missoulian*, 2006).

For about 30 minutes following the application of the TASER pulses, Thrasher showed no signs of medical distress, indicating that the electric current did not immediately affect his heart. Because the TASER device did not have an immediate effect and the medical examiner attributed the death to a preexisting heart condition, the device is excluded as the cause of death. Without additional information regarding any evidence the medical examiner may have found to conclude that the TASER device contributed to the death, the role of the device as a significant contributing factor is undetermined.

ERNESTO VALDEZ, JR.

AGE: 42
RACE/GENDER: HISPANIC MALE
AGENCY: PHOENIX (ARIZONA) POLICE DEPARTMENT
DATE OF INCIDENT: JULY 15, 2005
CAUSE OF DEATH: EXCITED DELIRIUM DUE TO COCAINE INTOXICATION
ROLE OF TASER DEVICE: EXCLUDED

Valdez broke into a restaurant after closing time, fought with an employee, and began throwing himself at the walls. The employees left the restaurant, locked Valdez inside, and called the police. When police arrived, they reported that Valdez exhibited bizarre behavior and incredible strength. He kept banging his head on the ground. During a struggle, officers shocked Valdez several times with TASER devices in the drive stun mode. Printouts from the devices showed eight discharges from one and three from the other, although it is unknown how many of the discharges made contact. Even then, Valdez continued to struggle. Officers placed Valdez in restraints and masked him to prevent him from spitting. Shortly thereafter, they noticed he was not breathing. Valdez died at the scene.

The medical examiner concluded that Valdez died of the complications of excited delirium. He noted significant levels of cocaine and methamphetamine in Valdez's blood and urine (see Anglen, 2005; Maricopa County (Arizona) Medical Examiner's Report 05–02639; Phoenix (Arizona) Police Report 2005–51340162; Amnesty International, 2006; Anglen, 2006).

Valdez continued to resist officers following the final application of the TASER pulses, indicating that the electric current did not have an immediate effect on his heart. Because the TASER device did not have an immediate effect and the medical examiner attributed the death to cocaine and methamphetamine-induced excited delirium, the device is excluded as the cause of death and as a significant contributing factor in the Valdez case.

CARLOS CASILLAS-FERNANDEZ

AGE: 31
RACE/GENDER: HISPANIC MALE
AGENCY: SANTA ROSA (CALIFORNIA) POLICE DEPARTMENT
DATE OF INCIDENT: JULY 16, 2005
CAUSE OF DEATH: EXCITED DELIRIUM DUE TO METHAMPHETAMINE INTOXICATION
ROLE OF TASER DEVICE: EXCLUDED

The police report in this incident was exempt from disclosure under California law, but media sources reported that officers responded to a call of a domestic disturbance regarding a man who was possibly under the influence of drugs, and who had been acting paranoid and delusional for the past three days. The officers contacted Casillas-Fernandez, who became uncooperative. He resisted the officers' attempts to take him into custody, and he attempted to assault the officers. While attempting to take Casillas-Fernandez into custody, officers used various control and compliance techniques including control holds, the use of a carotid restraint, several shocks with a TASER device, and chemical spray. Eventually it took five officers to restrain Casillas-Fernandez and place him in handcuffs. Once officers had restrained Casillas-Fernandez, they noticed he was having difficulty breathing. Officers immediately administered CPR and summoned emergency medical attention. Paramedics transported Casillas-Fernandez to a local hospital, where he died. The Sonoma County Sheriff's Office conducted the investigation.

The coroner reported that Casillas-Fernandez died of excited delirium due to methamphetamine intoxication occurring in the setting of physiologic stress with and restraint by police officers. He said that a carotid neckhold, chemical spray and a TASER discharge were without significant effect during the struggle. Casillas-Fernandez's underlying cardiac disease, which the coroner listed as mild cardiomegaly, coronary artery arteriosclerosis, and atrioventricular valve myxoid degeneration, while not profound, could have been a contributing factor (see Doyle, 2005; Sample, 2005; Sonoma County (California) Coroner Report 05–0847; *US States News*, 2005; Amnesty International, 2006; Anglen, 2006).

Fernandez continued to struggle with officers for several minutes following the application of the TASER pulses, indicating that the electric current did not immediately affect his heart, and the medical examiner specifically noted in the autopsy report that the TASER shock was without significant effect. Because the TASER device had no immediate effect and the medical examiner attributed the death to methamphetamine-induced excited delirium, the device is excluded as the cause of death and as a significant contributing factor in the Fernandez case.

MICHAEL LEON CRUTCHFIELD

AGE: 40
RACE/GENDER: AFRICAN-AMERICAN MALE
AGENCY: WEST PALM BEACH (FLORIDA) POLICE DEPARTMENT
DATE OF INCIDENT: JULY 17, 2005
CAUSE OF DEATH: ACUTE COCAINE TOXICITY, EXCITED DELIRIUM
ROLE OF TASER DEVICE: EXCLUDED

Crutchfield, who had a history of drug abuse, entered an assisted living center and began acting irrationally, grabbing and attacking elderly residents and screaming that someone was trying to kill him. Several employees of the center called 9-1-1. They reported that Crutchfield was acting crazed and that he was choking people. When the police arrived, they tried to speak with Crutchfield, but he would not respond to their questions. He kept saying that the police were trying to kill him. After a few minutes, Crutchfield tried to run away from the officers, so one officer fired a TASER device. Crutchfield went down, but started to struggle with officers as soon as the charge subsided. The officer eventually shocked Crutchfield three times before other officers could secure Crutchfield with handcuffs and leg restraints. A nurse from the facility checked on Crutchfield. She said he was breathing and had a pulse, but she suggested calling for medics. Soon after the paramedics loaded him onto a gurney, Crutchfield quit breathing. He fell unconscious and died at the scene a short time later.

The Florida state attorney's office issued a report stating that it was clear from the evidence that neither the officers nor the TASER discharge was responsible for Crutchfield's death. The medical examiner's office concluded that Crutchfield died of cocaine toxicity with excited delirium. Toxicology reports indicated a high concentration of cocaine in Crutchfield's blood and urine (see Barton, 2005; Krischer, 2005; Marra, 2005, July 20; Marra, 2005, December 1; Office of the Medical Examiner, Fifteenth District (Florida) Report ME 05–0800; West Palm Beach (Florida) Police Report 05–017011; Amnesty International, 2006; Anglen, 2006).

Crutchfield continued to struggle with officers following the final application of the TASER pulses, and a nurse, who examined him after the officers had secured him, said Crutchfield had a pulse and was breathing. Crutchfield did not show signs of medical distress until paramedics arrived several minutes later. Because the TASER device did not have an immediate effect and the medical examiner attributed the death to cocaine-induced excited delirium, the device is excluded as the cause of death and as a significant contributing factor in the Crutchfield case.

MAURICE CUNNINGHAM

AGE: 29
RACE/GENDER: AFRICAN-AMERICAN MALE
AGENCY: LANCASTER COUNTY (SOUTH CAROLINA) SHERIFF'S OFFICE
DATE OF INCIDENT: JULY 23, 2005
CAUSE OF DEATH: CARDIAC ARRHYTHMIA DUE TO TASER SHOCK
ROLE OF TASER DEVICE: CONFIRMED

Neither the Sheriff's Office report nor the autopsy report in this incident was available, but media accounts were very detailed. Cunningham escaped his jail cell when an officer tried to get a razor from him after he shaved. Cunningham jumped on the officer's back and stabbed her in the eye with a pencil. She got away from Cunningham, but he attacked a second officer and stabbed him in the eye with another pencil. A third officer rushed in to assist, and Cunningham tried to gouge his eyes with his fingers. A fourth officer shot Cunningham with a TASER device, shocking him five times. The jolts lasted seven seconds, eight seconds, six seconds, five seconds, and nine seconds, respectively. Only the last shock had any effect, knocking Cunningham to the floor, but he ripped the probes out and continued to fight. The fourth officer sprayed Cunningham with chemical spray. When Cunningham continued to fight, the third officer fired a second TASER device, hitting Cunningham in the left arm and left thigh. The deputy held the trigger for two minutes and forty-nine seconds. During those two minutes and forty-nine seconds, Cunningham collapsed, and an officer put handcuffs and leg irons on him. The officers dragged Cunningham to the front desk and called for medical assistance. EMS arrived and pronounced Cunningham dead.

Media sources reported that the coroner listed Cunningham's cause of death as cardiac arrhythmia due to TASER shocks. Pathologists found that Cunningham's heart suffered damage at a cellular level purportedly from the electrical current. The pathologists concluded that the probes that embedded in Cunningham's left thigh and left arm completed a circuit in his body that disrupted the electrical system that controls the heart, according to the autopsy report. Toxicology tests revealed that Cunningham had no illicit drugs or alcohol in his system.

The report also said that Cunningham, who had never attacked officers before, had a psychiatric disorder, possibly schizophrenia. The night before the attack, Cunningham complained that he was seeing snakes in his cell. Doctors had prescribed for Cunningham Aripiprazole and Risperidone, two drugs commonly used to treat psychiatric disorders. Jail records showed that Cunningham received the drugs as prescribed through the day of the incident. However, toxicology tests showed neither of the two drugs in his system (see Bell & Juarez, 2005; *United Press International*, 2005; Amnesty International, 2006; Anglen, 2006).

Assuming the media accounts are accurate, Cunningham collapsed during the final application of the TASER pulses, and the probes struck him in the left arm and left thigh, which could have established a circuit that potentially affected the rhythm of his heart. Additionally, the pathologists observed microscopic changes to Cunningham's heart that they believe was caused by the electric current. Consequently, the role of the TASER device as the cause of death in the Cunningham case is confirmed.

ELISEO MALDANADO

AGE: 33
RACE/GENDER: HISPANIC MALE
AGENCY: LOS ANGELES (CALIFORNIA) POLICE DEPARTMENT
DATE OF INCIDENT: JULY 25, 2005
CAUSE OF DEATH: UNKNOWN
ROLE OF TASER DEVICE: UNDETERMINED

Little is known about the Maldanado case because the police report was exempt from disclosure under California law. The Los Angeles County Medical Examiner's Office could not produce an autopsy report, and news reports were sketchy. However, news sources reported that officers responded to a call concerning a man who had attempted a theft and was fighting with some people. When officers arrived, they attempted verbally to calm Maldanado, whose behavior was extremely confrontational and irrational. Because of Maldanado's agitated state, the officers requested backup. After several attempts by officers to talk with him, Maldanado began to attack them. Officers used chemical spray, collapsible batons, and a TASER device to take Maldanado into custody. The officers were finally able to control and handcuff Maldanado after a ten-minute struggle. Maldanado and three officers suffered injuries during the altercation. Police summoned a rescue ambulance, and paramedics transported Maldanado to a local hospital. He failed to respond to medical treatment, and he died (see *City News Service*, 2005, July 26; Los Angeles Police Department, 2005, July 25).

Media sources do not report the results of an autopsy or a cause of death. Without additional information, the role of the TASER device in the Maldanado case is undetermined.

TERRENCE THOMAS

AGE: 35
RACE/GENDER: MALE (RACE NOT DETERMINABLE FROM THE AVAILABLE RECORDS)
AGENCY: NEW YORK (NEW YORK) POLICE DEPARTMENT
DATE OF INCIDENT: JULY 27, 2005
CAUSE OF DEATH: CARDIAC ARREST DUE TO ACUTE COCAINE INTOXICATION
ROLE OF TASER DEVICE: EXCLUDED

The police report in this incident was not available, but media sources reported that plainclothes narcotics detectives arrested Thomas when they discovered that the car he was riding in was stolen. In the vehicle, police found an open container of alcohol and a plastic bag filled with crack cocaine, which appeared to have been chewed, suggesting that someone in the car had tried to ingest the cocaine before officers could find it. After the detectives brought Thomas to the 105th Precinct stationhouse, they placed him in a cell. Thomas became ill and aggressive. Jailers called for an ambulance, but Thomas refused medical attention. He grew more combative, so an officer used a TASER device to subdue him. The paramedics transported Thomas, but he died before reaching the hospital.

The autopsy report was not subject to disclosure under New York law, but media sources reported that the medical examiner concluded that Thomas died of acute cocaine intoxication after swallowing crack cocaine. She said that the TASER device was neither the cause of death nor a significant contributing factor in the Thomas case (see *Associated Press*, 2005, July 28; Gendar & Lemire, 2005; Amnesty International, 2006; Anglen, 2006).

Thomas continued to struggle with jailers following the application of the TASER pulses, indicating that the electric current did not immediately affect his heart. Because the TASER device did not have an immediate effect and the medical examiner attributed the death to the effects of cocaine, the device is excluded as the cause of death and as a significant contributing factor in the Thomas case.

ERIC MICHAEL MAHONEY

AGE: 33
RACE/GENDER: WHITE MALE
AGENCY: FREMONT (CALIFORNIA) POLICE
    DEPARTMENT
DATE OF INCIDENT: JULY 29, 2005
DATE OF DEATH: AUGUST 3, 2005
CAUSE OF DEATH: ACUTE METHAMPHETAMINE
    AND AMPHETAMINE INTOXICATION
ROLE OF TASER DEVICE: EXCLUDED

The police report in this incident was exempt from disclosure under California law, but media sources reported that officers responded to a report of someone shooting a gun at a hotel. When an officer arrived, he came across Mahoney, who gave the officer false identification and then ran away. As Mahoney was climbing over a nearby wall in his attempt to get away, the officer shot him with a TASER device, shocking him between seven and nine times, although only half of the shots contacted his skin. Mahoney was wanted on two no-bail warrants, one for a parole violation and the other for burglary. Officers also found two knives and an ounce of methamphetamine at the scene, leading them to believe that Mahoney may have been under the influence of drugs. The TASER shocks appeared to have no effect, and it took three officers to finally subdue Mahoney and walk him, under his own power, to the police car. The officers took Mahoney to a local hospital for medical clearance before taking him to jail. About 20 minutes after arriving at the hospital, Mahoney created a disturbance, collapsed, and went into a coma. Five days later, his family removed him from life support and he died.

According to the coroner's report, at the time of his admission to the hospital, Mahoney had a blood methamphetamine level of 0.70 mg/L, which is well above the normal potentially toxic level of 0.2 to 0.6 mg/L. Mahoney also had amphetamine, diazepam and phenytoin in his blood. The coroner reported that there was no evidence that Mahoney had sustained any life-threatening injuries in his struggle with the officers, and he concluded that Mahoney died of acute methamphetamine and amphetamine intoxication (see Alameda County (California) Coroner's Report 2005–02253; Lee, 2005, August 6; Gaura, 2005; Amnesty International, 2006; Anglen, 2006).

Mahoney continued to struggle with officers following the application of the TASER pulses, indicating that the electric current did not immediately affect his heart. Because the TASER device did not have an immediate effect and the coroner attributed the death to the effects of methamphetamine and amphetamine, the device is excluded as cause of death and as a significant contributing factor in the Mahoney case.

BRIAN PATRICK O'NEAL

AGE: 32
RACE/GENDER: AFRICAN-AMERICAN MALE
AGENCY: SAN JOSE (CALIFORNIA) POLICE
    DEPARTMENT
DATE OF INCIDENT: AUGUST 1, 2005
CAUSE OF DEATH: CARDIAC ARRHYTHMIA DUE TO
    EXCITED DELIRIUM DUE TO METHAMPHETAMINE
    AND CANNABINOID INTOXICATION
CONTRIBUTING FACTORS: EXHAUSTIVE VIOLENT
    PHYSICAL STRUGGLE
ROLE OF TASER DEVICE: EXCLUDED

The police report in this incident was exempt from disclosure under California law, but media sources reported that a man working at his home computer was startled by O'Neal kicking open his condominium door shortly after 4 A.M. The resident confronted and fought with O'Neal. When O'Neal fled, the victim chased him to a nearby mini-market, where the two fought violently, both inside and outside the store. Police responded and separated the combatants, who they said were almost exhausted by the lengthy struggle. O'Neal then attempted to flee the scene and struggled with officers after they chased him down on foot. They shocked O'Neal with a TASER device, sprayed him with chemical spray, and struck him with a baton before they attempted to immobilize him in a body-length fabric corset intended to bind a suspect's arms and legs. At that point, O'Neal stopped breathing. Officers began CPR immediately, and O'Neal was taken by ambulance to a local hospital, where he later died.

The medical examiner concluded that O'Neal died of cardiac arrest due to physical exertion during methamphetamine and cannabinoid intoxication. He noted that toxicological analysis revealed a near toxic level of methamphetamine, a low level of amphetamine, and the presence of marijuana metabolites in O'Neal's body (see Gaura, 2005; Santa Clara County (California) Medical Examiner's Report 05–02726; Amnesty International, 2006; Anglen, 2006).

O'Neal continued to struggle with officers for a few minutes following the application of the TASER pulses, indicating that the electric current did not immediately affect his heart. Because the TASER device did not have an immediate effect and the medical examiner attributed the cardiac arrhythmia to methamphetamine-induced excited delirium, the device is excluded as the cause of death or as a significant contributing factor in the O'Neal case.

DWAYNE ZACHARY

AGE: 44
RACE/GENDER: AFRICAN-AMERICAN MALE
AGENCY: SACRAMENTO COUNTY (CALIFORNIA) SHERIFF'S OFFICE
DATE OF INCIDENT: AUGUST 4, 2005
CAUSE OF DEATH: SUDDEN CARDIAC ARREST WHILE BEING RESTRAINED PRONE AFTER PHYSICAL ALTERCATION WITH POLICE THAT INCLUDED USE OF TASER DEVICES DUE TO EXCITED DELIRIUM DUE TO ACUTE COCAINE AND MDMA (ECSTASY) INTOXICATION
CONTRIBUTING FACTORS: CHRONIC ILLICIT DRUG ABUSE
ROLE OF TASER DEVICE: EXCLUDED

The Sheriff's Office report in this incident was exempt from disclosure under California law, but media sources reported that dispatchers got a call from a woman who said Zachary had beaten her and was walking around his apartment naked. When three deputies went to the apartment, Zachary responded to them with sexual overtures and erratic behavior. Deputies searched for the caller but did not find her. The deputies left the apartment, considering the incident resolved, but Zachary, naked and sweating profusely, followed them into the parking lot. Deputies commanded him to stop, but Zachary ran back to the apartment. Fearing for the caller's welfare, deputies followed. In the apartment, Zachary began throwing furniture and a glass photo frame, injuring five officers. Emergency medical care was called for deputies during the fracas. Three deputies fired TASER devices at Zachary, with no effect. Officers eventually subdued Zachary by overpowering him. After Zachary was taken into custody, he went into distress and was taken to a local hospital, where he died.

The coroner ruled that Zachary died of sudden cardiac arrest while being restrained prone after physical altercation with police, that included use of TASER devices, attributed to excited delirium due to acute cocaine and Ecstasy intoxication. Although he mentioned in the cause of death the use of the TASER device, he does not explain how the TASER shocks contributed to the death. If, however, the coroner were to determine that the TASER device was a cause of death, he would have ruled the death a homicide. Instead, because he could not medically establish the physiological effects of the restraint, he chose to leave the manner of death undetermined (see Jewett, 2005, August 6; Sacramento County (California) Coroner's Report 05–04018; Amnesty International, 2006).

Zachary continued to struggle with officers after the application of the TASER pulses, indicating that the electric current did not have an immediate effect on his heart. Because the TASER device did not have an immediate effect and the medical examiner attributed the death to cocaine and Ecstasy-induced excited delirium, the device is excluded as the cause of death and as a significant contributing factor.

OLSON AGOODIE

AGE: 44
RACE/GENDER: AMERICAN INDIAN MALE
AGENCY: GLENDALE (ARIZONA) POLICE DEPARTMENT
DATE OF INCIDENT: AUGUST 5, 2005
CAUSE OF DEATH: METHAMPHETAMINE INTOXICATION
ROLE OF TASER DEVICE: EXCLUDED

In the parking lot of a convenience store, Agoodie was sitting in a car that did not belong to him. When officers asked him to get out of the car, Agoodie ignored them. When the officers reached into the vehicle to pull Agoodie out, he struggled. To remove him from the car, the officers shocked Agoodie in the arm and leg with TASER devices. The officers put Agoodie on the ground, but he refused to let them handcuff him. The officers shocked Agoodie again in the back to get him to place his arms behind his back. A printout from the TASER data port would later show four discharges between 2 and 16 seconds each. The officers suspected that Agoodie had taken illicit drugs, so they called for medical assistance to take him to a local hospital. Firefighters who responded had the officers remove the handcuffs, and they secured Agoodie to a gurney with Velcro restraints. During the transport, Agoodie began to vomit. At the hospital, Agoodie suffered a seizure and died.

The medical examiner concluded that Agoodie had died of methamphetamine intoxication, and he ruled the death an accident. He added that the TASER shocks were neither a cause of the death, nor were they a significant contributing factor (see Glendale (Arizona) Police Report 05–092969; Maricopa County (Arizona) Medical Examiner's Report 05–03018; Amnesty International, 2006; Anglen, 2006).

For several minutes following the application of the TASER pulses, Agoodie showed no signs of medical distress, indicating that the electric current did not have an immediate effect on his heart. Because the TASER device did not have an immediate effect and the medical examiner attributed the death to methamphetamine intoxication, the device is excluded as the cause of death and as a significant contributing factor.

FRANK GILMAN EDGERLY

AGE: 47
RACE/GENDER: WHITE MALE
AGENCY: PHOENIX (ARIZONA) POLICE DEPARTMENT
DATE OF INCIDENT: AUGUST 7, 2005
CAUSE OF DEATH: METHAMPHETAMINE INTOXICATION
ROLE OF TASER DEVICE: EXCLUDED

Edgerly entered a restaurant, went into the men's and women's restrooms, and caused extensive damage. After Edgerly attacked an employee who had tried to get him to leave, the employees of the restaurant called 9-1-1. When officers arrived, they found Edgerly locked in a stall in the women's restroom. The man displayed great strength, kicking and swinging his arms at three officers who were trying to arrest him. During the struggle, police shocked him with a TASER device five times for five seconds each to no apparent effect. After a lengthy struggle, the officers got Edgerly handcuffed and took him out of the restaurant. Once outside, the officers saw that the man was unconscious and had stopped breathing. Fire department paramedics transported Edgerly to a local hospital, where he was pronounced dead.

According to police reports, the medical examiner found methamphetamine and amphetamine in Edgerly's system. He concluded that Edgerly died of methamphetamine intoxication, and he ruled the death an accident (see Phoenix (Arizona) Police Report 2005–51505800; Amnesty International, 2006; Anglen, 2006).

For several minutes following the application of the TASER pulses, Edgerly continued to struggle with officers, indicating that the electric current did not have an immediate effect on his heart. Because the TASER device did not have an immediate effect and the medical examiner attributed the death to methamphetamine intoxication, the device is excluded as the cause of death and as a significant contributing factor.

ROBERT E. BOGGON

AGE: 65
RACE/GENDER: AFRICAN-AMERICAN MALE
AGENCY: ESCAMBIA COUNTY (FLORIDA) SHERIFF'S OFFICE
DATES OF INCIDENTS: AUGUST 25, 2005 AUGUST 26, 2005
DATE OF DEATH: AUGUST 29, 2005
CAUSE OF DEATH: COMBINED EFFECTS OF ARTHERIOSCLEROTIC AND HYPERTENSIVE CARDIOVASCULAR DISEASE AND PARANOID SCHIZOPHRENIA
CONTRIBUTING FACTORS: RESTRAINT IN A CHAIR AND HALOPERIDOL INJECTIONS
ROLE OF TASER DEVICE: EXCLUDED

On August 19, officers arrested Boggon following a disturbance at a local store. Family members said he suffered a "mental episode" and began acting strangely, knocking over boxes in the store. While in jail, Boggon continued to be uncooperative. Jailers shocked Boggon with TASER devices on two occasions, once on August 25 and twice on August 26. On the day of his death, August 29, Boggon threatened a jailer who had come into his cell to retrieve a dinner tray. The jailer sprayed Boggon with chemical spray then took him to the showers to wash away the contaminants. Following the shower, jailers placed Boggon in a restraint chair, and the jail nurse administered an injection of Lorazepam, and another of Haloperidol and Diphenhydramine hydrochloride, a major tranquilizer combination, at about 2:40 P.M. At 8:30 P.M., a jail nurse checked on Boggon. He was conscious and had a good pulse, but his blood pressure was a little high. At 10:30 P.M., when the nurse checked Boggon again, she found that he had died in his cell, still restrained in the chair.

A medical examiner concluded that Boggon died from the combined effects of arteriosclerotic and hypertensive cardiovascular disease and paranoid schizophrenia. She listed confinement to a restraint chair and injections of Haloperidol as contributory causes, but she did not mention the injections of Lorazepam, and Diphenhydramine hydrochloride. She noted that Boggon's death was the result of several factors; his paranoid schizophrenia with evidence of acute psychosis, the hours he spent in the chair straining against the restraints while exhibiting signs of excited delirium, Haloperidol injections, and his arteriosclerotic and hypertensive cardiovascular disease. She ruled the death a homicide (see Escambia County (Florida) Sheriff's Office Report 05–022396); Office of the Medical Examiner, First District (Florida) Report MLA05–678; Rasmussen, 2006).

For three days following the last application of the TASER pulses, Boggon showed no signs of medical distress, indicating that the electric current did not affect his heart. Because the TASER device did not have an immediate effect and the medical examiner attributed the death to the combined effects of cardiovascular disease and schizophrenia, the device is excluded as the cause of death and as a significant contributing factor in the Boggon case.

BRIAN LICHTENSTEIN

AGE: 31
RACE/GENDER: WHITE MALE
AGENCY: MARTIN COUNTY (FLORIDA) SHERIFF'S OFFICE
DATE OF INCIDENT: AUGUST 26, 2005

DATE OF DEATH: AUGUST 27, 2005
CAUSE OF DEATH: COCAINE TOXICITY
ROLE OF TASER DEVICE: EXCLUDED

Deputies responded to a 9-1-1 call of a man down. The caller said that he had seen a man with no clothes lying in the wet grass outside his house. When deputies and paramedics arrived, they found Lichtenstein still lying on the grass, but as they approached, he ran off into the woods. Deputies searching the woods could hear Lichtenstein screaming at them to leave him alone and let him die. They were to learn later that Lichtenstein had been on a three-day crack cocaine binge. When the deputies approached Lichtenstein in the woods, he jumped up at them. One deputy fired a TASER cartridge into Lichtenstein's back. Lichtenstein fell to the ground, but, as soon as the charge ended, he started to crawl away. The deputy activated the TASER device again, and again Lichtenstein went to the ground. As soon as the charge ended, Lichtenstein crawled into some bushes and refused to come out. The deputies had to crawl into the bushes to drag Lichtenstein out. Once the deputies had Lichtenstein under control, paramedics strapped him to a backboard and took him to a local hospital. Lichtenstein died the next morning.

Doctors at the hospital said Lichtenstein went into cardiac arrest on his way to the hospital and suffered two strokes. The medical examiner ruled that Lichtenstein died of cocaine toxicity. Because of the cocaine toxicity, Lichtenstein went into shock, and he suffered multiorgan failure. The medical examiner added that the TASER shocks had nothing to do with Lichtenstein's death (see Bender, 2005, August 28; Bender, 2005, August 30; Martin County (Florida) Sheriff's Office Report 05–11924; Office of the Medical Examiner, Nineteenth District (Florida) Report ME 2005–19–516; Amnesty International, 2006; Anglen, 2006).

Lichtenstein continued to resist officers for several minutes following the last application of the TASER pulses, indicating that the electric current did not immediately affect his heart. Because the TASER device did not have an immediate effect and the medical examiner attributed the death to the effects of cocaine, the device is excluded as the cause of death and as a significant contributing factor.

SHAWN A. NORMAN

AGE: 40
RACE/GENDER: WHITE MALE
AGENCY: ROSS COUNTY (OHIO) SHERIFF'S OFFICE
DATE OF INCIDENT: AUGUST 26, 2005
CAUSE OF DEATH: UNKNOWN
CONTRIBUTING FACTORS: COCAINE, MARIJUANA AND ANTIDEPRESSANTS IN HIS SYSTEM
ROLE OF TASER DEVICE: UNDETERMINED

Norman had been standing in the middle of the street yelling for someone to help him because someone was trying to kill him. He then walked into a house, picked up some keys from a table, and stole a truck. Nearby, he crashed the truck into a tree, jumping out of the vehicle just before the crash. When a deputy approached, responding to a 9-1-1 call, Norman's wife told him that her husband was in a nearby creek bed, unarmed, and that he was afraid that someone was going to shoot him. When the deputy found him, Norman disregarded orders to get on the ground. The deputy fired two shots with a TASER device, but they proved ineffective. Norman scuffled with the deputy and fell to the ground. At one point during the ensuing struggle, the deputy was shocked by his own device. Backup deputies arrived, but Norman ignored their commands and continued fighting. The backup deputy touch-stunned Norman in the back with his TASER device for roughly three seconds, prompting Norman to comply with commands, and the deputies handcuffed him. While they waited for an ambulance, the deputies checked for Norman's pulse but could not find one. Efforts to revive Norman with a defibrillator failed. Medical personnel used CPR and moved Norman to a local hospital, where he died.

The autopsy report was not available, but media sources reported that preliminary autopsy results from the coroner showed Norman had marijuana, cocaine, and antidepressants in his system. He also had preexisting medical conditions, such as an enlarged heart and arteriosclerosis (see Ross County Sheriff's Office Report 01–05–015334; *Associated Press*, 2005, August 30; Eckert, 2005; Amnesty International, 2006; Anglen, 2006).

Media sources did not report the final autopsy results. Without the final autopsy report, the role of the TASER device in the Norman case is undetermined.

DAVID ANTHONY CROSS

AGE: 44
RACE/GENDER: MALE (RACE NOT DETERMINABLE FROM THE AVAILABLE RECORDS)
AGENCY: SANTA CRUZ COUNTY (CALIFORNIA) SHERIFF'S OFFICE
DATE OF INCIDENT: SEPTEMBER 17, 2005
DATE OF DEATH: SEPTEMBER 18, 2005
CAUSE OF DEATH: ANOXIC ENCEPHALOPATHY DUE TO CARDIOPULMONARY ARREST AND CHEST COMPRESSION DURING RESTRAINT
CONTRIBUTING FACTORS: AMPHETAMINE USE, HYPERTENSIVE HEART DISEASE, OBESITY
ROLE OF TASER DEVICE: EXCLUDED

The Sheriff's Office report in this incident was exempt from disclosure under California law, but media sources reported that Santa Cruz police officers arrested Cross on a domestic violence charge, took him to the Santa Cruz County Jail, and placed him in the medical unit. About eight hours after he arrived, he began yelling and banging his head against his cell door. When he refused to stop, four deputies tried to restrain him. When they could not subdue Cross, one of the deputies shocked him with a TASER device. Once the deputies had him restrained, they realized that Cross had stopped breathing. Paramedics transported Cross to a local hospital, where he died the next day.

The autopsy report was not available, but media sources reported that the coroner concluded Cross lost the ability to breathe due to chest compression during the restraint. He went into cardiopulmonary arrest and then suffered anoxic encephalopathy. The coroner listed contributing causes of hypertensive heart disease and obesity. The examination also determined that Cross had amphetamine in his system. The coroner added that the TASER shocks did not contribute to Cross' death (see Smith, 2005; Amnesty International, 2006; Anglen, 2006).

Although Cross stopped breathing very shortly following the application of the TASER pulses, the medical examiner could find no evidence establishing a causal relationship, citing the effects of methamphetamine, hypertensive heart disease, and obesity. The TASER device is excluded as the cause of death and as a significant contributing factor.

TIMOTHY MICHAEL TORRES, JR.

AGE: 24
RACE/GENDER: Hispance Male
AGENCY: SACRAMENTO COUNTY (CALIFORNIA) SHERIFF'S OFFICE
DATE OF INCIDENT: SEPTEMBER 22, 2005
CAUSE OF DEATH: SUDDEN CARDIAC ARREST DURING PHYSICAL STRUGGLE THAT INCLUDED THE USE OF FORCEFUL RESTRAINT DUE TO EXCITED DELIRIUM DUE TO ACUTE METHAMPHETAMINE INTOXICATION
ROLE OF TASER DEVICE: EXCLUDED

The Sheriff's Office report in this incident was exempt from disclosure under California law, but media sources reported that deputies responded to a call that Torres had broken into his family's house and was holding his parents and teenage siblings at knifepoint. Torres's father had kicked him out of the house earlier that day. When deputies arrived, a fight ensued. Deputies struggled with Torres and used batons, handholds and a TASER device. After the first application of the device, Torres collapsed and struck the back of his head on the pavement. When the charge dissipated, Torres regained his feet and struggled for several more minutes. Two additional charges had no effect. Torres stopped breathing after deputies placed him in handcuffs. Paramedics transported Torres to a local hospital, where he died.

The coroner reported that Torres died from sudden cardiac arrest during physical struggle that included the use of forceful restraint. However, the coroner stated some concern that Torres could have asphyxiated while deputies held him in a prone position after wrestling the 290-pound man to the ground. The coroner concluded that, because Torres was able to return to his feet and continue fighting for several minutes after being shocked, the TASER device was not likely to have affected his heart. Forensic pathologists noted that Torres exhibited signs of excited delirium and had high levels of methamphetamine in his system when he encountered the deputies. The coroner reported that the level of methamphetamine present in Torrez's blood, coupled with his enlarged heart, would have been sufficient to result in Torrez's death (see Downing, 2005; Sacramento County (California) Coroner's Report 05–04868; Amnesty International, 2006; Anglen, 2006; Carreon, 2006).

Torrez continued to struggle with officers following the application of the TASER pulses, indicating that the electric current did not immediately affect his heart. Because the TASER device did not have an immediate effect and the medical examiner attributed the death to methamphetamine-induced excited delirium, the device is excluded as the cause of death and as a significant contributing factor.

PATRICK AARON LEE

AGE: 21
RACE/GENDER: WHITE MALE
AGENCY: NASHVILLE (TENNESSEE) METROPOLITAN POLICE DEPARTMENT
DATE OF INCIDENT: SEPTEMBER 24, 2005
CAUSE OF DEATH: EXCITED DELIRIUM
CONTRIBUTING FACTORS: CANNABIS, LSD, ENLARGED HEART
ROLE OF TASER DEVICE: EXCLUDED

Police received a call from a bar where Lee was creating a disturbance after employees had ejected him. When an officer confronted him, Lee reportedly stripped naked and began running around the parking lot. The officer shocked Lee with a TASER device, but Lee continued to resist and attempted to flee. Officers shocked Lee up to 19 times and hit him with batons and chemical spray before they were able to arrest him. Lee told officers he was high on drugs, and they found lysergic acid diethylamide (LSD) and marijuana in his belongings. He suffered a respiratory and cardiac arrest at the scene.

The autopsy report was not available, but media sources reported that the medical examiner listed Lee's cause of death as excited delirium. He said there was no evidence that any of the restraint techniques that police used on Lee, including the TASER shocks, directly caused his cardiac arrest and death. However, he added that the combined effects of multiple applications of the TASER device, chemical spray and physical force on someone in excited delirium are unclear and require additional research (see *Associated Press*, 2005, October 6; *Associated Press*, 2005, October 14; Nashville (Tennessee) Metropolitan Police Report 05-517462; Amnesty International, 2006; Anglen, 2006).

Lee continued to struggle with officers following the last application of the TASER pulses, indicating that the electric current did not immediately affect his heart. Because the TASER device did not have an immediate effect and the medical examiner attributed the cardiac arrest to excited delirium, the device is excluded as the cause of death and as a significant contributing factor in the Lee case.

MICHAEL LESEAN CLARK

AGE: 33
RACE/GENDER: AFRICAN-AMERICAN MALE
AGENCY: AUSTIN (TEXAS) POLICE DEPARTMENT
DATE OF INCIDENT: SEPTEMBER 26, 2005
CAUSE OF DEATH: CONSEQUENCES OF MASSIVE INTRAVASCULAR SICKLING ASSOCIATED WITH EXTREME PHYSICAL EXERTION DUE TO PHENCYCLIDINE AND COCAINE-INDUCED EXCITED DELIRIUM
ROLE OF TASER DEVICE: EXCLUDED

Police received a call of a fight between a man and a woman. When the first officer arrived, he spoke with Clark, who voluntarily climbed into the back seat of the police car. Later, when the officer tried to get Clark out of the back seat so he could handcuff him, Clark resisted. The officer sprayed

Clark with chemical spray, but Clark still refused to get out of the car. Several officers arrived as back-ups. Two officers fired TASER devices at Clark, shocking him three times, as other officers pulled him out of the car, handcuffed him, and applied leg restraints. During the fray, Clark bit one officer on the hand and injured another's shoulder. Once they had secured Clark, the officers called for emergency medical services to examine him. In the ambulance, Clark's condition deteriorated. The paramedics took Clark to a local hospital, where he died a short time later.

A report by the medical examiner ruled out the TASER shocks as the cause of Clark's death. She reported that Clark's death was a result of the consequences of massive intravascular sickling (sickle cell anemia) associated with extreme physical exertion due to phencyclidine (PCP) and cocaine-induced excited delirium. Toxicology tests showed cocaine, PCP, and marijuana in Clark's blood and urine (see Austin (Texas) Police Report 2005-2690925; Humphrey & Osborn, 2005; Stanley & Osborn, 2005; Travis County (Texas) Medical Examiner's Report ME-05-1805; Amnesty International, 2006; Anglen, 2006; Plohetski, 2006).

Because the TASER pulses did not have an immediate effect and the medical examiner attributed the death to sickle cell anemia and drug-induced excited delirium, the TASER device is excluded as the cause of death and as a significant contributing factor in the Clark case.

MARY ELEANOR MALONE JEFFRIES

AGE: 51
RACE/GENDER: AFRICAN-AMERICAN FEMALE
AGENCY: MARSHALL COUNTY (MISSISSIPPI) SHERIFF'S OFFICE
DATE OF INCIDENT: OCTOBER 1, 2005
CAUSE OF DEATH: UNKNOWN
ROLE OF TASER DEVICE: UNDETERMINED

A patrol officer observed Jeffries standing in the middle of the street naked and screaming at people. When the officer approached and asked what was wrong, Jeffries complained that someone was trying to get her. When the officer asked where her clothes were, she responded that the clothes did not belong to her. Jeffries had a strong smell of alcohol on her breath. When the officer asked her to step over to his vehicle, Jeffries ran into an alley. Jeffries kicked and fought with police officers while they attempted to handcuff her and get her into a patrol car. She bit one officer and spat on another. The officers subdued Jeffries and took her to the Marshall County Jail. At the jail, Jeffries refused to get out of the patrol car, so officers asked a jailer to bring out a restraint chair. A jailer applied a TASER device to Jeffries, placed her in the restraint chair, and carried her into the jail. While officers were booking Jeffries, they noticed that she was not doing well, and they called for medical assistance. Rescue personnel responded and took Jeffries to a local hospital, where she died about one hour later (see Holly Springs (Mississippi) Police Report 2005100002; Watson, 2005).

The autopsy report was not available, and media reports did not disclose a cause of death or final autopsy reports. Without additional information, the role of the TASER device in the Jeffries case is undetermined.

TIMOTHY GLEN MATHIS

AGE: 35
RACE/GENDER: WHITE MALE
AGENCY: LARIMER COUNTY (COLORADO) SHERIFF'S OFFICE
DATE OF INCIDENT: OCTOBER 3, 2005
DATE OF DEATH: OCTOBER 26, 2005
CAUSE OF DEATH: ACUTE PNEUMONIA AND BRAIN DAMAGE CAUSED BY CARDIAC ARREST
CONTRIBUTING FACTORS: EXCITED DELIRIUM, METHAMPHETAMINE
ROLE OF TASER DEVICE: EXCLUDED

Deputies received a call about a man who was bleeding and trying to break into motor homes. Mathis was incoherent and violent when sheriff's deputies responded, jumping up and down and covered in blood. He broke a window and stabbed himself in the leg with a stick. Mathis refused deputies' verbal commands, and he charged the officers with a brick in his hands. One deputy fired a TASER device, striking Mathis in the chest. The charge had no effect, and Mathis pulled out the probes. Another officer then fired his TASER device, striking Mathis in the back. Again, the device had no effect, and Mathis pulled out those probes. A deputy then started striking Mathis on the arm holding the brick, trying to get him to drop it. Those strikes caused Mathis to drop to his knees. The deputies then forced Mathis down to the ground and sat on his abdomen to hold him down while they handcuffed him. As they were arresting him, Mathis went into cardiac arrest. The deputies called for medical assistance and began CPR. Paramedics took Mathis to a local hospital, but he never recovered. He died in a hospice 23 days later.

The autopsy report was not available, but media sources reported that the coroner concluded Mathis had consumed methamphetamine and was exhibiting signs of excited delirium when police tried to restrain him. She noted that the TASER shocks had been ineffective. She ruled that the cause of death was acute pneumonia and brain damage caused by cardiac arrest (see *Associated Press*, 2005, November 10; Dickman, 2005; Larimer County (Colorado) Sheriff's Office Report 05–6367; Reed, 2005; Amnesty International, 2006; Anglen, 2006).

Mathis continued to struggle with officers following the final application of the TASER pulses, indicating that the electric current did not immediately affect his heart. His cardiac arrest occurred as deputies sat on him as they tried to put on the handcuffs. Because the TASER device did not have an immediate effect and the medical examiner attributed the cardiac arrest to the effects of methamphetamine-induced excited delirium, the device is excluded as the cause of death and as a significant contributing factor in the Mathis case.

DAVID MICHAEL CROUD

AGE: 29
RACE/GENDER: AMERICAN INDIAN MALE
AGENCY: DULUTH (MINNESOTA) POLICE
    DEPARTMENT
DATE OF INCIDENT: OCTOBER 12, 2005
DATE OF DEATH: OCTOBER 18, 2005
CAUSE OF DEATH: ANORXIC ENCEPHALOPATHY
    DUE TO CARDIOPULMONARY ARREST DUE TO
    ACUTE ALCOHOL INTOXICATION AND
    HALOPERIDOL ADMINISTRATION
ROLE OF TASER DEVICE: EXCLUDED

On October 12, Croud was in a casino, drunk, and harassing customers. When officers arrived, he became belligerent as they attempted to handcuff him. Officers pushed him against a wall, kneed him, and shocked him with a TASER device in drive stun mode, but it was not effective. Croud collapsed in the parking lot, and emergency medical personnel took him to a local hospital. Paramedics wheeled Croud into the emergency room in a wheelchair, still handcuffed. In the emergency room, he continued to struggle with medical staff. Officers held him down on a bed while nurses administered an injection of Haloperidol, a tranquilizer, and restrained him face down. For ten minutes, Croud continued to struggle against the restraints, so nurses gave him another injection. About ten minutes later, a doctor noticed that Croud was not breathing and was asystole. Medical staff resuscitated Croud and transferred him to the cardiac care unit. Croud went into a coma and never regained consciousness. He died six days later, on October 18.

The Minnesota Bureau of Criminal Apprehension conducted the investigation into Croud's death. The investigation would reveal that the TASER device was not charged properly and probably could not have delivered a shock. Investigators observed no marks on Croud's body that TASER shocks could have caused. The investigation revealed that Croud had a blood alcohol of 0.316 upon admission to the hospital, but toxicology tests showed no traces of abusive drugs, although Croud did have a prescription for Clonazepam.

The autopsy report is private under Minnesota law, but news reports said that the medical examiner determined Croud died of anoxic encephalopathy due to cardiopulmonary arrest due to acute alcohol intoxication and Haloperidol administration. He added that the TASER shocks had nothing to do with Croud's death (see *Associated Press*, 2005, October 20; *Associated Press*, 2005, November 2; Minnesota Department of Public Safety Report 2005-314; Bjerga, 2006; Metcalf, 2006).

For several minutes following the application of the TASER pulses, Croud showed no signs of medical distress. He continued to struggle with medical staff upon his arrival at the hospital, a clear indication that the electric current did not affect his heart. Because the TASER device did not have an immediate effect and the medical examiner attributed the cardiopulmonary arrest to the effects of alcohol intoxication and the administration of a tranquilizer, the device is excluded as the cause of death and as a significant contributing factor in the Croud case.

STEVEN MICHAEL CUNNINGHAM

AGE: 44
RACE/GENDER: WHITE MALE
AGENCY: FORT MYERS (FLORIDA) POLICE DEPARTMENT
DATE OF INCIDENT: OCTOBER 13, 2005
CAUSE OF DEATH: COCAINE TOXICITY WITH EXCITED DELIRIUM
CONTRIBUTING FACTORS: MENTAL ILLNESS
ROLE OF TASER DEVICE: EXCLUDED

The police report on this incident was not available, but media sources reported that Cunningham was at a mental health clinic when he began to struggle with staff members, who called the police. When officers arrived, Cunningham began fighting with an officer, who shocked him three times with a TASER device. Cunningham collapsed several minutes later in the parking lot of the mental health center. Paramedics transported him to a local hospital, where he died.

The medical examiner ruled that Cunningham died of cocaine toxicity with excited delirium. She noted that the officers had used the TASER device three or four times on Cunningham, but she made no reference to the shocks contributing to his death (see *Associated Press*, 2005b, October 14; Office of the Medical Examiner, Twenty-First District (Florida) Report 00786-2005; Amnesty International, 2006; Anglen, 2006).

For several minutes following the application of the TASER pulses, Cunningham showed no signs of medical distress, indicating that the electric current did not immediately affect his heart. Because the TASER device did not have an immediate effect and the medical examiner attributed the death to cocaine-induced excited delirium, the device is excluded as the cause of death and as a significant contributing factor in the Cunningham case.

JOSE MARAVILLA PEREZ, JR.

AGE: 33
RACE/GENDER: HISPANIC MALE
AGENCY: SAN LEANDRO (CALIFORNIA) POLICE DEPARTMENT
DATE OF INCIDENT: OCTOBER 20, 2005
CAUSE OF DEATH: METHAMPHETAMINE INTOXICATION ASSOCIATED WITH PHYSICAL EXERTION
CONTRIBUTING FACTORS: BROKEN VERTEBRA IN NECK AND 21 TASER APPLICATIONS
ROLE OF TASER DEVICE: EXCLUDED

Neither the police report nor the autopsy report was available in this incident. However, media sources reported that police confronted Perez at the home of a former girlfriend, who had a restraining order against him. When officers approached, Perez ran. Officers stopped Perez, who fought with officers attempting to arrest him. During the struggle, Perez was shocked with a TASER device in touch stun mode. Officers reported that he continued fighting until they were able to place him in hand and leg restraints. Perez was taken to jail, where he continued fighting with officers. He was again shocked with a TASER device. As officers attempted to take off the restraints, Perez stopped breathing. He was taken to a local hospital, where he died.

News accounts reported that an autopsy revealed Perez had a fractured neck and had received 21 TASER applications. However, the coroner listed the cause of death as methamphetamine intoxication associated with physical exertion (see Lee, 2005, October 26; Amnesty International, 2006; Anglen, 2006; Lee, 2006).

Perez continued to struggle with officers following the application of the TASER pulses, indicating that the electric current did not immediately affect his heart. Because the TASER device did not have an immediate effect and the coroner attributed the death to the effects of methamphetamine, the device is excluded as the cause of death and as a significant contributing factor in the Perez case.

MIGUEL SERRANO

AGE: 35
RACE/GENDER: HISPANIC MALE
AGENCY: NEW BRITAIN (CONNECTICUT) POLICE DEPARTMENT
DATE OF INCIDENT: OCTOBER 25, 2005
DATE OF DEATH: OCTOBER 31, 2005
CAUSE OF DEATH: ANOXIC ENCEPHALOPATHY
ROLE OF TASER DEVICE: EXCLUDED

The police and autopsy reports were exempt from disclosure under Connecticut law, but media sources reported that Serrano burst into an apartment house and began kicking on residents' doors. After unsuccessfully trying to force his way into one apartment, he went down to the basement and began smashing things. Officers tried to talk with Serrano, but he lunged at them. One officer shocked Serrano with a TASER device, handcuffed him, and carried him up the stairs. Because he seemed to be under the influence of alcohol or drugs or was suffering from an unknown condition, a police supervisor directed officers to have the man taken to a local hospital for an evaluation, but his condition deteriorated. He died a week later.

The medical examiner listed Serrano's cause of death as anoxic encephalopathy, or brain damage due to a lack of oxygen. He said Serrano was intoxicated on cocaine, and the combination of the drug and his altercation with police caused him to stop breathing and his heart to stop pumping shortly after the incident (see Munoz, Stacom, Goren, Fillo & Bachetti, 2005; Anglen, 2006; Munoz, 2006).

For several minutes following the application of the TASER pulses, Serrano showed no signs of medical distress. His condition only began to deteriorate after he arrived at the hospital. Because the TASER device did not have an immediate effect and the medical examiner attributed the death to the effects of cocaine and his physical struggle with officers, the device is excluded as the cause of death and as a significant contributing factor in the Serrano case.

CEDRIC STEMBERG-BARTON

AGE: 21
RACE/GENDER: AFRICAN-AMERICAN MALE
AGENCY: KING COUNTY (WASHINGTON) SHERIFF'S OFFICE
DATE OF INCIDENT: OCTOBER 25, 2005
CAUSE OF DEATH: ACUTE COCAINE TOXICITY
ROLE OF TASER DEVICE: EXCLUDED

Deputies stopped the car Stemberg-Barton and two other men were in because a computer check indicated the vehicle's owner had a misdemeanor arrest warrant. A deputy noticed that Stemberg-Barton had something in his mouth and tried to get him to spit it out. Another deputy arrived and tried to get Stemberg-Barton to spit out whatever he had in his mouth. During the struggle, the deputies applied chemical spray and touch stuns from a TASER device. They were unable to get the item in his mouth, and they placed Stemberg-Barton in custody, along with one of the other men in the car, for a firearms violation. A medical aid car arrived to clean Stemberg-Barton of the chemical spray, then left. Stemberg-Barton was coherent when the medics left, but about 45 minutes later, he had a seizure. He died about an hour later (see Castro, 2005; King County (Washington) Sheriff's Office Report 05–313722; *Seattle Post-Intelligencer*, 2005; Sullivan, 2005).

The autopsy report is private under Washington law, and news accounts do not report a cause of death or any autopsy results. However, a reporter for the *Seattle Post-Intelligencer* said that the coroner had ruled Stemberg-Barton died from acute cocaine toxicity (Castro, 2006). Because the TASER device did not have an immediate effect and the coroner attributed the death to the effects of cocaine, the device is excluded as the cause of death and as a significant contributing factor in the Stemberg-Barton case.

JOSHUA BROWN

AGE: 23
RACE/GENDER: WHITE MALE
AGENCY: LAFAYETTE PARISH (LOUSIANA) SHERIFF'S OFFICE
DATE OF INCIDENT: NOVEMBER 13, 2005
CAUSE OF DEATH: MULTIPLE BLUNT FORCE, SELF-INFLICTED INJURIES
ROLE OF TASER DEVICE: EXCLUDED

Police received a call concerning a man who had jumped head first from the window of his second floor apartment, naked, and was lying on the ground screaming. When deputies arrived, they found that Brown was on his feet, ramming his head repeatedly into a wooden fence, and yelling, "Stop hitting me." Brown broke through the fence and approached some bystanders. A deputy tried to speak with Brown, but Brown charged a paramedic. The deputy fired his TASER device, and Brown fell to the ground. The deputy tried to approach Brown, but Brown would try to grab him each time he got close. The deputy cycled the TASER device nine times trying to subdue Brown, but after each charge ended, Brown would try to get up. When he did finally get up, deputies rushed him, held him down, and secured him with handcuffs and leg restraints. Paramedics took Brown to a local hospital where his condition deteriorated. He died several hours later.

The coroner ruled that Brown died from multiple blunt force injuries that were self-inflicted. He noted that Brown had a skull fracture with associated cerebral edema. He ruled the death an accident (see Lafayette Parish (Louisiana) Coroner's Report 05–378; Lafayette Parish (Louisiana) Sheriff's Office Report 05–276017; Anglen, 2006).

Brown continued to struggle with officers following the final application of the TASER pulses, indicating that the electric current did not affect his heart. Because the TASER device did not have an immediate effect and the medical examiner attributed the death to self-inflicted injuries, including a skull fracture, the device is excluded as the cause of death and as a significant contributing factor in the Brown case.

JOSE ANGEL RIOS

AGE: 38
RACE/GENDER: HISPANIC MALE
AGENCY: SAN JOSE (CALIFORNIA) POLICE DEPARTMENT
DATE OF INCIDENT: NOVEMBER 17, 2005
CAUSE OF DEATH: CARDIOPULMONARY ARREST FOLLOWING VIOLENT STRUGGLE WITH POLICE IN INDIVIDUAL WITH ACUTE COCAINE INTOXICATION WITH PSYCHOSIS
CONTRIBUTING FACTORS: STATUS POST TASERING AND PEPPER SPRAYING; OBESITY; CARDIOVASCULAR DISEASE DUE TO CHRONIC COCAINE ABUSE
ROLE OF TASER DEVICE: EXCLUDED

The police report in this incident was exempt from disclosure under California law, but media sources reported that an off-duty officer observed Rios fighting with a woman over a 4-year-old child, and the woman was screaming for help. The off-duty officer identified himself, showed Rios his badge and demanded that he step away. Rios then began fighting with the officer and a neighbor who tried to help the officer. The off-duty officer used chemical spray on Rios, but he continued fighting. Other residents in the apartment complex began calling police and trying to stop Rios. After four other officers arrived, they stunned Rios with a TASER device at least twice, and they hit him with a baton. Rios continued to fight as the officers handcuffed him and placed him in an ambulance. Rios eventually lost consciousness, and he was dead on arrival at a local hospital.

The medical examiner found that Rios was obese, weighing about 330 pounds, and that he suffered from coronary artery stenosis and cardiac hypertrophy. Toxicology tests revealed a significant level of cocaine in Rios' system. The medical examiner concluded that Rios died of cardiopulmonary arrest following a violent struggle with the police that was due to acute cocaine intoxication with psychosis. He also noted as contributing factors the use of the TASER device and chemical spray, obesity, and cardiovascular disease due to chronic cocaine abuse. The medical examiner noted in his report that Rios exhibited symptoms of excited delirium. He also noted that, although no existing scientific studies of TASER shocks or chemical spray use have shown an association with sudden or delayed death, the effect of those methods on a person with excited delirium were unknown (see *Associated Press*, 2005, November 20; Benjamin, 2005; Santa Clara County (California) Medical Examiner's Report 05–04036; Amnesty International, 2006; Anglen, 2006).

The medical examiner noted that no studies connected TASER shocks with sudden or delayed death, and he admitted that the possible effects of the TASER device in these circumstances were unknown to him. However, Rios continued to struggle with officers following the application of the TASER pulses, indicating that the electric current did not immediately affect his heart. Moreover, the effects on the heart of sympathomimetic drugs, like cocaine, are well known. Consequently, because the TASER device did not have an immediate effect and the medical examiner attributed the death to the effects of cocaine, the device is excluded as the cause of death and as a significant contributing factor in the Rios case.

HANSEL CUNNINGHAM, III

AGE: 30
RACE/GENDER: MALE (RACE NOT DETERMINABLE FROM THE AVAILABLE RECORDS)
AGENCY: DES PLAINES (ILLIONIS) POLICE DEPARTMENT
DATE OF INCIDENT: NOVEMBER 20, 2005
CAUSE OF DEATH: SUFFOCATION
ROLE OF TASER DEVICE: EXCLUDED

Neither the police report nor the autopsy report was available in this incident, but media sources reported that police officers responded to a call about a disturbance at a group home. When police arrived, they found a caregiver covered with blood. The caregiver told officers that Cunningham had attacked him. His injuries were severe enough that he needed surgery to repair the damage. Officers tried to arrest Cunningham, but he broke free. An officer tried to use his TASER device to subdue Cunningham, but it had no effect. The officers used chemical spray on Cunningham, and, when that did not work, they tackled him, held him down, and handcuffed him. The officers noticed Cunningham was having trouble breathing, and they called for the paramedics who were treating the caregiver. The paramedics sedated Cunningham, and he lost consciousness. They transported Cunningham to a local hospital, where he died.

News accounts said that the coroner ruled Cunningham suffocated after officers held him face down on the ground while they subdued him. There was no mention of the TASER shocks as a contributing factor (see Boykin & Malone, 2005; Main, 2005, November 22; Amnesty International, 2006; Anglen, 2006; *Associated Press*, 2006, February 7).

Cunningham continued to struggle with officers following the application of the TASER pulses, indicating that the electric current did not affect his heart. Because the TASER device did not have an immediate effect and the coroner attributed the death to suffocation, the device is excluded as the cause of death and as a significant contributing factor in the Cunningham case.

BARNEY LEE GREEN

AGE: 38
RACE/GENDER: WHITE MALE
AGENCY: PASADENA (TEXAS) POLICE DEPARTMENT
DATE OF INCIDENT: NOVEMBER 21, 2005
CAUSE OF DEATH: ACUTE COCAINE TOXICITY
ROLE OF TASER DEVICE: EXCLUDED

Police pulled Green over for a routine traffic stop. As the officer who made the stop approached Green's car, he saw that the driver was chewing vigorously and trying to wash down what he was chewing with a drink of water. The officer told him to spit it out, but Green refused. The officer ordered Green, who had just weeks before been released from prison on drug possession charges, to place his hands on the steering wheel. When Green repeatedly refused, the officer sprayed him in the face with chemical spray and then stunned him twice in the shoulder with a TASER device. The officer got Green out of his car and handcuffed him, but he was unable to recover what was in Green's mouth. The officer ordered him to spit out whatever he was chewing, but Green refused. The officer again stunned him with the TASER device. The officers eventually loaded Green into a patrol car without recovering whatever he had been trying to swallow. As they were driving to the jail, Green announced, "I'm going to die," and began shaking and convulsing. The officer radioed for an ambulance to meet them at the jail. By the time they arrived at the jail, Green was slumped over the seat, unconscious. The ambulance took Green to a local hospital. The medical staff placed him on life support, but he died later that day.

The medical examiner noted high levels of cocaine and cocaine metabolites in Green's system. He also noted the presence of cocaine in Green's stomach. He concluded that Green died of acute cocaine toxicity, and he ruled the death an accident (see Harris County Medical Examiner's Report ML2005–3303; Pasadena (Texas) Police Department, 2005; *Fort Worth Weekly*, 2006).

Green continued to resist the officer following the final application of the TASER pulses, indicating that the electric current did not immediately affect his heart. Because the TASER device did not have an immediate effect and the medical examiner attributed the death to the effects of cocaine, the device is excluded as the cause of death and as a significant contributing factor in the Green case.

TYLER MARSHALL SHAW

AGE: 19
RACE/GENDER: WHITE MALE
AGENCY: ASOTIN COUNTY (WASHINGTON) SHERIFF'S OFFICE

DATE OF INCIDENT: NOVEMBER 25, 2005
CAUSE OF DEATH: ARRHYTHMIA FOLLOWING MULTIPLE BLUNT FORCE INJURIES AND USE OF ELECTRO MUSCULAR INCAPACITATION DEVICES DURING A STATE OF EXCITED DELIRIUM
ROLE OF TASER DEVICE: EXCLUDED

Shaw, an inmate at the jail in a juvenile cell, suddenly began beating his head repeatedly against the cement walls and floor of his cell and throwing himself against the steel door. Shaw had previously been diagnosed with bipolar disorder. He received his medications while in the jail, but he had a history of refusing to take them. When jailers moved in to subdue him, Shaw charged them. Jailers shocked him with four TASER devices, perhaps as many as 21 times, but the charges seemed to have no effect. They also struck him with a baton, but the baton did not subdue Shaw. Shaw broke away from the jailers and ran to the public access door. When jailers cornered him, Shaw began hitting his head on the floor. The jailers put Shaw in a restraining chair. They realized moments later that Shaw was not breathing. The jailers called for medical assistance, but paramedics were unable to revive him. Shaw died at the scene. Washington State Patrol detectives assisted with the investigation.

The medical examiner's report was not available, because it is exempt from public records disclosure under Washington law. However, a news report, quoting a Sheriff's office spokesperson, said that Shaw died of an arrhythmia following blunt force injuries and use of electromuscular incapacitation devices during a state of excited delirium. A second medical examiner, who reviewed the autopsy, concluded that Shaw died from excited delirium with restraint stress. He concluded that Shaw's death was an accident that occurred because his abnormal mental state led to physical agitation with the application of restraint measures (see Asotin County Sheriff's Office Report 05A07692; Patrick, 2005; Washington State Patrol Report 05–012929; Sandaine, 2006, June 9; Sandaine, 2006, October 12).

Assuming that the media accounts of the medical examiners' conclusions are accurate, there is a problem with the first medical examiner's conclusions. The first medical examiner stated that the arrhythmia followed blunt force injuries and use of the TASER device, but noting that the arrhythmia followed the use of the TASER device did not establish a causal relationship. Shaw's continued resistance following the application of the TASER pulses indicated that the electric current did not affect his heart. Because the TASER device did not have an immediate effect and both medical examiners attributed the death to the effects of excited delirium, the device is excluded as the cause of death and as a significant contributing factor.

TRACY RENE SHIPPY

AGE: 35
RACE/GENDER: WHITE FEMALE
AGENCY: LEE COUNTY (FLORIDA) SHERIFF'S OFFICE
DATE OF INCIDENT: NOVEMBER 26, 2005
CAUSE OF DEATH: ACUTE COCAINE TOXICITY WITH EXCITED DELIRIUM
ROLE OF TASER DEVICE: EXCLUDED

The Sheriff's Office report in this incident was not available, but media sources reported that Shippy entered a store and asked employees to call 9-1-1 because she had been in a fight. She then began knocking over and smashing the store's displays. A deputy tackled Shippy and put her in handcuffs, then began moving her toward a squad car with another deputy, who shocked Shippy once on her shoulder with a TASER after she began kicking the deputies. Shippy tried to kick out one of the windows in the back of the vehicle before the deputies moved her to a more a secure car with bars over its back-seat windows. Paramedics evaluated her, but, about 15 minutes later, a deputy noticed that Shippy had become unresponsive and her face was turning blue. Paramedics took Shippy to a local hospital, where she died.

The medical examiner ruled that Shippy died of acute cocaine toxicity with excited delirium. The toxicology test detected elevated levels of cocaine in Shippy's blood, but the test also detected the presence of therapeutic levels of Alprazolam, also known as Xanax. The Medical Examiner ruled the death an accident (see *Associated Press*, 2005, November 27; *Associated Press*, 2005, November 29; Office of the Medical Examiner, Twenty-First District (Florida) Report 00901-2005; Amnesty International, 2006; Anglen, 2006).

For several minutes following the application of the TASER pulses, Shippy showed no signs of medical distress, even after paramedics had examined her. Because the TASER device had no immediate effect and the medical examiner attributed the death to the effects of cocaine, the device is excluded as the cause of death and as a significant contributing factor in the Shippy case.

KEVIN DEWAYNE WRIGHT

AGE: 39
RACE/GENDER: AMERICAN INDIAN MALE
AGENCY: LONGVIEW (WASHINGTON) POLICE DEPARTMENT
DATE OF INCIDENT: NOVEMBER 30, 2005
CAUSE OF DEATH: UNKNOWN
ROLE OF TASER DEVICE: EXCLUDED

An employee of the Cowlitz Indian Tribe office called 9-1-1 and reported that Wright had become combative, had smashed a chair, and had taken a fighting stance with employees. When officers arrived, a mental health counselor told them that Wright was out of control. The officers and the counselor tried to speak with Wright, who had a history of mental illness and methamphetamine use, and they tried to convince Wright to get medical help. Wright refused and grew more aggressive and threatening. He refused to leave the room he was in, so police entered to remove him. When Wright began fighting, officers shocked him five times with TASER devices. They handcuffed him and called in paramedics, who were already on scene. Paramedics sedated Wright with Haloperidol. Shortly after he was sedated, Wright stopped breathing. The paramedics transported Wright to a local hospital, where he died.

The investigation revealed that doctors had prescribed methocarbamol, hydrocodone, Wellbutrin XL® and Bupropin ER® for Wright. Preliminary toxicology tests on Wright's blood were positive for opiates and amphetamines. The medical examiner told investigators that the TASER shocks had nothing to do with Wright's death, but the cause of death has not been disclosed (see Cowlitz County (Washington) Sheriff's Office Report A05-16648; Main, 2005, November 22; Anglen, 2006).

Wright continued to resist officers following the application of the TASER pulses, indicating that the electric current did not immediately affect his heart. Because the TASER device did not have an immediate effect and the medical examiner ruled out the device as a potential cause, the TASER device is excluded as the cause of death and as a significant contributing factor in the Wright case.

JEFFREY DEAN EARNHARDT

AGE: 47
RACE/GENDER: WHITE MALE
AGENCY: ORANGE COUNTY (FLORIDA) SHERIFF'S OFFICE
DATE OF INCIDENT: DECEMBER 1, 2005
CAUSE OF DEATH: EXCITED DELIRIUM WITH METHAMPHETAMINE
ROLE OF TASER DEVICE: EXCLUDED

Deputies responded to a call of a disturbance and found Earnhardt screaming, naked, and running in traffic. When deputies approached him, he was angry, sweating, and his skin was hot to the touch. Earnhardt refused to obey commands and get out of the road. He rushed and struck one of the deputies, who responded with two jolts from a TASER device, which did not affect him. When more deputies arrived, they tackled Earnhardt and restrained him, striking him with batons. Once they had him handcuffed, they applied hobbles. During the struggle, Earnhardt stopped breathing. Paramedics transported Earnhardt to a local hospital, where he died. Earnhardt had a history of methamphetamine abuse, and, according to family members, he had been acting as if he were intoxicated for more than a day.

The medical examiner noted that Earnhardt had a body temperature of 102.6 degrees Fahrenheit upon arrival at the hospital, and that he exhibited muscle rigidity. He concluded that Earnhardt died of agitated delirium induced by toxic levels of methamphetamine, with bizarre behavior, hyperthermia, and collapse. He noted that law enforcement officers had applied a TASER device on Earnhardt to keep him out of the highway. The medical examiner specifically commented that the death was the result of the effects of drug ingestion, and he ruled the death accidental (see *Associated Press*, 2005, December 1; Office of the Medical Examiner, Ninth District (Florida) Report ME 2005-001475; Orange County (Florida) Sheriff's Office Report 05-106191; Amnesty International, 2006; Anglen, 2006).

Earnhardt continued to resist officers for several minutes following the application of the TASER pulses, indicating that the electric current did not immediately affect his heart. Because the TASER device did not have an immediate effect and the medical examiner attributed the death to the effects of methamphetamine intoxication, the device is excluded as a cause of death and as a significant contributing factor in the Earnhardt case.

MICHAEL STANLEY TOLOSKO

AGE: 31
RACE/GENDER: WHITE MALE
AGENCY: SONOMA COUNTY (CALIFORNIA) SHERIFF'S OFFICE
DATE OF INCIDENT: DECEMBER 7, 2005
CAUSE OF DEATH: CARDIORESPIRATORY ARREST DUE TO AGITATED PSYCHOSIS DUE TO PSYCHIATRIC ILLNESS
ROLE OF TASER DEVICE: EXCLUDED

Neither the police report nor the autopsy report were available in this incident, but media sources reported that police went to Tolosko's home after his mother called 9-1-1 to ask for help with her son, who had a history of mental illness. He had stopped taking his medication and had locked himself in his room. The officers heard a disturbance in the home that sounded like someone destroying things, and they had difficulty getting the home's occupants to come to the door. Officers entered and found

Tolosko in his room, which had been converted from a garage. When they opened the door, Tolosko attacked them. The officers used a TASER device and handcuffed Tolosko, but he continued to struggle. Officers then applied a restraint cord. Shortly after officers gained control of Tolosko, he became unconscious and stopped breathing. He was taken to a local hospital, where he died.

News accounts reported that the coroner ruled Tolosko died of cardiorespiratory arrest due to agitated psychosis. There was no mention of the TASER device (see Seltzer, 2005; *Vallejo Times Herald*, 2005; Amnesty International, 2006; Anglen, 2006; Egelko, 2007).

Tolosko continued to resist officers for a few minutes following the application of the TASER pulses, indicating that the electric current did not immediately affect his heart. Because the TASER device did not have an immediate effect and the coroner attributed the cardiorespiratory arrest to agitated psychosis, the device is excluded as the cause of death and as a significant contributing factor in the Tolosko case.

HOWARD STARR

AGE: 32
RACE/GENDER: AFRICAN-AMERICAN MALE
AGENCY: FLORENCE COUNTY (SOUTH CAROLINA) SHERIFF'S OFFICE
DATE OF INCIDENT: DECEMBER 17, 2005
CAUSE OF DEATH: UNKNOWN
ROLE OF TASER DEVICE: UNDETERMINED

Deputies chased Starr and another man they suspected of using a stolen van to break into a convenience store during a burglary. The van hit a patrol car during the chase and the men tried to run away. One of the deputies shocked Starr with a TASER device after he refused to stop running. Starr fell face first with his hands under his chest. The deputies kicked Starr's leg a couple of times, demanding to see his hands. When Starr did not comply, the deputy took his hands and cuffed them behind his back. The officers then rolled Starr over and noticed he was not breathing. After not finding a pulse, the deputies started CPR and called for paramedics. Starr was taken to a local hospital, where he died (see *Associated Press*, 2005, December 20; *Associated Press*, 2005, December 23; Florence County (South Carolina) Sheriff's Office Report 05–12–521; Amnesty International, 2006; Anglen, 2006).

The autopsy report was not available, and media sources did not report a cause of death or autopsy findings. Without additional information, the role of the TASER device in the Starr case is undetermined.

DAVID L. MOSS, JR.

AGE: 26
RACE/GENDER: AFRICAN-AMERICAN MALE
AGENCY: OMAHA (NEBRASKA) POLICE DEPARTMENT
DATE OF INCIDENT: DECEMBER 29, 2005
CAUSE OF DEATH: VENTRICULAR FIBRILLATION/ ARRHYTHMIA, SECONDARY TO EXCITED DELIRIUM SYNDROME
CONTRIBUTING FACTORS: PHENCYCLIDINE INTOXICATION
ROLE OF TASER DEVICE: EXCLUDED

Officers received a call to check on a suspicious person. A woman told police she walked outside and found a man, later identified as Moss, salivating and growling like a dog. Moss walked into her car a couple of times, ran underneath a porch, and grabbed a stick. The woman's son-in-law confronted the man before calling 9-1-1. When officers approached, Moss refused to obey their commands. The officers used a TASER device on Moss, but it was ineffective. The officers eventually were able to handcuff Moss and put leg restraints on him. Officers kept their hands on Moss' shoulders to keep him under control as they waited for an ambulance to arrive. Moss was on his stomach, head up and breathing, and he continued to tense up. Moss appeared to go into cardiac arrest as paramedics placed him in an ambulance. Medics performed CPR at the scene and transported Moss to a local hospital, where he died.

The autopsy report was not available, but according to news reports, the medical examiner found that Moss had a toxic level of phencyclidine (PCP) in his body. Moss' official cause of death was listed as ventricular fibrillation/arrhythmia, secondary to excited delirium syndrome, which was caused by his use of PCP (see Cole & Zagurski, 2005; Omaha (Nebraska) Police Report RB# 83421-F; Amnesty International, 2006; Anglen, 2006; Safranek, 2006).

For several minutes following the application of the TASER pulses, Moss displayed no signs of medical distress, indicating that the electric current did not immediately affect his heart. Because the TASER device did not have an immediate effect and the medical examiner attributed the death to the effects of PCP, the device is excluded as the cause of death and as a significant contributing factor in the Moss case.

Chapter 4

ANALYSIS OF TASER CASE STUDY DATA

It is a capital mistake to theorize before one has data.
Sir Arthur Conan Doyle, "Scandal in Bohemia"

Before one can theorize on the role of TASER electronic control devices as they relate to in-custody deaths, one must first separate evidence from conjecture and analyze the collected data. Many critics assume the role of the TASER devices based simply on the number of deaths, or on a misunderstanding of how the devices work, or on speculation of potential problems with the use of electromuscular disruption technology. One must analyze the credible evidence–what medical experts know about sudden death, the technical operations of conducted energy weapons, the physiological effects of the TASER devices, and the facts of each of each case–to determine the true role of the TASER devices. The question, then, is "What does the credible evidence tell us about the TASER device?"

The TASER device has been through four generations, and because the technology of the weapons has changed, it is necessary to analyze the data in two groups. The first and second generation weapons relied mainly on pain compliance technology. The third and fourth generation weapons relied on electromuscular disruption technology. Incidents before 2000 occurred before the introduction of third generation technology, so those cases constitute the first group for analysis. Cases from 2000 to the end of 2005 occurred following introduction of electromuscular disruption technology and constitute the second group for analysis.

GROUP ONE ANALYSIS

Available records indicate that, from 1983 through 1999, 42 deaths followed the application of first and second generation weapons, the TASER TF–76, the AIR TASER 34000, and the TASERTRON. Of those 42 cases, the only records available for analysis in 20 were news accounts. In 17 cases, the available records consisted of news accounts and an autopsy report or death certificate. In only four cases were the police reports and autopsy available along with news reports. In one case, the only record available was the autopsy.

Of the 42 cases, 35 occurred in California, which is not surprising when one remembers that TASER was originally a California company that marketed mainly in California during its early years. Only two cases involved women. One was an African-American woman aged 29, and the other was a 62-year-old woman whose race and ethnicity were not discernable from the available records. The other 40 cases were men. One was a 25 year-old American Indian, 12 were African-American males between the ages of 24 and 39, 11 were Hispanic males between the ages of 24 and 40, and 13 were white males between the ages of 25 and 47. In three cases, the races and ethnicity of the men were not discernable from the available records. Their ages ranged from 25 to 48. For all 40 men, the mean age was 33.4 years with a standard deviation of 6.4. ANOVA calculations indicated there was no statistically significant difference in the ages between the men's racial/ethnic groups.

Of the 42 group one cases, at least 36 of the deceased suffered one or more predisposing factors for sudden death. In six of those cases, the deceased suffered two or more predisposing factors. Autopsy reports were not available for the other six cases, but media sources did not mention predisposing factors for sudden death in those cases. Of the 42 cases, 29 had documented drug use. Four of the 29 drug use cases also had a preexisting heart disease, and two of those suffered from a mental illness. Three other deceased, in addition to the four who also had drug use, had a preexisting heart disease. One of those three cardiac disease cases also suffered from a mental illness. Additionally, two others suffered from a mental illness absent any other predisposing risk factor. In 34 cases, the deceased was described as having engaged in bizarre behavior, such as being nude, talking to nonexistent people, screaming and yelling incoherently, hyperactive physical behavior, or the indiscriminate destruction of property. See Table 10, Group One Predisposing Factors for Sudden Death.

Table 10
GROUP ONE PREDISPOSING FACTORS OF SUDDEN DEATH

Gender/Ethnicity	Drug Use	Heart Disease	Mental Illness	Obesity	Bizarre Behavior
Male (40)	28	7	5	1	33
African-American	7	3	2	1	10
American Indian	-	1	-	-	1
Hispanic	8	1	-	-	9
White	10	1	2	-	10
Unknown	3	1	1	-	3
Female (2)	1	-	-	-	1
African-American	1	-	-	-	-
Unknown	-	-	-	-	1

Table 11
GROUP ONE NARCOTICS USE

Gender/Ethnicity	Cocaine	PCP	Methamphetamine	Morphine	Marijuana
Male (40)	15	7	5	2	1
African-American	6	1	-	-	-
Hispanic	3	4	2	1	-
White	5	2	2	1	1
Unknown	1	-	1	-	-
Female (2)	1	-	-	-	-
African-American	1	-	-	-	-
Unknown	-	-	-	-	-

Illicit drugs were the cause of death or a significant contributing factor in 27 of the 42 group one cases. In three of those cases, the deceased ingested more than one drug. Cocaine was observed in 16 cases, phencyclidine (PCP) in seven, methamphetamine in five, morphine or morphine derivatives in two, and marijuana in one. See Table 11, Group One Narcotics Use.

In 37 cases when data were available, the number of shocks from a TASER device ranged from 0 to 9. Newspapers presented one case as being related to the discharge of a TASER device, but tests on the weapon later proved that the device was inoperative and could not have delivered a shock. In five cases, the data were not available. The average number of applications in the 36 cases when the TASER device was discharged was 2.0. Of the 42 group one cases, the TASER pulse was ineffective in subduing the target 30 times, or 71.4 percent. In only three cases did the deceased collapse immediately following application of the device. In three cases, the time between application of the TASER device and collapse is not discernable from the available data. In the other 36 cases, the deceased did not collapse within five to fifteen seconds following application of the TASER devices, an indication that the current from the TASER pulses did not affect their hearts' rhythm.

After reviewing the evidence presented in the case studies in Chapter 3, the role of the TASER electronic control devices as a cause of death or as a significant contributing factor was listed either as excluded, doubtful, possible, confirmed, or undetermined. As the cause of death, the TASER device

Table 12
GROUP ONE CASE STUDY DISPOSITIONS

Taser	Gender/Ethnicity	Excluded	Doubtful	Possible	Confirmed	Undetermined
Cause of Death	**Male (40)**	31	1	2	0	6
	African-American	9	1	1	-	-
	American Indian	1	-	-	-	-
	Hispanic	9	-	-	-	2
	White	10	-	1	-	-
	Unknown	2	-	-	-	4
	Female (2)	2	-	-	-	-
	African-American	1	-	-	-	-
	Unknown	1	-	-	-	-
Significant Contributing Factor	**Male (40)**	28	4	2	0	6
	African-American	8	2	1	-	-
	American Indian	1	-	-	-	-
	Hispanic	8	1	-	-	2
	White	9	1	1	-	-
	Unknown	2	-	-	-	4
	Female (2)	2	0	0	0	0
	African-American	1	-	-	-	-
	Unknown	1	-	-	-	1

was excluded in 33 cases, doubtful in one case, possible in two cases and undetermined in six cases. In no case was a TASER device confirmed as a cause of death. As a significant contributing factor, the TASER devices were excluded in 30 cases, doubtful in four cases, possible in two cases and undetermined in six cases. In no case was the TASER device confirmed as a significant contributing factor. See Table 12, Group One Case Study Dispositions.

GROUP TWO ANALYSIS

Available records indicate that, from 2000 through 2005, 171 deaths followed the application of third and fourth generation weapons, the ADVANCED TASER M26 and the TASER X26. Of those 171 cases, the only records available for analysis in 30 were news accounts. In 48 cases, the available records consisted of news accounts and either the autopsy report or a death certificate. In 42 cases, news accounts and the law enforcement report was available. In 51 cases the police reports, the autopsy report, and news accounts were available.

Seven cases involved women, three of whom were African-American between the ages of 35 to 51, and four of whom were white females ages 33 to 46. The mean age was 39 years with a standard deviation of 6.3. The other 164 cases were men. Three were American Indian males between the ages of 29 and 39, 58 were African-American males between the ages of 18 and 65, 26 were Hispanic males between the ages of 21 and 40, and 69 were white males between the ages of 17 and 59. In nine cases, the races and ethnicities of the men were not discernable from the available records. Their ages ranged from 30 to 55. For all 164 men, the mean age was 36.1 years with a standard deviation of 8.9. ANOVA calculations indicated there was no statistically significant difference in the ages between the men's racial/ethnic groups.

Of the 171 group two cases, at least 151 of the deceased suffered one or more predisposing factors for sudden death. In 69 of those cases, the deceased suffered two or more factors. In two cases, the autopsy report did not list a predisposing factor for sudden death. In the other 18 cases, the autopsy reports were not available, but media sources did not mention predisposing factors for sudden death in those cases. In 124 of the 171 cases, or 72.5 percent, the deceased had documented drug use. In 36 cases, the deceased had a documented mental illness. In 65 cases, the deceased suffered from preexisting cardiovascular disease, 17 suffered from obesity, five from alcohol abuse, five from head trauma,

Table 13
GROUP TWO PREDISPOSING FACTORS OF SUDDEN DEATH

Gender/Ethnicity	Drug Use	Heart Disease	Mental Illness	Obesity	Bizarre Behavior
Male (164)	120	61	33	16	117
African-American	42	24	9	5	36
American Indian	1	-	2	-	3
Hispanic	24	8	3	3	19
White	46	27	16	8	54
Unknown	7	2	3	-	5
Female (7)	4	4	3	1	7
African-American	1	1	-	1	3
Unknown	3	3	3	-	4

Table 14
GROUP TWO NARCOTICS USE

Gender/Ethnicity	Cocaine	PCP	Methamphetamine	Ecstasy	Marijuana
Male (164)	76	4	31	4	18
African-American	34	4	3	2	12
American Indian	-	-	1	-	-
Hispanic	14	-	6	-	-
White	23	-	18	2	6
Unknown	5	-	3	-	-
Female (7)	3	-	1	-	-
African-American	1	-	-	-	-
Unknown	2	-	1	-	-

and three from diabetes. In 124 cases, or 72.5 percent, the deceased was described as having engaged in bizarre behavior, such as being nude, talking to nonexistent people, screaming and yelling incoherently, hyperactive physical behavior, or the indiscriminate destruction of property. See Table 13, Group Two Predisposing Factors for Sudden Death.

Illicit drugs were the cause of death or a significant contributing factor in 110 of the 170 group two cases. In 25 of those cases, the deceased ingested more than one drug. Cocaine was observed in 76 cases, methamphetamine in 31, marijuana in 18, phencyclidine (PCP) in four, Ecstasy in four, and morphine or morphine derivatives in two. See Table 14, Group Two Narcotics Use.

In 153 cases when data were available, the number of shocks from the TASER devices ranged from 0 to 21. Newspapers presented one case as being related to the discharge of a TASER device, but tests proved that the device was not properly charged and could not have delivered a shock. In 17 cases, the data were not available. The average number of applications in the 152 cases when the TASER device was discharged was 3.7. Of the 171 group two cases, the TASER electronic control device was ineffective in subduing the target in 101 cases, or 59.1 percent. In only 19 cases did the deceased collapse immediately following application of the TASER device. In 16 cases, the time between application of the TASER device and collapse is not discernable from the available data. In the other 136 cases, the deceased did not collapse within five to fifteen seconds following application of the TASER pulse, an indication that the current from the TASER device did not affect their hearts' rhythm.

After reviewing the evidence presented in the case studies in Chapter 3, the role of the TASER device as a cause of death or as a significant contributing factor was listed either as excluded, doubtful, possible, confirmed, or undetermined. As the cause of death, the TASER devices were excluded in 149 cases, doubtful in ten cases, undetermined in 11 cases, and confirmed in one case, that of Maurice Cunningham. As a significant contributing factor, the TASER devices were excluded in 136 cases, doubtful in 12 cases, possible in six cases, and undetermined in 16 cases. In one case, that of Jerry W. Pickens, the TASER was confirmed as a significant contributing factor. See Table 15, Group Two Case Study Dispositions.

SUMMARY

Of the 213 case studies presented in Chapter 3, TASER devices were excluded as the cause of death in 182 cases, or 84.5 percent. The role of the TASER device was doubtful in 11 cases, possible in two cases, and undetermined in 20 cases. As a cause of death, the TASER electronic control device was confirmed in only one case, the death of Maurice Cunningham. As a significant contributing factor, the TASER devices were excluded in 166 cases, or 97.1 percent. The role of TASER devices were doubtful in 16 cases, possible in eight cases, and undetermined in 22 cases. In only one case was the TASER device confirmed as a significant contributing factor, the death of Jerry W. Pickens. Thus, from 1983 through 2005, there were only two deaths that could be said to be related to the use of a TASER electronic control device. Those two deaths represent less than one percent of the number of deaths that critics of TASER technology attribute to it.

Table 15
GROUP TWO CASE STUDY DISPOSITIONS

Taser	Gender/Ethnicity	Excluded	Doubtful	Possible	Confirmed	Undetermined
Cause of Death	**Male (164)**	144	10	0	1	9
	African-American	50	4	-	1	3
	American Indian	3	-	-	-	-
	Hispanic	23	2	-	-	1
	White	62	3	-	-	3
	Unknown	6	1	-	-	2
	Female (7)	5	-	-	-	2
	African-American	2	-	-	-	1
	Unknown	3	-	-	-	1
Significant Contributing Factor	**Male (164)**	131	12	6	1	14
	African-American	48	4	3	-	3
	American Indian	3	-	-	-	-
	Hispanic	21	3	-	-	2
	White	54	4	2	1	7
	Unknown	5	1	1	-	2
	Female (7)	5	-	-	-	2
	African-American	2	-	-	-	1
	Unknown	3	-	-	-	1

Appendix A

UNEXPECTED DEATHS FOLLOWING APPLICATION OF TASER DEVICES - 2006

Name	Police Agency	State	Date of Incident
Roberto Gonzales[1]	Waukegan Police Department	TX	January 3, 2006
Steven Hooker	Kalamazoo Police Department	MI	January 4, 2006
Matthew Dunlevy[2]	Laguna Beach Police Department	CA	January 5, 2006
Carlos Claros-Castro	Davidson County Sheriff's Office	NC	January 7, 2006
Daryl Dwayne Kelley	Harris County Sheriff's Office	TX	January 13, 2006
Shimeka Rena Lewis	Jefferson County Sheriff's Office	TX	January 16, 2006
Daniel Rivera Tamez[3]	Harlingen Police Department	TX	January 18, 2006
Nicholas Ryan Hanson	Jackson County Sheriff's Office	OR	January 22, 2006
Troy Anthony Rigby[4]	Broward County Sheriff's Office	FL	January 23, 2006
Jaime Coronel[5]	Monterey County Sheriff's Office	CA	January 24, 2006
Murray Bush	Jefferson Parish Sheriff's Office	LA	January 25, 2006
Jorge Luis Trujillo-Hernandez[6]	San Jose Police Department	CA	January 25, 2006
Karl W. Marshall	Kansas City Police Department	MO	January 28, 2006
Benites Salmon Sichiro	Spokane County Sheriff's Office	WA	January 29, 2006
Jessie Lee Williams, Jr.[7]	Harrison County Sheriff's Office	MS	February 4, 2006
Darvel Emile Smith	Louisiana State Police	LA	February 13, 2006
Gary Bartley[8]	St. Tammany Parish Sheriff's Office	LA	February 18, 2006
Samuel Hair[9]	Fort Pierce Police Department	FL	February 21, 2006
Christopher Levert McCargo[10]	Bradley County Sheriff's Office	TN	February 24, 2006
Twan Tran[11]	Port Arthur Police Department	TX	February 28, 2006
Melvin Anthony Jordan[12]	Norman Police Department	OK	March 3, 2006
Robert R. Hamilton	St. John's County Sheriff's Office	FL	March 7, 2006
Cedric Davis[13]	Merced County Sheriff's Office	CA	March 10, 2006
Otto Zehm[14]	Spokane Police Department	WA	March 18, 2006
Timothy W. Grant	Portland Police Bureau	OR	March 20, 2006
Theodore Rosenbury[15]	Washington County Sheriff's Office	MD	March 24, 2006
Thomas Clint Tipton	Clearwater Police Department	FL	April 5, 2006

Nicholas R. Mamino, Jr.	Collinsville Police Department	IL	April 15, 2006
Billy Ray Cook	Bladen County Sheriff's Office	NC	April 16, 2006
Juan Manuel Nunez[16]	Lubbock Police Department	TX	April 16, 2006
Alvin Itula	Salt Lake City Police Department	UT	April 22, 2006
Emily Marie Delafield	Green Cove Springs Police Department	FL	April 24, 2006
Jose Romero	Dallas Police Department	TX	April 24, 2006
Curtis Lee Smith	New Holland Borough Police Department	PA	April 25, 2006
Jeremy Davis[17]	Bellmead Police Department	TX	April 29, 2006
Brian Craig Carlile[18]	Pittsburgh Bureau of Police	PA	May 1, 2006
Kenneth Cleveland	Ashtabula County Sheriff's Office	OH	May 7, 2006
Brian Nash	Los Angeles Police Department	CA	May 25, 2006
Unknown	Phoenix Police Department	AZ	June 5, 2006
Felipe Fragoso Herrera[19]	Las Vegas Police Department	NV	June 6, 2006
Vicki Avila[20]	Orange County Sheriff's Office	CA	June 10, 2006
Jerry Preyer	Escambia County Sheriff's Office	FL	June 13, 2006
Jason Troy Dockery	Cookville Police Department	TN	June 18, 2006
Kenneth Eagleton[21]	Harris County Sheriff's Office	TX	June 18, 2006
Joseph Stockdale	Indianapolis Police Department	IN	June 21, 2006
Jermall Williams	South Bend Police Department	IN	July 2, 2006
Robert L. Sisneros, Jr.	Evans Police Department	CO	July 3, 2006
Michael Deon Babers[22]	St. Petersburg Police Department	FL	July 6, 2006
Christopher Tull[23]	Cincinnati Police Department	OH	July 8, 2006
Nickolos Cyrus	Mukwonago Police Department	WI	July 9, 2006
Jesus Negron	New Britain Police Department	CT	July 11, 2006
Raymond Mitchell	Riverside County Sheriff's Office	CA	July 19, 2006
Shannon Lane Johnson	Chatham County Sheriff's Office	NC	July 23, 2006
George Victor Holder	Salina Police Department	KS	July 30, 2006
Anthony Jones[24]	Merced Police Department	CA	August 1, 2006
Ryan Michael Wilson	Lafayette Police Department	CO	August 4, 2006
Curry McCrimmon	Melbourne Police Department	FL	August 8, 2006
James Nunez	Santa Ana Police Department	CA	August 8, 2006
Glen Thomas	Indian River County Sheriff's Office	FL	August 9, 2006
Raul M. Gallegos-Reyes	Arapahoe County Sheriff's Office	CO	August 17, 2006
Kenyata H. Allen	Mobile Police Department	AL	August 18, 2006
Noah Lopez[25]	Fort Worth Police Department	TX	August 18, 2006
Mark D. McCullaugh, Jr.	Summit County Sheriff's Office	OH	August 20, 2006
Timothy Picard	Woonsocket Police Department	RI	August 20, 2006
Mark L. Lee	Rochester Police Department	NY	August 25, 2006
Terry Wayne Robinson	Clermont County Sheriff's Office	OH	August 26, 2006
Juan Soto, Jr.	Liberal Police Department	KS	August 30, 2006

Jesus Mejia	Los Angeles Police Department	CA	September 04, 2006
Larry Noles	Louisville Police Department	KY	September 05, 2006
Perry Simmons	Montgomery Police Department	AL	September 08, 2006
Laborian Detron Simmons	Marion County Sheriff's Office	FL	September 13, 2006
John Philip Chasse, Jr.	Portland Police Bureau	OR	September 17, 2006
Marcus Roach-Burrus	Neenah Police Department	WI	September 18, 2006
Joseph M. Kinney	Madison Township Police Department	OH	September 29, 2006
John David Johnson, III	Clay County Sheriff's Office	FL	September 30, 2006
Vardan Kasilyan	Las Vegas Police Department	NV	September 30, 2006
Kip Black	North Charleston Police Department	SC	October 1, 2006
Michael Lee Templeton[26]	Craighead County Sheriff's Office	AK	October 4, 2006
Herman Carroll	Harris County Sheriff's Office	TX	October 6, 2006
Armando Ibarra[27]	Greenacres Police Department	FL	October 8, 2006
Gerald Raymond Guimond	Patagonia City Marshal	AZ	October 9, 2006
James Arthur Simons	Lincoln Park Police Department	MI	October 9, 2006
Nicholas Brown	Milford Police Department	CT	October 19, 2006
James Lewis	Las Vegas Police Department	NV	October 19, 2006
Eddie Charles Ham, Jr.	Montgomery Police Department	AL	October 22, 2006
Roger Holyfield	Jerseyville Police Department	IL	October 29, 2006
Jeremy Lee Bell Foos	Delaware Police Department	OH	October 31, 2006
Rosendo Gaytan[28]	Rockdale Police Department	TX	November 1, 2006
Curtis M. Sloan	Georgetown Police Department	IL	November 1, 2006
Matthew Barnett[29]	East Norriton Police Department	PA	November 7, 2006
William Jobe[30]	Federal Way Police Department	WA	November 9, 2006
Weizhong Wang	Los Angeles County Sheriff's Office	CA	November 10, 2006
Darren Faulkner	DeSoto County Sheriff's Office	MS	November 14, 2006
Timothy Wayne Newton	Rocky Mount Police Department	NC	November 14, 2006
Briant K. Parks	Columbus Police Department	OH	December 3, 2006
Terrill Patrick Enard	Lafayette Police Department	LA	December 17, 2006
Daniel Walter Quick	Butte County Sheriff's Office	CA	December 30, 2006

1. Gonzales died the following day, January 4, 2006.
2. Dunlevy died the following day, January 6, 2006.
3. Tamez died the following day, January 19, 2006.
4. Rigby suffered a heart attack while in jail on January 26, 2006. He died February 6, 2006, after being removed from life support.
5. Coronel died January 31, 2006, after being removed from life support.
6. Trujillo-Hernandez died the following day, January 26, 2006.
7. Williams died February 6, 2006, after being removed from life support.
8. Bartley died the following day, February 19, 2006.

9. Hair died February 24, 2006, after being removed from life support.
10. McCargo died January 30, 2007, having lapsed into a coma on February 24, 2006.
11. Tran died March 12, 2006, after being removed from life support.
12. Jordan died the following day, March 4, 2006.
13. Davis died March 17, 2006.
14. Zehm died March 20, 2006, after being removed from life support.
15. Rosenbury was pronounced dead the following day, March 25, 2006.
16. Nunez was pronounced dead the following day, April 17, 2006.
17. Davis died May 1, 2006.
18. Carlile died May 21, 2006, having never left the hospital since the incident.
19. Herrera died the following day, June 7, 2006.
20. Avila died on June 11, 2006.
21. Eagleton died June 21, 2006.
22. Babers died the following day, July 7, 2006.
23. Tull died the following day, July 9, 2006.
24. Jones died the following day, August 2, 2006.
25. Lopez died August 23, 2006, having lapsed into a coma on August 18, 2006.
26. Templeton died the following day, October 5, 2006.
27. Ibarra died the following day, October 9, 2006.
28. Gaytan died November 5, 2006, having never left the hospital since the incident.
29. Barnett died Novemer 10, 2006.
30. Jobe died November 11, 2006.

Appendix B

UNEXPECTED DEATHS BY JURISDICTION – 1983 THROUGH 2006

Jurisdiction	Cases
Alabama	
Birmingham Police Department	1
Mobile Police Department	2
Montgomery Police Department	2
Arkansas	
Craighead County Sheriff's Office	1
Little Rock Police Department	1
Arizona	
Glendale Police Department	1
Maricopa County Sheriff's Office	1
Mesa Police Department	1
Patagonia City Marshall	1
Phoenix Police Department	7
California	
Bakersfield Police Department	1
Bell/Cudahy Police Department	1
Brea Police Department	1
Butte County Sheriff's Office	1
California Department of Corrections	2
California Highway Patrol	1
Clearlake Police Department	1
Escondido Police Department	1
Fairfield Police Department	2
Fontana Police Department	1
Fremont Police Department	1

Jurisdiction	Cases
Fresno Police Department	1
Fullerton Police Department	1
Gardena Police Department	1
Inyo County Sheriff's Office	1
Laguna Beach Police Department	1
Livingston Police Department	1
Los Angeles County Sheriff's Office	9
Los Angeles Police Department	21
Merced County Sheriff's Office	2
Merced Police Department	1
Monterey County Sheriff's Office	1
Orange County Sheriff's Office	1
Pacifica Police Department	1
Pomona Police Department	3
Riverside County Sheriff's Office	1
Sacramento County Sheriff's Office	6
Sacramento Police Department	1
Salinas Police Department	1
San Bernardino County Sheriff's Office	1
San Diego Police Department	3
San Jose Police Department	3
San Leandro Police Department	1
Santa Ana Police Department	2
Santa Clara County Sheriff's Office	1
Santa Clara Police Department	1
Santa Cruz County Sheriff's Office	1
Santa Rosa Police Department	1
Seaside Police Department	1
Sonoma County Sheriff's Office	1
Tustin Police Department	1
Vallejo Police Department	1
Ventura Police Department	1

Jurisdiction	Cases	Jurisdiction	Cases
Colorado		St. John's County Sheriff's Office	2
		St. Petersburg Police Department	1
Arapahoe County Sheriff's Office	1	Volusia County Sheriff's Office	1
Denver Police Department	1	West Palm Beach Police Department	1
Evans Police Department	1		
Glendale Police Department	1	**Georgia**	
Lafayette Police Department	1		
Larimer County Sheriff's Office	1	Atlanta Police Department	1
Pueblo Police Department	1	Burke County Sheriff's Office	1
		Fulton County Sheriff's Office	1
Connecticut		Gwinnett County Sheriff's Office	2
		Houston County Sheriff's Office	2
Milford Police Department	1	LaGrange Police Department	1
New Britain Police Department	2	Whitfield County Sheriff's Office	1
Florida		**Illinois**	
Alachua County Sheriff's Office	1	Chicago Police Department	1
Broward County Sheriff's Office	2	Collinsville Police Department	1
Clay County Sheriff's Office	1	Columbia Police Department	1
Clearwater Police Department	1	Des Plaines Police Department	1
Collier County Sheriff's Office	1	Georgetown Police Department	1
Columbia County Sheriff's Office	1	Jerseyville Police Department	1
Coral Springs Police Department	1	Madison Police Department	1
DeLand Police Department	1	Waukegan Police Department	1
Delray Beach Police Department	1		
Division of the State Fire Marshal	1	**Indiana**	
Escambia County Sheriff's Office	4		
Fort Myers Police Department	1	Indianapolis Police Department	2
Fort Pierce Police Department	1	Johnson County Sheriff's Office	1
Green Cove Springs Police Department	1	Monroe County Sheriff's Office	1
Greenacres Police Department	1	South Bend Police Department	1
Hillsborough County Sheriff's Office	1	Whiteland Police Department	1
Hollywood Police Department	2		
Indian River County Sheriff's Office	1	**Kansas**	
Lee County Sheriff's Office	2		
Marion County Sheriff's Office	1	Liberal Police Department	1
Martin County Sheriff's Office	1	Salina Police Department	1
Melbourne Police Department	1		
Miami Police Department	2	**Kentucky**	
Miami-Dade Police Department	1		
Nassau County Sheriff's Office	1	Kenton County Police Department	1
Okaloosa County Sheriff's Office	1	Louisville Police Department	1
Orange County Sheriff's Office	4		
Orlando Police Department	1	**Louisiana**	
Pembroke Pines Police Department	1		
Polk County Sheriff's Office	1	Jefferson Parish Sheriff's Office	4
Putnam County Sheriff's Office	1	Lafayette Parish Sheriff's Office	1

Jurisdiction	Cases
Lafayette Police Department	2
Louisiana State Police	1
St. Tammany Parish Sheriff's Office	1

Maryland

Jurisdiction	Cases
Montgomery County Police Department	1
Prince George's County Sheriff's Office	1
Washington County Sheriff's Office	1

Michigan

Jurisdiction	Cases
Carrollton Township Police Department	1
Kalamazoo Police Department	1
Lincoln Park Police Department	1
Livingston County Sheriff's Office	1
Saginaw County Sheriff's Office	1
Zilwaukee Police Department	1

Minnesota

Jurisdiction	Cases
Duluth Police Department	1
Minneapolis Police Department	2
St. Paul Police Department	1

Mississippi

Jurisdiction	Cases
DeSoto County Sheriff's Office	1
Harrison County Sheriff's Office	1
Marshall County Sheriff's Office	1

Missouri

Jurisdiction	Cases
Joplin Police Department	1
Kansas City Police Department	1
Springfield Police Department	1
St. Louis County Police Department	1

Montana

Jurisdiction	Cases
Butte-Silver Bow County Sheriff's Office	1

North Carolina

Jurisdiction	Cases
Bladen County Sheriff's Office	1
Cabarrus County Sheriff's Office	1
Chatham County Sheriff's Office	1
Davidson County Sheriff's Office	1

Jurisdiction	Cases
Rocky Mount Police Department	1

Nebraska

Jurisdiction	Cases
Omaha Police Department	1

Nevada

Jurisdiction	Cases
Las Vegas Police Department	6
Sparks Police Department	1
Washoe County Sheriff's Office	1

New Mexico

Jurisdiction	Cases
Albuquerque Police Department	2

New York

Jurisdiction	Cases
New York Police Department	3
Rochester Police Department	1
Southampton Village Police Department	1
Suffolk County Police Department	1

Ohio

Jurisdiction	Cases
Akron Police Department	1
Ashtabula County Sheriff's Office	1
Canton Police Department	1
Cincinnati Police Department	3
Clermont County Sheriff's Office	1
Columbus Police Department	1
Dayton Police Department	1
Delaware Police Department	1
Hamilton Police Department	1
Lucas County Sheriff's Department	1
Madison Township Police Department	1
Ross County Sheriff's Office	1
Springfield Township Police Department	1
Summit County Sheriff's Office	1
Toledo Police Department	1
Union Township Police Department	1

Oklahoma

Jurisdiction	Cases
Carter County Sheriff's Office	1
Chickasha Police Department	1
Norman Police Department	1
Oklahoma City Police Department	3

Jurisdiction	Cases	Jurisdiction	Cases
Oregon		Bellmead Police Department	1
		Dallas Police Department	1
Jackson County Sheriff's Office	1	Euless Police Department	1
Portland Police Bureau	2	Fort Worth Police Department	4
		Harlingen Police Department	1
Pennsylvania		Harris County Constable's Office Pct. 1	1
		Harris County Sheriff's Office	3
Bushkill Township Police Department	1	Jefferson County Sheriff's Office	1
East Norriton Police Department	1	Johnson County Sheriff's Office	1
New Holland Borough Police Department	1	Lubbock Police Department	1
Philadelphia Police Department	1	Pasadena Police Department	1
Pittsburgh Bureau of Police	1	Port Arthur Police Department	1
Township of Spring Police Department	1	Rockdale Police Department	1
		Tarrant County Sheriff's Office	1
Rhode Island		Waco Police Department	1
Woonsocket Police Department	1	**Utah**	
South Carolina		Heber Police Department	1
		Salt Lake City Police Department	1
Anderson County Sheriff's Office	1		
Florence County Sheriff's Office	1	**Washington**	
Lancaster County Sheriff's Office	1		
North Charleston Police Department	1	Asotin County Sheriff's Office	1
		Auburn Police Department	1
Tennessee		Federal Way Police Department	1
		King County Sheriff's Office	1
Bradley County Sheriff's Office	1	Kitsap County Sheriff's Office	1
Cookville Police Department	1	Longview Police Department	1
Metropolitan Nashville Police Department	1	Olympia Police Department	1
Nashville Police Department	1	Spokane County Sheriff's Office	1
		Spokane Police Department	1
Texas		**Wisconsin**	
Amarillo Police Department	2		
Arlington Police Department	1	Mukwonago Police Department	1
Austin Police Department	2	Neenah Police Department	1
Beaumont Police Department	1		

Appendix C

LIST OF FATAL SHOOTINGS – 1983 THROUGH 2006

Name	Department	State	Date
Thomas Bubenhofer	Cincinnati Police Department	OH	02/06/1987
Eliberto Saldana	Los Angeles Police Department	CA	05/19/1987
Name Not Disclosed	New York Police Department	NY	04/11/1989
Name Not Disclosed	Los Angeles Police Department	CA	09/29/1989
Allan Eberhardt	San Pedro Police Department	CA	07/21/1991
Keith Hamilton	Los Angeles County Sheriff's Office	CA	08/13/1991
Melodee Ann Belew	Oceanside Police Department	CA	03/31/1993
Robert Hutchinson	Des Moines Police Department	IA	05/16/1994
Harvey Price	Cincinnati Police Department	OH	02/01/1995
Mitchell Edey	New York Police Department	NY	02/08/1995
Eduardo Gonzales Santana	Stanislaus County Sheriff's Office	CA	02/01/1996
William Lee Fowler, Jr.	Fresno Police Department	CA	03/12/1997
Doron Lifton	Fairfield Police Department	CA	09/03/1997
Darryl Hood	Los Angeles Police Department	CA	11/15/1997
Kevin Cerbelli	New York Police Department	NY	10/25/1998
Frank La	Huntington Beach Police Department	CA	11/11/1998
Donald Venerable, Jr.	Sacramento Police Department	CA	02/09/1999
Richard Rodriguez	Los Angeles Police Department	CA	04/03/2000
Glen Eric Peterson	Ontario Police Department	CA	10/25/2000
John Jones, Jr.	Houston Police Department	TX	03/12/2001
Joseph Evans	Gardena Police Department	CA	04/01/2001
Gabindo Benjamin Flores	San Diego Police Department	CA	04/28/2001
Michael Wellman	Escondido Police Department	CA	02/09/2002
Shawn Jerel Maxwell	Seattle Police Department	WA	02/18/2002
Abu Kassim Jeilani	Minneapolis Police Department	MN	03/10/2002
Cecil Rudolph Menifield	Los Angeles Police Department	CA	07/06/2002
Kedrian Edwards	New York Police Department	NY	07/29/2002
Ronald A. Jackson	Akron Police Department	OH	01/08/2003
Christopher Jay Yerbey	Bradley County Sheriff's Office	TN	03/02/2003
Kendra S. James	Portland Police Department	OR	05/07/2003
Calvin Clarence Barrett	Olmsted County Sheriff's Office	MN	05/22/2003
Lawrence Licano	Pamona Police Department	CA	07/06/2003
Oscar Curtis Tilghman	Fresno Police Department	CA	07/17/2003

Kevin Laron Smith	Vallejo Police Department	CA	08/02/2003
Mario Albert Madrigal, Jr.	Mesa Police Department	AZ	08/25/2003
Mary Ann Minchew	Mesa Police Department	AZ	09/06/2003
Jamaal Bonner	Aurora Police Department	CO	12/03/2003
Gary Elsholtz	Ft. Worth Police Department	TX	12/07/2003
Cephus Smith	Baltimore Police Department	MD	12/08/2003
Glenn David	Fremont Police Department	CA	02/18/2004
Julio Morais	Philadelphia Police Department	PA	02/19/2004
Gabriel Angel Garcia	Bakersfield Police Department	CA	02/21/2004
Roy Curtis Gomez	Sacramento Police Department	CA	03/06/2004
Name Not Disclosed	Las Vegas Police Department	NV	04/07/2004
Name Not Disclosed	Los Angeles County Sheriff's Office	CA	04/16/2004
Luis Fabian Delgado	Los Angeles County Sheriff's Office	CA	05/04/2004
Deborah Ann Kerr	Port Huron Police Department	MI	05/13/2004
Robert Anthony Carillo	Salinas Police Department	CA	05/16/2004
Rob VanBuren	Plymouth Township Police Department	MI	07/22/2004
Johnny Nakao	San Jose Police Department	CA	08/12/2004
Richard Grant Compton	Birmingham Police Department	AL	09/03/2004
Maurice Gosserand	Jefferson Parish Sheriff's Office	LA	09/25/2004
Zaim Bojcic	San Jose Police Department	CA	09/26/2004
Aaron Shaw	Covington County Sheriff's Department	AL	10/28/2004
Joey Ellis	Dallas Police Department	TX	11/12/2004
Booker T. Carloss	Oakland Police Department	CA	12/04/2004
Martin Louie Castro-Soriano	University of California - Davis Police	CA	12/14/2004
Aquileo Jiminez-Duran	Las Vegas Police Department	NV	12/16/2004
David Higgins	Las Vegas Police Department	NV	12/17/2004
Brian Charles Smith	Placentia Police Department	CA	12/22/2004
Placido Torres	Las Vegas Police Department	NV	12/30/2004
Name Not Disclosed	San Mateo County Sheriff's Office	CA	01/02/2005
Matthew Lee Arthur	Levy County Sheriff's Office	FL	02/13/2005
William David Basham	Temple Terrance Police Department	FL	02/25/2005
Bradley J. Spiker	Caseyville Police Department	IL	03/26/2005
Ronald Love Oxford	Coos County Sheriff's Office	OR	03/30/2005
Daniel C. Barnes	Ohio State Highway Patrol	OH	04/09/2005
Anthony William McElroy	Corona Police Department	CA	05/05/2005
Leopoldo Tijerina	Las Vegas Police Department	NV	05/15/2005
Gregory T. Sanderson	Tampa Police Department	FL	05/19/2005
Kevin Michael Fortney	San Bernardino Police Department	CA	05/21/2005
Alberto T. Rodriguez-Sobrino	Orange County Sheriff's Office	CA	05/22/2005
Alvie Allen Dorsey	San Bernardino County Sheriff's Office	CA	05/25/2005
Samuel Martinez	San Jose Police Department	CA	05/26/2005
Name Not Disclosed	Los Angeles County Sheriff's Office	CA	05/29/2005
Ricky Mills	Orange County Sheriff's Office	FL	06/14/2005
Michael A. Meluzzi	Sarasota Police Department	FL	07/08/2005
Roosevelt T. Summerville	Forest Park Police Department	OH	08/06/2005
Steven Loren Fountain	Campbell Police Department	CA	09/07/2005
Fouad Kaady	Sandy Police Department	OR	09/08/2005
Juan Ramon Aguilar	Houston Police Department	TX	09/09/2005
Marcus Quinn Berton Tesei	Los Angeles Police Department	CA	09/24/2005
Jose Fernandez	Seminole County Sheriff's Office	FL	10/12/2005

Name	Department	State	Date
Charles W. Wittland	Wilmington Police Department	DE	10/20/2005
Godfrey John, Jr.	Mendocino County Sheriff's Office	CA	10/23/2005
James Raymond Bothwell	Council bluff Police Department	IA	10/25/2005
Mark William Nelson	Redmond Police Department	WA	11/06/2005
Gustavo Pena Izaguirre	Stockton Police Department	CA	11/13/2005
Name Not Disclosed	Phoenix Police Department	AZ	11/19/2005
John Hayes	Tallahassee Police Department	FL	11/26/2005
Steven Hamlin	Pittsburgh Police Department	PA	12/07/2005
Isaac J. Mouton	Lafayette Police Department	LA	12/16/2005
Clinton A. Carey	Clackamas County Sheriff's Office	OR	12/28/2005
Santana Baca	Las Vegas Police Department	NV	01/02/2006
Dennis Lamar Young	Portland Police Department	OR	01/04/2006
Deven Lepierro	Sacramento County Sheriff's Office	CA	01/06/2006
Donald R. Yates	Pinellas County Sheriff's Office	FL	01/12/2006
Name Not Disclosed	Long Beach Police Department	CA	02/14/2006
Chester Washington	Palm Beach County Sheriff's Office	FL	02/18/2006
Wayne Scantlebury	Sacramento Police Department	CA	03/05/2006
Marilou Forrest	Palm Beach County Sheriff's Office	FL	03/27/2006
Lee Deante Brown	Riverside Police Department	CA	04/03/2006
Oscar Garcia	Stanislaus County Sheriff's Office	CA	04/11/2006
Victor Montero-Diaz	Madison Police Department	WI	04/12/2006
Phillip Conatser	North Las Vegas Police Department	NV	06/09/2006
John Thomas Clarke	Charlotte County Sheriff's Office	FL	06/16/2006
Gene Velasquez	Monterey County Sheriff's Office	CA	06/26/2006
Sasha Yuksel	New Hampshire State Police	NH	07/24/2006
Jaime Pineda	Palm Springs Police Department	FL	08/06/2006
Kristen J. Moore	Sierra Vista Police Department	AZ	08/29/2006
Jamie Alvarez	Little Rock Police Department	AK	09/02/2006
Robert J. Garcia	Santa Fe Police Department	NM	09/03/2006
Stanley Wong	San Mateo Police Department	CA	09/04/2006
Mark E. Taylor	Aurora Police Department	CO	09/18/2006
Trent Buckins	Lake Charles Police Department	LA	10/12/2006
Daniel J. Rice	Bowling Green Police Department	OH	10/16/2006
Scott William Boyko	Springfield Police Department	OR	10/21/2006
Jordan Laird Case	Washington County Sheriff's Office	OR	10/22/2006
Youwus Vilpre	Southeast Regional Fugitive Task Force	GA	10/25/2006
Bernet Vaba	Stockton Police Department	CA	11/01/2006
Derek J. Hale	Wilmington Police Department	DE	11/06/2006
Name Not Disclosed	Los Angeles County Sheriff's Office	CA	11/10/2006
Nick Gordon	Lowndes County Sheriff's Office	MS	11/18/2006
Joseph Hanrath	South Bend Police Department	IN	11/20/2006
Tyler J. Lowrey	Boise Police Department	ID	11/20/2006
John Hevener	Bernalillo County Sheriff's Office	NM	11/22/2006
Trevor New	Kalispell Police Department	MT	11/23/2006
Ty Louis Alexander	San Jose Police Department	CA	11/25/2006
Jonni Kiyoshi Honda	Eureka Police Department	CA	12/08/2006
Walter Douglas	Jefferson Parish Sheriff's Office	LA	12/12/2006
Bertram Hickman	Wilmington Police Department	DE	12/15/2006

GLOSSARY

Acetylcholine. (Often abbreviated ACh.) An ester of choline that is the neurotransmitter at somatic neuromuscular junctions, the entire parasympathetic nervous system, sympathetic preganglionic fibers (cholinergic fibers), and at some synapses in the central nervous system.

Acetylcholinesterase. An enzyme that stops the action of acetylcholine, present in various body tissues, including muscles, nerve cells, and red blood cells.

Acidosis. An actual or relative increase in the acidity of blood due to an accumulation of acids, as in diabetic acidosis or renal disease, or an excessive loss of bicarbonate, as in renal disease.

Action Potential. The change in electrical potential of nerve or muscle fiber when it is stimulated; depolarization followed by repolarization.

Acute Exhaustive Mania. See Neuroleptic Malignant Syndrome.

Agitated Delirium. See Excited Delirium.

Alprazolam. (Also known by the trade name Xanax.) A benzodiazepine and antianxiety agent, administered orally to treat anxiety and manage panic attacks.

Amitriptyline Hydrochloride. A tricyclic antidepressant administered orally or intramuscularly.

Ampere. A unit of electric current or the amount of electric charge.

Anoxia. The extreme form of hypoxia where there is no oxygen present.

Anoxic Encephalopathy. A hypoxic condition in which there is no supply of oxygen to the brain, or parts of the brain, despite adequate blood flow.

Anticholinergics. Members of a class of pharmaceutical compounds which serve to reduce the effects mediated by acetylcholine in the central nervous system and peripheral nervous system.

Aripiprazole. One of the most recent of the atypical antipsychotic medications to be approved for the treatment of schizophrenia.

Arteriosclerosis. A disease of the arterial vessels marked by thickening, hardening, and loss of elasticity in the arterial walls.

Asphyxia. Condition caused by insufficient intake of oxygen.

Aspiration Pneumonia. Pneumonia caused by inhalation of gastric contents, food, or other substances.

Asthma. A disease caused by increased responsiveness of the tracheobronchial tree to various stimuli, which results in episodic narrowing and inflammation of the airways.

Asystole. Cardiac standstill; absence of electrical activity and contractions of the heart evidenced on the surface electrocardiogram as an isoelectric (flat) line during cardiac arrest.

Atherosclerosis. The most common form of arteriosclerosis, marked by cholesterol-lipid-calcium deposits in the walls of arteries.

Atrial Flutter. A cardiac dysrhythmia marked by rapid regular atrial beating of about 300 beats per minute, and usually a regular ventricular response.

Autonomic Nervous System. The part of the nervous system that controls involuntary bodily functions. It is inappropriately named because, rather than being truly autonomic, it is intimately responsive to changes in somatic activities.

Atrioventricular Bundle of His. The only direct muscular connection between the atria and the ventricles.

Atrioventricular Valve Myxoid Degeneration. A tumor of the heart affecting the function of the valve between the atria and ventricles.

Bell's Mania. See Excited Delirium.

Benzodiazepines. Any of a group of chemically similar psychotropic drugs with potent hypnotic and sedative action; used predominantly as antianxiety and sleep-inducing drugs.

Benzoylecgonine. The major metabolite of cocaine, it is formed by hydrolysis of cocaine in the liver.

Benztropine Mesylate. (Also known by the trade name Cogentin.) An antiparasympathomimetic agent usually used with other drugs in treating Parkinsonism.

Bipolar Disorder. (Previously known as manic-depressive psychosis.) A psychological disorder marked by manic and depressive episodes. Bipolar disorders are divided into four main categories: bipolar I, bipolar II, cyclothymia, and nonspecified disorders. Mania is the essential feature of bipolar I, whereas recurrent moods of both mania and depression mark bipolar II.

Body Mass Index. *See* Quetelet Index.

Bradycardia. A slow heartbeat marked by a pulse rate below 60 beats per minute in an adult.

Bupropin. *See* Bupropion.

Bupropion. (Also known by trade names Wellbutrin and Zyban.) An antidepressant medication that is also moderately effective in aiding smoking cessation.

Cannabinoids. A group of secondary metabolites found in the cannabis plant, which are responsible for the plant's peculiar pharmacological effects.

Cardiac Action Potential. A specialized action potential in the heart, with unique properties necessary for function of the electrical conduction system of the heart.

Cardiac Arrest. The abrupt cessation of normal circulation of the blood due to failure of the heart to contract effectively during systole.

Cardiac Arrhythmia. Irregular heart action caused by physiological or pathological disturbances in the discharge of cardiac impulses from the sinoatrial node or their transmission through conductive tissue of the heart. Cardiac dysrhythmia is technically more correct, as arrhythmia would imply that there is no rhythm, but this term is not used frequently.

Cardiac Dysrhythmia. *See* Cardiac Arrhythmia.

Cardiac Hypertrophy. An enlargement of the heart caused by an increased size of the myocardium. It may be caused by exercise, valvular stenosis, and many other conditions. The myocardium increases in size by enlargement of each cell, not by an increase in number of cells.

Cardiomegaly. A medical condition wherein the heart is enlarged. Cardiomegaly is generally categorized as due to dilation or due to ventricular hypertrophy.

Cardiopulmonary Arrest. The combination of cardiac arrest and respiratory arrest.

Cardiotoxicity. The occurrence of heart muscle damage.

Carisoprodol. A drug that relaxes muscles by acting on the central nervous system.

Catecholamines. One of many biologically active amines, including metanephrine, dopamine, epinephrine, and norepinephrine, derived from the amino acid tyrosine. They have a marked effect on the nervous and cardiovascular systems, metabolic rate, temperature, and smooth muscle.

Cerebral Edema. Swelling of the brain due to an increase in its water content, caused by a variety of conditions including increased permeability of brain capillary endothelial cells, swelling of brain cells associated with hypoxia or water intoxication, trauma to the skull, and interstitial edema resulting from obstructive hydrocephalus.

Cerebrovascular Disease. Any disease by which the arteries in the brain, or are connected to the brain, are defective.

Cirrhosis. A chronic liver disease characterized pathologically by liver scarring with loss of normal hepatic architecture and areas of ineffective regeneration.

Clonazepam. A benzodiazepine used to treat anxiety, panic, and seizure disorders.

Clonus. Spasmodic alternation of muscular contractions between antagonistic muscle groups caused by a hyperactive stretch reflex from an upper motor neuron lesion.

Clozapine. The first of the atypical antipsychotic medications to be developed, normally used for treatment-resistant schizophrenia and for reducing the risk of suicidal behavior in patients with schizophrenia.

Cocaethylene. The ethyl ester of benzoylecgonine, formed in the body when cocaine and alcohol have been taken simultaneously.

Cocaine. A crystalline tropane alkaloid obtained from the leaves of the coca plant, cocaine is a stimulant of the central nervous system.

Cogentin. See Benztropine Mesylate.

Congestive Heart Failure. (Also called Congestive Cardiac Failure.) Inability of the heart to circulate blood effectively enough to meet the body's metabolic needs. Heart failure may affect the left ventricle, right ventricle, or both. It may result from impaired ejection of blood from the heart during systole or from impaired relaxation of the heart during diastole.

Coulomb. The amount of electric charge carried by a current of one ampere flowing for one second.

Conducted Energy Weapon. Defensive system that deploys an electromuscular disruption charge that affects the sensory and motor functions of the somatic nervous system.

Creatine Kinase. (Also known as Phosphocreatine Kinase or Creatine Phosphokinase.) An enzyme expressed by various tissue types, its function is the catalysis of the conversion of creatine to phosphocreatine, consuming adenosine triphosphate and generating adenosine diphosphate and the reverse reaction. In tissues that consume adenosine triphosphate rapidly, especially skeletal muscle, but also brain and smooth muscle, phosphocreatine serves as an energy reservoir for the rapid regeneration of adenosine triphosphate, the major source of energy in biochemical reactions.

Darvocet. *See* Propoxyphene.

Dehydration. The clinical consequences of negative fluid balance.

Delirium. An acute, reversible state of agitated confusion marked by disorientation without drowsiness, hallucinations or delusions, difficulty in focusing attention, inability to rest or sleep, and emotional, physical, and autonomic overactivity. Delirium itself is not a disease, but rather a clinical syndrome that results from an underlying disease or new problem with mentation.

Delirium Tremens. The most severe expression of alcohol withdrawal syndrome, marked by visual, auditory, or tactile hallucinations, extreme disorientation, restlessness, and hyperactivity of the autonomic nervous system evidenced by such findings as pupillary dilation, fever, tachycardia, hypertension, and profuse sweating.

Delta 9 Carboxy THC. *See* Tetrahydrocannabinol.

Delta 9 THC. *See* Tetrahydrocannabinol.

Dextromethorphan. An antitussive drug that is found in many over-the-counter cold and cough preparations, usually in the form of dextromethorphan hydrobromide, but commonly used as a recreational drug.

Diabetes Mellitus. A metabolic disorder characterized by hyperglycemia, or high glucose blood sugar.

Diazepam. A benzodiazepine derivative possessing anxiolytic, anticonvulsant, sedative, skeletal muscle relaxant, and amnestic properties, making it a useful drug for treating anxiety, insomnia, seizures, alcohol withdrawal, and muscle spasms.

Diffuse Axonal Injury. One of the most common and devastating types of brain injury wherein damage occurs over a more widespread area than in focal brain injury, and refers to extensive lesions in white matter tracts.

Dilantin. *See* Phenytoin Sodium.

Diphenhydramine Hydrochloride. An over-the-counter antihistamine and sedative.

Disseminated Intravascular Coagulation. A pathological process where blood starts to coagulate throughout the whole body.

Dopaminergic. Caused by dopamine, or concerning tissues that are influenced by dopamine.

Doxepin. A tricyclic antidepressant.

Ecgonine. An organic chemical found naturally in coca leaves with a close structural relation to cocaine, it is both a metabolite and a precursor.

Ecstasy. *See* Methylenedioxymethamphetamine.

Edema. Swelling of any organ or tissue due to accumulation of excess lymph fluid, without an increase of the number of cells in the affected tissue.

Electrocardiogram. (EKG, abbreviated from the German *Elektrokardiogramm*.) A graphic produced by an electrocardiograph, which records the electrical activity of the heart over time.

Electro-Muscular Disruption Technology. An alternative in less-lethal technology that uses pulses of electricity to incapacitate suspects.

Emphysema. A chronic lung disease characterized by loss of elasticity of the lung tissue, destruction of structures supporting the alveoli, and destruction of capillaries feeding the alveoli.

End-plate Potential. The postsynaptic potential induced at the neuromuscular junction by the opening of the Nicotinic acetylcholine receptor.

Encephalopathy. A term for any diffuse disease of the brain that alters brain function or structure.

Enteroischemia. A restriction of the blood flow to the intestines.

Ephedrine. A sympathomimetic amine, similar in structure to the synthetic derivatives amphetamine and methamphetamine, and commonly used as a stimulant, appetite suppressant, concentraton aid, decongestant and to treat hypotension associated with regional anaesthesia.

Epinephrine. (Also called Adrenaline.) A catecholamine, a sympathomimetic monoamine released by the adrenal gland.

Ethanol. The alcohol found in alcoholic beverages.

Excited Delirium. Characterized by an acute onset of bizarre and violent behavior and may be accompanied by combativeness, hyperactivity, unexpected strength, paranoid delusions, incoherent shouting, hallucinations, and hyperthermia. Underlying causes of excited delirium include bipolar disorder, chronic schizophrenia, intoxication with sympathomimetics, cocaine intoxication, alcohol withdrawal, and head trauma.

Fluoxetine Hydrochloride. (Marketed as Prozac in the USA.) An antidepressant drug used in the treatment of depression, body dysmorphic disorder, obsessive-compulsive disorder, bulimia nervosa, premenstrual dysphoric disorder and panic disorder.

Fluphenazine. A typical antipsychotic drug used for the

treatment of psychoses such as schizophrenia and acute manic phases of bipolar depression.

Ganglioglioma. Rare, benign brain tumors arising from ganglia-type cells occurring mainly in the temporal lobe of the brain.

Haldol. See Haloperidol.

Haloperidol. A conventional butyrophenone antipsychotic drug.

Hepatic Steatosis. A reversible condition where large vacuoles of lipid accumulate in the cells of the liver, and caused by various diseases, such as chronic alcoholism and obesity.

Hepatitis. A gastroenterological disease, featuring inflammation of the liver.

Hydrocodone. A semi-synthetic opioid derived from two of the naturally occurring opiates, codeine and thebaine, commonly used to relieve pain.

Hyoid Bone. A bone in the human neck, not articulated to any other bone. It is supported by the muscles of the neck and in turn supports the root of the tongue.

Hyperammonemia. A metabolic disturbance characterised by an excess of ammonia in the blood, a condition that may lead to encephalopathy and death.

Hypercholesterolemia. The presence of high levels of cholesterol in the blood, which is not a disease but a metabolic derangement that can be secondary to many diseases and can contribute to many forms of disease, most notably cardiovascular disease.

Hyperkalemia. An elevated blood level of potassium, which, when extreme, is considered a medical emergency due to the risk of potentially fatal arrhythmias.

Hypertension. A medical condition where the blood pressure is chronically elevated.

Hyperthermia. An acute condition that occurs when the body produces or absorbs more heat than it can dissipate.

Hyperthyroidism. A disease caused by excessive levels of thyroid hormone in the body.

Hypertrophic Cardiomyopathy. A genetic disorder caused by various mutations in genes that cause the heart muscle to thicken, which can obstruct blood flow and prevent the heart from functioning properly.

Hypoglycemia. A pathologic state produced by a lower than normal amount of glucose in the blood.

Hyponatremia. An electrolyte disturbance that exists when the sodium level in the plasma falls too low.

Hypothyroidism. The disease state caused by insufficient production of thyroid hormone by the thyroid gland.

Hypoxia. The reduction of oxygen in tissues below levels which are considered to be normal.

Infarction. When an artery is blocked by some obstruction, such as a blood clot, causing necrosis, the accidental death of living cells or tissue.

Instantaneous Death. Death with immediate collapse without preceding symptoms.

Interventricular Septum. The myocardial wall between the ventricles of the heart.

Intraparenchymal Hemorrhage. Bleeding within the brain tissue.

Ischemia. Medical term for hypoxia where there is a restriction in blood supply, generally due to factors in the blood vessels.

Joule. The work required to move an electric charge of one coulomb through an electrical potential difference of one volt.

Lethal Catatonia. See Excited Delirium.

LSD. See Lysergic Acid Diethylamide.

Lysergic Acid Diethylamide. (Commonly called LSD or LSD-25.) A semisynthetic psychedelic drug.

Lithium Carbonate. A chemical compound that is used as a mood stabilizer in psychiatric treatment of manic states and bipolar disorder.

Lorazepam. A benzodiazepine derivative, it is a drug classified as a sedative-hypnotic, anxiolytic and anticonvulsant.

Loxapine. A typical antipsychotic medication, used primarily in the treatment of schizophrenia.

Malignant Hyperthermia. A genetic disease marked by skeletal muscle dysfunction after exposure to some anesthetics or other stressors when body temperatures may climb above 105 degrees Fahrenheit.

Mania. (See also Bipolar Disorder.) A form of psychosis characterized by exalted feelings, delusions of grandeur, elevation of mood, psychomotor overactivity, and overproduction of ideas.

Manic-Depressive Psychosis. See Bipolar Disorder.

Marijuana. (Also Marihuana.) Produced from parts of the cannabis plant, primarily the cured flowers and gathered trichomes of the female plant. The major active chemical compound is _9-tetrahydrocannabinol, commonly referred to as THC, which has psychoactive effects when consumed, usually by smoking or ingestion.

Meprobamate. A carbamate derivative which is used as an anxiolytic, or antianxiety, drug.

Metabolic Acidosis. A state in which the blood pH is low (less than 7.35) due to increased production of hydrogen ions by the body or the inability of the body to form bicarbonate ($HCO3-$) in the kidney.

Methamphetamine. (Pharmaceutically referred to as methylamphetamine or desoxyephedrine.) A psychostimulant drug used primarily for recreational purposes, but is sometimes prescribed for ADHD and narcolepsy.

Methocarbamol. A central muscle relaxant for skeletal muscles, used to treat spasms.

Methylenedioxyamphetamine. (MDA.) A psychedelic hallucinogenic drug and empathogen/entactogen of the phenethylamine family.

Methylenedioxymethamphetamine. (Commonly known as Ecstasy or MDMA.) A synthetic entactogen of the phenethylamine family, whose primary effect is believed to be the stimulation of secretion as well as inhibition of re-uptake of large amounts of serotonin as well as dopamine and norepinephrine in the brain, inducing a general sense of openness, empathy, energy, euphoria, and well-being.

Mirtazapine. An antidepressant used for the treatment of mild to severe depression.

Mitral Valve Prolapse. A heart valve condition marked by the displacement of an abnormally thickened mitral valve leaflet into the left atrium during systole.

Morphine. An extremely powerful opiate analgesic drug and the principal active agent in opium.

Myocardial Contraction Band Necrosis. Microscopic structures in the heart muscle that appear like bands of dead tissue and are formed by overcontraction of the muscle cells.

Myocardial Fibrosis. The replacement of the muscle cells of the heart with noncontractile scar tissue.

Myocardial Hypertrophy. An increase in the size and mass of the heart that ultimately leads to congestive heart failure and sudden death.

Myocardial Infarction. A disease state that occurs when the blood supply to a part of the heart is interrupted.

Myocarditis. An inflammation of the myocardium, the muscular part of the heart.

Neuroleptic Drug. Antipsychotic drugs used to treat psychoses.

Neuroleptic Malignant Syndrome. A life-threatening, neurological disorder most often caused by an adverse reaction to neuroleptic or antipsychotic drugs.

Neuromuscular Junction. (Also known as the Myoneural Junction.) The synapse or junction of the axon terminal of a motoneuron with the motor end plate, the highly-excitable region of muscle fiber plasma membrane responsible for initiation of action potentials across the muscle's surface, ultimately causing the muscle to contract.

Neurons. Electrically excitable cells in the nervous system that function to process and transmit information.

Neurotransmitters. Chemicals that are used to relay, amplify and modulate electrical signals between a neuron and another cell.

Norepinephrine. A catecholamine and a phenethylamine released from the medulla of the adrenal glands, which, with epinephrine, affects the fight-or-flight response, activating the sympathetic nervous system to directly increase heart rate, release energy from glucose and glycogen, and increase muscle readiness.

Nortriptyline Hydrochloride. A second generation tricyclic antidepressant, sometimes used for chronic pain modification.

Obesity. An unhealthy accumulation of body fat.

Olanzapine. (Also known by the trade name Zyprexa.) An atypical antipsychotic agent used to treat psychosis and schizophrenia that controls both the positive symptoms of schizophrenia, such as delusions and hallucinations, and the negative symptoms, such as passivity and social isolation.

Osteoporosis. A disease of bone in which the bone mineral density is reduced, bone microarchitecture is disrupted, and the amount and variety of noncollagenous proteins in bone is changed, making the bones more susceptible to fracture.

Palpitation. An awareness of the beating of the heart, whether it is too slow, too fast, irregular, or at its normal frequency; brought on by overexertion, adrenaline, alcohol, disease, or drugs, or as a symptom of panic disorder.

Peripheral Nervous System. The portion of the nervous system outside the central nervous system.

Phencyclidine. (A contraction of the chemical name phenylcyclohexylpiperidine, abbreviated PCP.) A dissociative drug formerly used as an anesthetic agent, exhibiting hallucinogenic and neurotoxic effects.

Phenytoin Sodium. (Marketed as Dilantin in the USA.) A commonly used antiepileptic medication.

Positional Asphyxia. See Restraint Asphyxia.

Postural Asphyxia. See Restraint Asphyxia.

Postural Hypotension. (Also known as Orthostatic Hypotension.) A sudden fall in blood pressure that occurs when a person assumes a standing position.

Promethazine. A first-generation H1 receptor antagonist antihistamine and antiemetic medication, commonly used as an antiallergic medication.

Propoxyphene. (Also known by the trade name Darvocet) An analgesic in the opioid category used to treat mild to moderate pain.

Prozac. See Fluoxetine Hydrochloride.

Psychosis. A generic psychiatric term for a mental state in which thought and perception are severely impaired.

Pulmonary Edema. Swelling and/or fluid accumulation in the lungs that can lead to impaired gas exchange and may cause respiratory failure.

Pulmonary Emphysema. A type of chronic obstructive lung disease often caused by exposure to toxic chemicals or long-term exposure to tobacco smoke.

Pulmonary Interstitial Pneumonitis. A lung disease wherein the walls of the air sacs may become inflamed, and the tissue (interstitium) that lines and supports the sacs becomes increasingly scarred.

Pulmonary Thromboembolism. A blockage of an artery in the lungs, which occurs when a blood clot, generally a venous thrombus, becomes dislodged from its site of formation and embolizes to the arterial blood supply of one of the lungs.

Pulse Oximetry. A noninvasive method that allows health care providers to monitor the oxygenation of a patient's blood.

Quetelet Index. (Also known as Body Mass Index.) The individual's body mass divided by the square of the height, and almost always expressed in the unit kg/m^2.

Refractory Period. A period during which a cell is incapable of repeating a particular action, or, more precisely, the time it takes for an excitable membrane to be ready for a second stimulus once it returns to its resting state following an excitation.

Respiratory Arrest. The cessation of the normal tidal flow of the lungs due to paralysis of the diaphragm, collapse of the lung, or any number of respiratory failures.

Restraint Asphyxia. A form of asphyxia that occurs when someone's position prevents them from breathing adequately.

Resting Potential. Membrane potential that would be maintained if there were no action potentials, synaptic potentials, or other active changes in the membrane potential.

Rhabdomyolysis. The breakdown of skeletal muscle due to injury, either mechanical, physical or chemical, the principal result being acute renal failure due to accumulation of muscle breakdown products in the bloodstream, several of which are injurious to the kidney.

Risperidone. An atypical antipsychotic medication most often used to treat delusional psychosis, including schizophrenia, but may also be used to treat some forms of bipolar disorder, psychotic depression and Tourette syndrome.

Schizophrenia. A psychiatric diagnosis that describes a mental disorder characterized by impairments in the perception or expression of reality and/or by significant social or occupational dysfunction.

Sensory Nervous System. The part of the nervous system responsible for processing sensory information.

Sequela. (Plural - Sequelae) A pathological condition resulting from a disease, injury, or other trauma.

Sickle Cell Trait. Describes the way a person can inherit some of the genes of sickle cell disease, but not develop symptoms.

Sickle Cell Disease. A blood disorder in which the body produces an abnormal type of the oxygen-carrying substance hemoglobin in the red blood cells.

Sick Sinus Syndrome. Also called bradycardia-tachycardia syndrome, a group of abnormal heartbeats, or arrhythmias, presumably caused by a malfunction of the sinus node, the heart's natural pacemaker.

Synapse. Specialized junction through which cells of the nervous system signal to one another and to nonneuronal cells such as muscles or glands.

Sinus Rhythm. A term used in medicine to describe the normal beating of the heart, as measured by an electrocardiogram.

Somatic Nervous System. That part of the peripheral nervous system associated with the voluntary control of body movements through the action of skeletal muscles and reception of external stimuli.

Steatohepatitis. A type of liver disease, characterized by inflammation of the liver with concurrent fat accumulation in liver, classically seen in alcoholics, but also frequently found in people with diabetes and obesity.

Steatosis. A condition, which is a marker of sublethal cellular injury, characterized by the presence of abnormally large quantities of fat within a cell.

Stenosis. An abnormal narrowing in a blood vessel or other tubular organ or structure.

Stroke. An acute neurologic injury in which the blood supply to a part of the brain is interrupted.

Subarachnoid Hemorrhage. Bleeding between the brain tissue and the skull.

Subdural Hematoma. A form of traumatic brain injury in which blood collects between the dura, the outer protective covering of the brain, and the arachnoid, the middle layer of the meninges.

Sudden Death. Death that is nonviolent or nontraumatic, that is unexpected, that is witnessed, and that is instantaneous or occurs within a few minutes of an abrupt change in a previous clinical state.

Sympathetic Nervous System. Part of the autonomic nervous system that activates what is often termed the fight or flight response.

Sympathomimetics. A class of drugs whose effects mimic those of a stimulated sympathetic nervous system, such as increased cardiac output, dilated bronchioles, and constricted blood vessels. Sympathomimetics include cocaine and methamphetamine.

Tachycardia. A rapid beating of the heart. By convention the term refers to heart rates greater than 100 beats per minute in an adult.

TASER. Acronym for Thomas A. Swift's Electric Rifle, a registered trademark of TASER International.

Temazepam. A powerful hypnotic drug, a benzodiazepine derivative, that possesses powerful anxiolytic, anticonvulsant, amnestic, sedative and skeletal muscle relaxant properties.

Tetanus. A medical condition characterized by a prolonged contraction of skeletal muscle fibers.

Tetrahydrocannabinol. (Also known as THC, Δ^9-THC, Δ^9-tetrahydrocannabinol, Δ^1-tetrahydrocannabinol, or dronabinol.) The main psychoactive substance found in the cannabis plant.

Thrombogenic. Capable of or likely to produce a blood clot.

Thrombosis. The formation of a clot, or thrombus, inside a blood vessel, obstructing the flow of blood through the circulatory system.

Tracheal Stenosis. A narrowing of the windpipe.

Tracheotomy. (Also known as Tracheostomy.) A surgical procedure performed on the neck to open a direct airway through an incision in the trachea, or windpipe.

Tricyclic Antidepressants. A class of antidepressant drugs first used in the 1950s, named after the drugs' molecular structure, which contains three rings of atoms.

Troponin. A complex of three proteins that is integral to muscle contraction in skeletal and cardiac muscle, but not smooth muscle.

Vasospasm. A condition in which blood vessels spasm, leading to vasoconstriction, which can lead to tissue ischemia and necrosis.

Ventricular Fibrillation. A cardiac condition that consists of a lack of coordination of the contraction of the muscle tissue of the large chambers of the heart that eventually leads to the heart stopping altogether.

Volt. The amount of force required to send one ampere of current through a resistance of one ohm.

Wellbutrin. *See* Bupropion.

Xanax. *See* Alprazolam.

Zyban. *See* Bupropion.

Zyprexa. *See* Olanzapine.

BIBLIOGRAPHY

Abbey, K. (2003, December 28). Shootings by APD get close look. *Amarillo Globe-News* (Amarillo, TX).

ACLU of Northern California. (2005, August 16). ACLU seeks details in another Sacramento TASER shooting; Family says Sacramento man was TASERed while committing suicide. Press release. Retrieved June 30, 2006, from http://www.aclunc.org/pressrel/050816-tasers.html.

Aidem, P. (2006, May 16). County pays $75,000 over inmate death; settlement avoids risky court fight. *Daily News of Los Angeles* (Los Angeles, CA), p. SC1.

Akron (Ohio) Police Report 05-000475: Death of Dennis S. Hyde; January 5, 2005.

Alabama Department of Forensic Sciences Report of Autopsy 01A-02MB05430: Death of Clever Craig, Jr.; June 28, 2002.

Alachua County (Florida) Sheriff's Report 01-007808: Death of Mark Lorenzo Burkett; June 13, 2001.

Alameda County (California) Coroner's Report 2005-02253: Death of Eric Michael Mahoney; August 3, 2005.

Albuquerque Journal. (2003, March 18). Drugs, alcohol in man's system. Author (Albuquerque, NM), p. D2.

Albuquerque (New Mexico) Police Report 03-53426: Death of Christopher Smith; March 16, 2003.

Aljentera, C. (2005. January 17). ACLU includes Seaside, Calif., police department in criticism of TASER usage. *Monterey County Herald* (Monterey, CA).

Amarillo (Texas) Police Report 2003-00036037: Death of Corey Calvin Clark; April 16, 2003.

Amarillo (Texas) Police Report 2003-00080149: Death of Troy Dale Nowell; August 4, 2003.

Amnesty International. (2004, November). *Excessive and lethal force? Amnesty International's Concerns About Deaths and Ill-treatment Involving Police Use of TASERs.* Author. Retrieved May 26, 2006, from http://www.amnestyusa.org/countries/usa/Taser_report.pdf.

Amnesty International. (2006, May). *Amnesty International's Continuing Concerns About TASER Use.* Author. Retrieved May 26, 2006, from http://www.amnestyusa.org/news/document.do?id=ENGAMR510302006.

Anderson County (South Carolina) Coroner Report OA-04-0000143: Death of William Malcolm Teasley; August 16, 2004.

Anderson, M. (2005, March 16). Use of TASERs gets a closer look. *Chattanooga Times Free Press* (Chattanooga, TN), p. B1.

Anderson, M. (2006, February 15). Whitfield County denies lawsuit claims against sheriff's deputies. *Chattanooga Times Free Press* (Chattanooga, TN), p. NG4.

Anderson, T. (2005, September 27). Deputies won't be charged in death; D.A. report cites circumstances. *Daily News of Los Angeles* (Los Angeles, CA), p. N4.

Anglen, R. (2005, August 9). TASER's role in 2 deaths examined. *Arizona Republic.* Retrieved July 4, 2006, from http://www.yourlawyer.com/articles/read/10464.

Anglen, R. (2006, January 5). 167 cases of death following stun-gun use. *Arizona Republic.* Retrieved May 26, 2006, from http://www.azcentral.com/specials/special43/articles/1224taserlist 24-ON.html and http://www.azcentral.com/specials/special43/articles/0527taserlist100-200.html.

Anwar, Y. (2005, August 16). ACLU seeks details in another Sacramento TASER shooting; family says Sacramento man was TASERed while committing suicide. ACLU of Northern California, press release. Retrieved June 13, 2006, from http://www.aclunc.org/pressrel/050816-tasers.html.

Arapahoe County (Colorado) Coroner's Report 03A278: Death of Glenn Richard Leybe; September 29, 2003.

Arlington (Texas) Police Report 02-97548: Death of Ronald Wright; January 7, 2003.

Arkansas State Crime Laboratory Medical Examiner Division Report ME-322-04: Death of Robert Harold Allen; April 17, 2004.

Asotin County Sheriff's Office Report 05A07692: Death of Tyler Marshall Shaw; November 25, 2005.

Associated Press. (1984, September 21). NAACP calls for federal investigation in Waynesboro death. Author (Dateline: Atlanta, GA).

Associated Press. (1996, June 6). Inmate's death after scuffle ruled an accident. Author (Dateline: Phoenix, AZ).

Associated Press. (1996, October 17). Inmate predicted his death. Author (Dateline: Phoenix, AZ).

Associated Press. (2001, December 17). Man dies after police shoot him with stun gun. Author (Dateline: Hamilton, OH).

Associated Press. (2002, June 29). Police used TASER gun on man who died. Author (Dateline: Mobile, AL).

Associated Press. (2003, October 1). Man dies after officer uses stun gun. Author (Dateline: Glendale, CO).

Associated Press. (2004, March 8). Police say cocaine, not TASER, killed Unadilla man. Author (Dateline: Warner Robins, GA).

Associated Press. (2004, June 2). Man dies after Orlando police shock him with TASER gun. Author (Dateline: Orlando, FL).

Associated Press. (2004, June 12). Man ID'd who died during Reno struggle with officers. Author (Dateline: Reno, NV).

Associated Press. (2004, June 25). Man dies after being restrained by police. Author (Dateline: Bethlehem, PA).

Associated Press. (2004, July 3). Man dies after arrest by Okaloosa sheriff's office. Author (Dateline: Destin, FL).

Associated Press. (2004, July 23). Defense expert says TASER not likely a factor in inmate's death. Author (Dateline: Bloomington, IN).

Associated Press. (2004, July 31). Probe to determine if man's death linked to police stun gun. Author (Dateline: Mesa, AZ).

Associated Press. (2004, August 6). Report: Alabama man's death sixth linked to TASER. Author (Dateline: Phoenix, AZ).

Associated Press. (2004, August 7). Cocaine found in man who died after TASER shooting in Okaloosa. Author (Dateline: Fort Walton Beach, FL).

Associated Press. (2004, August 12). Man dies of injuries in Joplin explosion. Author (Dateline: Joplin, MO).

Associated Press. (2004, August 17). SLED investigating death at Anderson County jail. Author (Dateline: Anderson, SC).

Associated Press. (2004, August 19). Pathologist says TASER contributed to jailed man's death. Author (Dateline: Anderson, SC).

Associated Press. (2004, September 26). Rookie officer killed investigating break-in. Author (Dateline: El Paso, TX).

Associated Press. (2004, October 6). Police: Stun gun used during arrest of man who died. Author (Dateline: Lafayette, LA).

Associated Press. (2004, November 3). Suspect dies in custody after being stunned by TASER. Author (Dateline: LaGrange, GA).

Associated Press. (2004, November 19). Coroner: Inmate's death due to cocaine, alcohol, not TASER. Author (Dateline: Lafayette, LA).

Associated Press. (2004, December 13). Prosecutor: No charges against police in TASER death. Author (Dateline: Easton, PA).

Associated Press. (2004, December 21). Minister dies after attack by man calling him a "devil." Author (Dateline: Whiteland, IN).

Associated Press. (2004, December 22). Attorney plans insanity defense for man accused of killing minister. Author (Dateline: Whiteland, IN).

Associated Press. (2004, December 30). Man accused of killing pastor dies. Author (Dateline: Indianapolis, IN).

Associated Press. (2004, December 30). Police investigating death of Fort Meyers teen subdued by TASER. Author (Dateline: Naples, FL).

Associated Press. (2005, January 5). Coroner probes role of stun gun in suspect's death. Author (Dateline: Indianapolis, IN).

Associated Press. (2005, January 6). Coroner: Cocaine, not TASER, killed man zapped by Hollywood Police. Author (Dateline: Hollywood, FL).

Associated Press. (2005, January 8). Home intruder dies after Escambia deputies shock him with TASER. Author (Dateline: Pensacola, FL).

Associated Press. (2005, January 12). Autopsy: Shocks from stun gun may have acted as trigger in man's death. Author (Dateline: Mesa, AZ).

Associated Press. (2005, February 2). Man dies after being shocked with stun gun nine times. Author (Dateline: Toledo, OH).

Associated Press. (2005, February 17). County in Ohio suspends use of TASERs following death. Author (Dateline: Toledo, OH).

Associated Press. (2005, March 8). DeLand burglary suspect dies after police use TASER stun gun. Author (Dateline: DeLand, FL).

Associated Press. (2005, March 22). Kitsap County hit with $2 million claim over TASER-related death. Author (Dateline: Port Orchard, WA).

Associated Press. (2005, April 4). Man dies after police shock him with TASER gun. Author (Dateline: Fort Worth, TX).

Associated Press. (2005, April 27). Officer who used

TASER on man suspended. Author (Dateline: Fort Worth, TX).
Associated Press. (2005, May 4). Drug overdose killed man who died after TASER shock. Author (Dateline: Midland, TX).
Associated Press. (2005a, May 6). Inmate dies in Jefferson Parish prison. Author (Dateline: New Orleans. LA).
Associated Press. (2005b, May 6). Man dies in Miami-Dade jail after police use TASER stun gun. Author (Dateline: Miami, FL).
Associated Press. (2005, May 14). Man dies after being shot with TASER. Author (Dateline: Batavia, OH).
Associated Press. (2005, May 23). Man dies after being subdued with stun gun, pepper spray. Author (Dateline: Reading, PA).
Associated Press. (2005, May 28). Nashville drug suspect dies after police shock him with stun gun. Author (Dateline: Nashville, TN).
Associated Press. (2005, June 9). Drugs ruled cause of death of man hit twice with a TASER. Author (Dateline: Naples, FL).
Associated Press. (2005, June 13). Autopsy links death of Fort Lauderdale man to cocaine, not TASER. Author (Dateline: West Park, FL).
Associated Press. (2005, June 15a). Man dies after falling from window, running into traffic. Author (Dateline: Canton, OH).
Associated Press. (2005, June 15b). Waco man dies after confrontation with police. Author (Dateline: Waco, TX).
Associated Press. (2005, July 8). Cocaine ruled cause of death of man hit three times with TASER. Author (Dateline: DeLand, FL).
Associated Press. (2005, July 13). Teenager dies after being shot by TASER gun. Author (Dateline: Euless, TX).
Associated Press. (2005, July 15). Butte man shocked with stun gun dies. Author (Dateline: Butte: MT).
Associated Press. (2005, July 17). Recent Texas deaths after people were shocked with TASERs. Author (Dateline: Fort Worth, TX).
Associated Press. (2005, July 23). Officers suspended for use of force on man who died in custody. Author (Dateline: Canton, OH).
Associated Press. (2005, July 28). Drug suspect dies after being shot with TASER in holding cell. Author (Dateline: New York, NY).
Associated Press. (2005, August 13). Lawsuit alleges TASER misrepresented safety of stun gun. Author (Dateline: Phoenix, AZ).
Associated Press. (2005, August 24). News in brief from eastern Pennsylvania. Author (Dateline: Reading, PA).
Associated Press. (2005, August 28). Coroner rules death of man in police custody a homicide. Author (Dateline: Canton, OH).
Associated Press. (2005, August 30). Man dies after being shocked by TASER, had drugs in system. Author (Dateline: Laurelville, OH).
Associated Press. (2005, October 6). Federal authorities asked to investigate stun gun-related death. Author (Dateline: Nashville, TN).
Associated Press. (2005, October 14a). Man's autopsy report reveals no direct link to TASERs. Author (Dateline: Nashville, TN).
Associated Press. (2005, October 14b). Former Tennessee man dies after TASER shock in Florida. Author (Dateline: Fort Meyers, FL).
Associated Press. (2005, October 20). American Indian man dies after fight with Duluth police. Author (Dateline: Duluth, MN).
Associated Press. (2005, November 2). Examiner rules death of man under arrest was accidental. Author (Dateline: Duluth, MN).
Associated Press. (2005, November 10). TASER death ruled homicide; No charges will be filed. Author (Dateline: Fort Collins, CO).
Associated Press. (2005, November 20). Fresno man dies after police shoot him with TASER in San Jose. Author (Dateline: San Jose, CA).
Associated Press. (2005, November 27). Woman dies after Lee County deputies shock her with stun gun. Author (Dateline: Fort Myers, FL).
Associated Press. (2005, November 29). Former Iowa woman dies after Taser shock. Author (Dateline: Waterloo, IA).
Associated Press. (2005, December 1). Winter Garden man dies after being shocked by TASER. Author (Dateline: Orlando, FL).
Associated Press. (2005, December 20). Florence man dies after being shocked with TASER during arrest. Author (Dateline: Florence, SC).
Associated Press. (2005, December 23). Federal authorities investigating Taser death during arrest. Author (Dateline: Florence, SC).
Associated Press. (2006, January 11). Medical Examiner: TASER contributed to Lee County inmate's death. Author (Dateline: Bonita Springs, FL).
Associated Press. (2006, February 7). Medical examiner: Autistic man died from suffocation, not TASER. Author (Dateline: Des Plaines, IL).
Associated Press. (2006, June 8). Judge rules stun gun not responsible for man's death. Author (Dateline: Saginaw, MI.).

Atlanta Journal-Constitution. (2004, April 26). Law & order. Author (Atlanta, GA), p. D3.

Auburn (Washington) Police Report 04-079192: Death of Willie Smith, III; July 11, 2004.

Austin (Texas) Police Report 2004-1680065: Death of Abel Ortega Perez; June 16, 2004.

Austin (Texas) Police Report 2005-2690925: Death of Michael Lesean Clark; September 26, 2005.

Autrey, J. (2005, October 3). Woman died in jail after JPS refused admission. *Fort Worth Star-Telegram* (Fort Worth, TX).

Baeder, B. (2005, February 5). Man's family tries to cope with death at Pitchess. *Daily News of Los Angeles* (Los Angeles, CA), p. SC1.

Bailey, B. & Salner, R. (1989, May 25). Inmate dies after "stun" shots. *San Jose Mercury News* (San Jose, CA), p. 1A.

Barton, A. (2005, August 14). Drugs shadow TASER deaths: 17 of the 27 people who died in Florida after being shocked with a TASER were high on cocaine, officials say. *Palm Beach Post* (West Palm Beach, FL), p. 1A.

Battershill, P., Naughton, B., Laur, D., Panton, K., Massine, M., & Anthony, R. (2005, June). *TASER Technology Review Final Report.* Victoria, BC: Office of the Police Complaint Commissioner.

Bell, N. & Juarez, A. (2005, October 5). Inmate shocked for nearly 3 minutes; man who stabbed deputy was killed by TASER, autopsy says. *Charlotte Observer* (Charlotte, NC), p. 1A.

Bender, M. (2005, August 28). Man dies after Martin deputy fires TASER at him; suspect suffers 2 strokes on way to hospital. *Palm Beach Post* (Palm Beach, FL), p. 1C.

Bender, M. (2005, August 30). Cocaine blamed in man's death. *Palm Beach Post* (Palm Beach, FL), p. 1B.

Benjamin, M. (2005, November 21). Fresno man dies after fight with officers; San Jose police used TASERs to try to break up a fight between man and his wife. *Fresno Bee* (Fresno, CA), p. B4.

Berenson, A. (2004, July 18). As police use of TASERs soars, questions over safety emerge. *New York Times* (New York, NY), p. A1.

Berger, L. (1991, July 3). Man dies after police use TASER. *Los Angeles Times* (Los Angeles, CA), p. 1.

Berks County (Pennsylvania) Coroner's Investigation Report 05-0336: Death of Lee Marvin Kimmel; May 23, 2005.

Bernstein, L. (1987, December 23). Autopsy fails to fix blame in TASER gun death. *Los Angeles Times* (Los Angeles, CA), p. 2.

Bird, P. (2005, January 19). Death of killer, 40, is ruled natural: Coroner: man who choked minister suffered complications from an irregular heartbeat. *Indianapolis Star* (Indianapolis, IN), p. B2.

Birmingham News. (2005, July 8). Metro briefs. Author (Birmingham, AL), p. 3B.

Bjerga, D. (2006, October 25). Unpublished e-mail message to the author.

Bleetman, A. & Steyn, R. (2003, April 27). *The Advanced TASER: A Medical Review.* Birmingham, UK: Birmingham Heartlands Hospital.

Botonis, G. (2005, January 6). Probe begun in death after TASER employed. *Daily News of Los Angeles* (Los Angeles, CA), p. AV1.

Botonis, G. (2005, March 9). Sheriff's custody fatality probed; man suspected of dealing drugs stunned with TASER. *Daily News of Los Angeles* (Los Angeles, CA), p. AV1.

Boyd, D. (2005, June 29). Woman shot with Taser dies after being arrested. *Fort Worth Star-Telegram* (Fort Worth, TX), p. B3.

Boyer, B. (2002, March 8). Autopsy shows drugs, not stun gun, killed man. *Philadelphia Inquirer* (Philadelphia, PA), p. B2.

Boykin, A. & Malone, T. (2005, November 22). Autistic man dies in struggle; police used TASER, pepper spray to try to subdue him. *Chicago Daily Herald* (Chicago, IL), p. 1.

Brannon, A. (2004, July 7). Pensacola, Fla., group seeks investigation of TASER's use in detainee's death. *Columbia Daily Tribune* (Columbia, MO).

Broward County (Florida) Medical Examiner's Report 01-1664: Death of Steven Vasquez; December 21, 2001.

Broward County (Florida) Medical Examiner's Report 02-0117: Death of Vincent Delostia; January 27, 2002.

Broward County (Florida) Medical Examiner's Report 04-1798: Death of Kevin Downing; December 15, 2004.

Broward County (Florida) Medical Examiner Report 05-0924: Death of Horace Owens; June 11, 2005.

Broward County (Florida) Sheriff's Office Report WP05-06-00467: Death of Horace Owens; June 11, 2005.

Bruce, B. (2005). *Six Month TASER Study.* Columbus, OH: Columbus Police Department.

Burgess, R. (2005, June 9). Man sues police for TASER use; Federal-court plaintiff claims he was stunned 17 times, scarred. *Advocate* (Baton Rouge, LA), p. 1B.

Burkeman, O. & Borger, J. (2001, August 2). Shock tactics: They deliver a charge so great that people who have been shot remember nothing. In America they have been used to tackle behaviour as unthreatening as "hugging a lamppost." So should we be glad that stun guns are coming to Britain? Oliver Burkeman and Julian Borger report. *The Guardian* (London, England), p. 2.

Burress, C. (2005, February 22). Man hit by police stun guns dies 2 days later; TASER use, links to deaths controversial. *San Francisco Chronicle* (San Francisco, CA), p. B5.

Business Wire. (2002, April 2). Medical Examiner finds TASER was not cause of death during in-custody death of Hollywood man. Author (Dateline: Scottsdale, AZ).

Butte-Silver Bow County (Montana) Sheriff's Office Report CR05-03479: Death of Otis Thrasher; July 15, 2005.

Cabarrus County (North Carolina) Sheriff's Office Report 2004-006203: Death of Anthony Lee McDonald; August 13, 2004.

Canham, M. (2004, December 21). Chief: TASER didn't cause man's death; autopsy findings: The family of Douglas Meldrum disputes that, saying the medical examiner hasn't ruled it out; TASER disputed as the cause of man's death. *Salt Lake Tribune* (Salt Lake City, UT), p. C1.

Carreon, C. (2006, February 2). TASER didn't cause death, coroner says. *Sacramento Bee* (Sacramento, CA), p. B3.

Castro, H. (2005, November 2). Deputies back on duty a week after man's death. *Seattle Post-Intelligencer* (Seattle, WA).

Castro, H. (2006, March 25). TASERs didn't cause death; cocaine, struggle with officers to blame. *Seattle Post-Intelligencer* (Seattle, WA), p. B2.

Castro, H. (2006, December 27). Unpublished e-mail to the author.

Chan, T., Neuman, T., Clausen, J., Eisele, J. & Vilke, G. (2004, September). Weight force during prone restraint and respiratory function. *American Journal of Forensic Medicine and Pathology, 25*,(3), pp. 185-189.

Chanen, D. (2003, August 29). Man chased by police, dies later; After van crash, he fled on foot, died at hospital. *Star Tribune* (Minneapolis, MN), p. 3B.

Chanen, D. (2004, February 10). TASER safety questioned after man's heart attack; the guns use a small enough jolt to be considered safe, their manufacturer says. *Star Tribune* (Minneapolis, MN), p. 1B.

Chiang, W. (2006). Amphetamines. In Flomenbaum, N., Goldfrank, L., Hoffman, R., Howland, M., Lewin, N. & Nelson, L. (Eds.), *Goldfrank's Toxicologic Emergencies, 8th Edition.* New York, NY: The McGraw-Hill Companies. Retrieved January 10, 2007 from http://online.statref.com/document.aspx?fxid=68&docid=652.

Chickasha (Oklahoma) Police Report 05003259: Death of James Edward Hudson; January 28, 2005.

Cincinnati (Ohio) Police Homicide Unit (1994). Summary of investigation; death in police custody of Lee Grant Griffin.

Cincinnati (Ohio) Police Report 50501162: Death of Shirley Andrews; March 3, 2005.

Cincinnati (Ohio) Police Use of Force Form 2005-62523.1: Shirley Andrews, March 3, 2005.

City News Service. (1997, July 4). LAPD. Author (Dateline: Los Angeles, CA).

City News Service. (1998, March 10). Death probe. Author (Dateline: Los Angeles, CA).

City News Service. (2003, October 8). TASER death. Author (Dateline: Yorba Linda, CA).

City News Service. (2005, February 15). SD TASER death. Author (Dateline: San Diego, CA).

City News Service. (2005, May 23). Suspect dies. Author (Dateline: Tustin, CA).

City News Service. (2005, July 26). Suspect dies. Author (Dateline: Los Angeles, CA).

City News Service. (2005, August 26). Questions linger in death of schizophrenic man shot with police stun gun. Author (Dateline: San Diego, CA).

Coco, T. & Klasner, A. (2004). Drug-induced rhabdomyolisis. *Current Opinion in Pediatrics, 16*, 206-210.

Colavecchio-Van Sickler, S. (2004, May 25). Suspect hit with TASER dies after being subdued. *St. Petersburg Times* (St. Petersburg, FL), p. 1B.

Cole, K. & Zagurski, K. (2005, December 31). Death after TASER zap confounds family. *Omaha World-Herald* (Omaha, NE), p. 1A.

Collier County (Florida) Sheriff's Office Report 04000 40700: Death of Christopher Dearlo Hernandez; December 28, 2004.

Columbia County (Florida) Sheriff's Office Report 2005-009094: Death of Milton Woolfolk, Jr.; March 11, 2005

Connor, G. (2006, August). Essential elements in TASER™ policy and procedure. *Law & Order, 54*, 8, pg. 87-90.

Conway, M. (2005, April 5). Autopsy set today in TASER subduing; Delphi man, 32, battled with four officers trying to arrest him on warrants. *Modesto Bee* (Modesto, CA), p. A1.

Cook, T. & Campbell, D. (1979). *Quasi-Experimentation: Design and Analysis Issues for Field Settings.* Boston, MA: Houghton Mifflin Company.

Cordele Dispatch. (2003, December 11). Man dies after

being shocked twice with police TASER gun. Author (Cordele, GA).

County of Los Angeles (California) Coroner's Report 1983-09964: Death of Vincent Alvarez; August 10, 1983.

County of Los Angeles (California) Coroner's Report 1985-04921: Death of Cornelius Garland Smith; April 11, 1985.

County of Los Angeles (California) Coroner's Report 1986-06105: Death of Death of Anthony Manwell Williams, III; May 9, 1986.

County of Los Angeles (California) Coroner's Report 1987-06037: Death of Miguel Contreras; July 22, 1987.

County of Los Angeles (California) Coroner's Report 1987-11625: Death of Stewart Alan Vigil; December 4, 1987.

County of Los Angeles (California) Coroner's Report 1988-09032: Death of Charles Eugene Miles; September 10, 1988.

County of Los Angeles (California) Coroner's Report 1991-03620: Death of Douglas L. Danville; April 21, 1991.

County of Los Angeles (California) Coroner's Report 1991-05975: Death of Douglas Charles; July 1, 1991.

County of Los Angeles (California) Coroner's Report 1992-08402: Death of David Martinez; September 14, 1992.

County of Los Angeles (California) Coroner's Report 1994-00214: Death of Daniel Gizowski; January 6, 1994.

County of Los Angeles (California) Coroner's Report 1996-00138: Death of Byron Williams; January 5, 1996.

County of Los Angeles (California) Coroner's Report 1996-09471: Death of Andrew Hunt, Jr.; December 27, 1996.

County of Los Angeles (California) Coroner's Report 2002-04388: Death of Eddie Rene Alvarado; June 10, 2002.

County of Los Angeles (California) Coroner's Report 2002-05448: Death of Johnny Lozoya; July 19, 2002.

County of Los Angeles (California) Coroner's Report 2004-02590: Death of Phillip LeBlanc; April 1, 2004.

County of Los Angeles (California) Coroner's Report 2004-08657: Death of Jessie Robert Tapia; November 15, 2004.

County of Los Angeles (California) Coroner's Report 2005-00453: Death of Jerry John Moreno; January 10, 2005.

County of San Diego (California) Coroner's Report 87-2253: Death of Mario Gastelum; November 2, 1987.

County of San Diego (California) Medical Examiner's Report 05-00304: Death of Robert Fidalgo Camba; February 12, 2005.

County of San Diego (California) Medical Examiner's Report 05-01049: Death of Nazario Javier Camba; May 28, 2005.

County of San Diego (California) Medical Examiner's Report 05-01049: Death of Nazario Javier Solorio; June 2, 2005.

County of Santa Clara (California) Coroner's Report 89-144-003: Death of Jeffrey Michael Leonti; May 24, 1989.

County of Summit (Ohio) Medical Examiner's Report N-008-05: Death of Dennis S. Hyde; January 5, 2005.

County of Summit (Ohio) Medical Examiner's Report N-220-05: Death of Richard Thomas Holcomb; May 28, 2005.

County of Ventura (California) Coroner's Report 246-90: Death of Duane Jay Johnson; February 13, 1990.

Cowlitz County (Washington) Sheriff's Office Report A05-16648: Death of Kevin Dewayne Wright; November 30-2005.

Crecente, B. (2004, September 16). TASERed man died of drug mixture; autopsy cites cocaine and antidepressant as the cause of death. *Rocky Mountain News* (Denver, CO), p. 6A.

Curreri, F. (2004, February 24). Man's death angers relatives. *Las Vegas Review-Journal* (Las Vegas, NV), p. 1B.

Curtiss, A. (1992, September 15). Man dies after slashing self, fighting police: Arrest: Distraught suspect resists officers while cutting body, authorities say. TASER darts and a baton blow to the chest are used to subdue him. *Los Angeles Times* (Los Angeles, CA), p. 1.

Dallas County (Texas) Medical Examiner Report JP1922-05: Death of Robert Earl Williams; June 14, 2005.

Davies, D. (2004, December 24). Man dies after Delray Police use TASER on him. *Palm Beach Post* (West Palm Beach, FL), p. 1B.

Davies, D. (2005, March 16). Report: Drugs killed suspect, not TASERs. *Palm Beach Post* (West Palm Beach, FL), p. 1B.

Davis, J. (2004, August 21). Man dies after jolt from TASERs; Fresno police tried to subdue him during a domestic disturbance. *Fresno Bee* (Fresno, CA), p. A1.

Davis, L. (1998, August 5). The Fairfield Wives; Dr. John Parkinson, a civic and religious leader in the perfectly suburban town of Fairfield, told women they needed pelvic exams. Long exams. Several times a week. For years. And they believed him. *SF Weekly* (San Francisco, CA).

Davison, N. & Lewer, N. (2005, May). *Bradford Non-*

Lethal Weapons Research Project: Research Report No. 7. Bradford, England: Centre for Conflict Resolution, Department of Peace Studies, University of Bradford.

Dayton (Ohio) Police Report 0406300555: Death of Eric Bernard Christmas; June 30, 2004.

DeFalco, B. (2005, May 4). Father questions police use of stun gun on son who died in Arizona. *Associated Press.* (Dateline: Phoenix, AZ).

Defence Scientific Advisory Council, Sub-committee on the Medical Implications of Less-lethal Weapons. (2004). *Second Statement on the Medical Implications of Less-lethal Weapons (DOMILL).* London, England: Secretary of State for the Home Department.

Deland (Florida) Police Report 0DL050001569: Death of Willie Michael Towns; March 6, 2005.

Delray Beach (Florida) Police Report 04-35821: Death of Timothy Bolander; December 23, 2004.

Denver (Colorado) Coroner's Report 2004-3347: Death of Richard Karlo; August 19, 2004.

Denver (Colorado) Police Report 2004-38306: Death of Richard Karlo; August 19, 2004.

Department of California Highway Patrol Report 05553 UN: Arrest of Jeanne Marie Hamilton; December 22, 2004.

Dickman, P. (2005, November 10). Coroner: Deputies cause of man's death. *Daily Reporter-Herald* (Loveland, CO), p. A3.

DiMarco, J. (2003). Sudden cardiac death. In M. Crawford (Ed.), *Current Diagnosis & Treatment in Cardiology, 2nd Edition.* New York, NY: The McGraw Hill Companies. Retrieved January 10, 2007, from http://online.statref.com/document.aspx?fxid=19&docid=218.

Dobner, J. & Choate, A. (2004, December 19). Autopsy is performed on man zapped by TASER. *Deseret Morning News* (Salt Lake City, UT).

Dotson, R. (1984, September 26). Inquest declares Gardner died of natural causes. *The True Citizen* (Waynesboro, GA), p. 6A.

Downing, J. (2005, September 23). Man shot with TASER dies; Latest in string of fatalities follows tussle with officers. *Sacramento Bee* (Sacramento, CA), p. B1.

Doyle, J. (2005, July 19). Weekend death revives TASER dispute; critics say police overuse stun guns, which may be more lethal than advertised. *San Francisco Chronicle,* (San Francisco, CA), p. B3.

Driehaus, B. (1999, March 4). Inmate death reports at odds on stun gun. *Cincinnati Post* (Cincinnati, OH), p. 13A.

Duchschere, K. (2003, August 9). Police looking into transient's death. *Star Tribune* (Minneapolis, MN), p. 4B.

Duggan, P. (1992, November 3). Bone splinter in heart killed P.G. woman; Deputy sheriff subdued her with rubber bullet to chest. *Washington Times* (Washington, D.C.), p. B3.

Eckert, K. (2005, August 30). Drugs blamed in man's death; Ross County sheriff says deputy's use of TASER was called for. *Columbus Dispatch* (Columbus, OH), p. 4D.

Egelko, B. (2007, January 4). Deputies cleared in TASER-related death. *San Francisco Chronicle* (San Francisco, CA), p. B3.

Eiserer, T. (2003, January 7). Police: Stun gun used as last resort; man "never stopped threatening to jump," Arlington officer says. *Dallas Morning News* (Dallas, TX).

Epperson, J. (2006, March 28). As fatalities rise, a recent report cites police for overuse of TASER stun guns. *Daily Oklahoman* (Oklahoma City, OK).

Escambia County (Florida) Sheriff's Office Report 05-000663: Death of Carl Nathaniel Trotter; January 8, 2005.

Escambia County (Florida) Sheriff's Office Report 05-022396: Death of Robert E. Boggon; August 29, 2005.

Estrada, H. & Gustafson, P. (2004, June 10). St. Paul man dies after struggling with police. *Star Tribune* (Minneapolis, MN), p. 1B.

Euless (Texas) Police Report 0500043238: Death of Kevin Ray Omas; July 12, 2005.

Fairfield (California) Police Report 99-13233: Death of David Torres Flores; September 28, 1999.

Farkas, K. (2005, January 6). The "devil" died, never got upstairs; bleeding burglar unfazed by stun gun. *Plain Dealer* (Cleveland, OH), p. B3.

Farnham, F & Kennedy, H. (1997) Acute excited states and sudden death: Much journalism, little evidence. *British Medical Journal, 315*, pp. 1107-8.

Financial Wire. (2006, June 7). 20th TASER case ruling says device is not cause of death involving acidosis. Author (Dateline Saginaw, MI).

Fish, R. and Geddes, L. (September 2001). Effects of stun guns and TASERS. *Lancet, 358*, p. 687.

Flannery, G. (2005, March 9). Janitorial wages and other cruel realities. *City Beat* (Cincinnati, OH). Retrieved June 25, 2006 from http://www.citybeat.com/2005-03-09/porkopolis.shtml.

Fletcher, E. & Jewett, C. (2004, November 9). Man shot by stun guns dies; deputies were subduing the Elk Grove resident, who was acting erratically. *Sacramento Bee* (Sacramento, CA). p. B1.

Fletcher, G., Flipse, T. & Oken, K. (2004). Exercise and the cardiovascular system. In Fuster, V., Alexander, R. & O'Rourke, R. (Eds.), *Hurst's the Heart, 11th Edition.*

New York, NY: The McGraw-Hill Companies. Retrieved January 10, 2007 from http://online.statref.com/document.aspx?fxid=67&docid=1952.

Florence County (South Carolina) Sheriff's Office Report 05-12-521: Death of Howard Starr; December 17, 2005.

Florida Department of Law Enforcement Report PE-27-0018: Death of David Sean Lewandowski; June 26, 2003.

Florida Division of State Fire Marshal's Report 04-3821: Arrest of Byron W. Black; November 23, 2004.

Florida Times-Union. (2005, June 18). Autopsy: Man was in state of delirium; Preliminary results find TASERed Palatka man's condition caused by drugs. Author (Jacksonville, FL), p. B3.

Floyd, J. (2006, April 28). Police brutality or "excited delirium?" *Dallas Morning News* (Dallas, TX).

Ford, A. (1992, July 21). The troubled L.A. County Sheriff's Department: Inmate's death raises questions about deputies' conduct: Law enforcement: Other prisoners say they saw deputies beat him before shooting him with an electronic TASER gun. Officials say the man was violent and under the influence of drugs. *Los Angeles Times* (Los Angeles, CA), p. 19.

Fort Worth (Texas) Police Report 04132367: Death of Robert Guerrero; November 2, 2004.

Fort Worth (Texas) Police Report 05038655: Death of Eric J. Hammock; April 30, 2005.

Fort Worth (Texas) Police Report 05075522: Death of Carolyn J. Daniels; June 24, 2005.

Fort Worth Weekly. (2006, March 8). A stunning toll: TASER-related deaths and questionable uses of the weapon are mounting in Texas. Author (Fort Worth, TX).

Frankel, T. (2004). Questions linger for parents on report by coroner. *St. Louis Post-Dispatch* (St. Louis, MO), p. A1.

Frazier, M. (2005, April 22). Kin of dead man sues over stun gun. *Arkansas Democrat-Gazette* (Little Rock, AK).

Fresno Bee. (2001, May 11). Burglary suspect held. Author (Fresno, CA), p. B2.

Fresno Bee. (2004, August 28). Cocaine overdose cited in man's death Fresno musician had been subdued by police firing TASERs. Author (Fresno, CA), p. B2.

Fresno County (California) Coroner's Report 04-08.172: Death of Michael Lewis Sanders; August 20, 2004.

Fulton County (Georgia) Medical Examiner's Report 04-0974: Death of Daryl Lavon Smith; May 30, 2004.

Fulton County (Georgia) Sheriff's Office Report 04M059770: Death of Daryl Lavon Smith; May 30, 2004.

Gafni, M. (2005, January 6). Autopsy reveals TASER use. *Vallejo Times-Herald* (Vallejo, CA).

Gafni, M. (2005, February 2). Pathologist expounds on autopsy. *Vallejo Times-Herald* (Vallejo, CA).

Ganong, W. (2005). *Review of Medical Physiology, 22nd Ed.* New York, NY: Lange Medical Books.

Garcia et al. v. City of Fullerton et al., 2002 Cal. App. Unpublished. LEXIS 8245.

Gathright, A. (2005, January 4). Man dies after police shock him with TASER: Relatives had requested help calming him down. *San Francisco Chronicle* (San Francisco, CA), p. B1.

Gathright, A. (2005, March 18). Police stun guns had role in man's death. *San Francisco Chronicle* (San Francisco, CA), p. B4.

Gathright, A. (2005, June 25). Multiple stun-gun jolts found justified. *San Francisco Chronicle*, (San Francisco, CA) p. B2.

Gathright, A. (2005, February 24). Police chief defends use of stun guns; he blames man's death on drugs, not 10 jolts from officers. *San Francisco Chronicle* (San Francisco, CA), p. B4.

Gaura, M. (2005, August 3). Suspect dies, another hospitalized after separate stun gun incidents. *San Francisco Chronicle* (San Francisco, CA), p. B4.

Geary, F. (2004, June 26). Coroner's inquest: Jurors rule TASER a factor in death. *Las Vegas Review-Journal* (Las Vegas, NV), p. 2B.

Geary, F. (2004, August 4). Police defend use of TASERs. *Las Vegas Review-Journal* (Las Vegas, NV), p. 1A.

Geary, F. (2004, October 23). Inquest: TASER death excusable. *Las Vegas Review-Journal* (Las Vegas, NV), p. 1B.

Geary, F. (2005, July 9). LV officers cleared in death. *Las Vegas Review-Journal* (Las Vegas, NV), p. 1B.

Gendar, A. & Lemire, J. (2005, August 27). Crack, not TASER, killed ex-con. *Daily News* (New York, NY), p. 14.

Georgia Bureau of Investigation Report A84-1892: Death of Larry Donnell Gardner; August 17, 1984.

Georgia Bureau of Investigation Record of Medical Examiner Report 2003-4005152: Death of Curtis Lamar Lawson; November 9, 2003.

Georgia Bureau of Investigation Record of Medical Examiner Report 2004-1028709: Death of Greshmond Gray; November 2, 2004.

Georgia Bureau of Investigation Record of Medical Examiner Report 2004-4001604: Death of Melvin Samuel; April 16, 2004.

Georgia Bureau of Investigation Record of Medical

Examiner Report 2005-7002735: Melinda Kaye Neal; June 22, 2005.

Gibbons, T. (2002, February 14). Northeast man dies after police use stun gun in arrest; Officers said he was vocal after being hit; he had cocaine in his system, but whether it was a factor is still unclear. *Philadelphia Inquirer* (Philadelphia, PA), p. B1.

Ginsburg, M. (2005, October 7). Bay area; ACLU demands limits on police use of stun guns. *San Francisco Chronicle* (San Francisco, CA), p. B4.

Glendale (Arizona) Police Report 05-092969: Death of Olsen Agoodie; August 5, 2005.

Goodrich, R. (2004, December 22). Officials await lab tests in death of prisoner. *St. Louis Post-Dispatch* (St. Louis, MO), p. C2.

Goodrich, R. (2005, March 17). Autopsy shows man died from cocaine, not stun gun hit. *St. Louis Post-Dispatch* (St. Louis, MO), p. B1.

Goodyear, C. (1999, October 1). Former officer dies in custody/Fairfield cops used stun device, spray after called to home. *San Francisco Chronicle* (San Francisco, CA), p. A24.

Gordon, S. (1992, April 12). Prison death probed: Family files claim against the state. *Modesto Bee* (Modesto, CA), p. A1.

Green County (Missouri) Medical Examiner Report A03-097: Death of Timothy Roy Sleet; June 9, 2003.

Greenwood, J. & Bernard, P. (2004, August 28). Man dies hour after being shot with TASER; Examiner says device probably not the cause. *Tampa Tribune* (Tampa, FL), p. 2

Grieco, L. (2004, July 2). Coroner postpones ruling on TASER-linked death; office awaits results of toxicology tests. *Dayton Daily News* (Dayton, OH), p. A1.

Gwinnett County (Georgia) Medical Examiner's Office Report 04G-0402: Death of Frederick Jerome Williams; May 27, 2004.

Gwinnett County (Georgia) Sheriff's Office Use of Force Report on Frederick Jerome Williams; May 27, 2004.

Haegeli, L., Sterns, L., Adam, D. & Leather, R. (2005). Effect of a TASER shot to the chest of a patient with an implantable defibrillator. *Heart Rhythm, 3*, 3, pp. 339-341.

Hamilton County (OH) Coroner's Report 122314: Death of LeeGrand Griffin; June 5, 1994.

Hamilton County (Ohio) Coroner's Office Autopsy Case Number CC05-01487: Death of Vernon Anthony Young; May 13, 2005.

Hand, E. (2004, August 15). Much-debated TASERs grow in use with police in state. *Arkansas Democrat-Gazette* (Little Rock, AK).

Harris County Medical Examiner's Office Report ML 2005-0552: Death of Joel Dawn Casey; February 18, 2005.

Harris County Medical Examiner's Report ML2005-3303: Death of Barney Lee Green; November 21, 2005.

Hart, D. (1986, June 12). Canoga Park man died after being shot with TASER. *Daily News of Los Angeles* (Los Angeles, CA), p. 9.

Healy, P. (2004, September 21). Relatives sue stun gun maker and police over a man's death. *New York Times* (New York, NY), p. B6.

Heard, K. & Hillen, M. (2006, October 6). 50-year-old man dies after stun-gun shock: Deputy used the device during scuffle. *Arkansas Democrat-Gazette* (Little Rock, AR).

Hick, J., Smith, S. & Lynch, M. (1999). Metabolic acidosis in restraint associated cardiac arrest: A case series. *Academic Emergency Medicine, 6*, pp. 239-43.

Hillsborough County (Florida) Medical Examiner Report 04-03066 A: Death of Henry John Lattarulo; May 22, 2004.

Hillsborough County (Florida) Sheriff's Office Report 2004-50360: Death of Henry John Lattarulo; May 22, 2004.

Ho, J. (2005, August). Sudden in-custody death. *Police, 29*, 8, pp. 47+.

Ho, J., Dawes, D., Bultman, L., Thacker, J., Skinner, J, Johnson, M. & Miner, J. (2007, February 5). Respiratory effect of prolonged electrical weapon application on human volunteers. *Academic Emergency Medicine*, published online before print. Retrieved on February 19, 2007 from http://www.aemj.org/cgi/contentabstract/j.aem.2006.11.016v1.

Ho, J., Miner, J., Heegaard, W. & Reardon, R. (2005). Deaths in American police custody: A 12-month surveillance study. Poster presentation. University of Minnesota Medical School.

Ho, J., Miner, J., Lakireddy, D., Bultman, L. & Heegaard, W. (2006, January). *Cardiovascular and Physiologic Effects of Conducted Electrical Weapon Discharge in Resting Adults. Society for Academic Emergency Medicine, 13*, pp. 589-595.

Holly Springs (Mississippi) Police Report 2005100002: Death of Mary Eleanor Malone Jeffries; October 1, 2005.

Hollywood (Florida) Police Report 02-015065: Death of Vincent Delostia: January 27, 2002.

Hollywood (Florida) Police Report 04-117086: Death of Kevin Downing; December 15, 2004.

Horne, T. & Spalding, T. (2005, March 5). Coroner probes stun gun's possible role in death; official says suspect hit by police TASER also had ingested quantity of drugs. *Indianapolis Star* (Indianapolis, IN), p. B1.

Houston Chronicle. (2002, August 13). Autopsy shows cocaine was in Porter's system. Author (Houston, TX).

Houston Chronicle. (2005, February 20). Hepatitis test awaited in TASER death. Author (Houston, TX), p. B6.

Hovey, H. (2003, March 17). Man dies day after struggle with police; cause unknown. *Albuquerque Tribune* (Albuquerque, NM), p. A2.

Howard, T. (2004, March 30). Man under arrest dies at hospital; autopsy fails to show a cause of death. *St. Louis Post-Dispatch* (St. Louis, MO), p. B1.

Howard, T. (2004, August 5). Detainee's death is ruled an accident. *St. Louis Post-Dispatch* (St. Louis, MO), p. B1.

Hughes, J. (2005, February 15). Death of man, after use of stun gun, won't deter police using TASER. *San Diego Union-Tribune* (San Diego, CA), p. B2.

Hume, E. (2004, December 25). Questions in jail inmate's death; man shocked twice by TASER had bad heart, his family says. *Sacramento Bee* (Sacramento, CA), p. B1.

Hume, E. (2005, February 19). TASER controversy spawns legislation but stun gun is ruled out as cause of Elk Grove man's death. *Sacramento Bee* (Sacramento, CA), p. A1.

Humphrey, K. & Osborn, C. (2005, September 28). Man who died in custody had multiple TASER injuries; Austin police say 9 officers involved in arrest; 2 of them gave suspect shocks during struggle. *Austin American-Statesman* (Austin, TX), p. A1.

Huntley, S. (2003, October 1). TASER eyed in death. Glendale police chief denies that shock caused man to die. *Rocky Mountain News* (Denver, CO), p. 6A.

Huntley, S. (2003, October 28). ACLU wants cops' stun-gun policy revised. *Rocky Mountain News* (Denver, CO), p. 8A.

Hyde, J. (2005, April 9). Was TASER a factor in man's death? Deseret Morning News (Salt Lake City, UT).

Iacoboni, D. (2005, May 29). Akron man dies; had been stunned with a TASER gun. *Plain Dealer* (Cleveland, OH), p. B2.

Indianapolis (Indiana) Police Report 05-0030179: Death of Mark Young; March 10, 2005.

International Association of Chiefs of Police. (2006). *Electro-Muscular Disruption Technology: A Nine-Step Strategy for Effective Deployment*. Alexandria, VA: Author.

James, G. (1993, December 3). Bronx man dies after struggle with officers. *New York Times* (New York, NY), p. B3.

Jauchem, J. (2004, November 16). *Effectiveness & Safety of Electro-Muscular Incapacitating Devices*. Brooks Air Force Base, TX: Air Force Research Laboratory.

Jauchem, J., Sherry, C., Fines, D. & Cook, M. (2006). Acidosis, lactate, electrolytes, muscle enzymes, and other factors in the blood of *Sus scrofa* following repeated TASER exposures. *Forensic Science International, 161*, pp. 20-30.

Jefferson County (Alabama) Coroner's Report 05-0746: Death of Rockey Bryson; July 7, 2005.

Jefferson Parish (Louisiana) Sheriff's Office Report F-03847-04: Death of Jerry W. Pickens; May 31, 2004.

Jefferson Parish (Louisiana) Sheriff's Office Report E-06413-05: Death of Lawrence Berry; May 6, 2005.

Jewett, C. (2005, April 14). Coroner: TASER wasn't a factor in man's death. *Sacramento Bee* (Sacramento, CA), p. B1.

Jewett, C. (2005, June 7). Excess force alleged in death; but police say a stun gun was the best way to subdue the capital man. *Sacramento Bee* (Sacramento, CA), p. B1.

Jewett, C. (2005, August 6). Suspect hit with TASERs, then dies; authorities say he was combative and assaulted deputies. *Sacramento Bee* (Sacramento, CA), p. A1.

Johnson County (Texas) Sheriff's Office Report S04325 93: Death of Samuel Ramon Wakefield; September 12. 2004.

Johnson County (Texas) Medical Examiner Report WA 04-27: Death of Samuel Ramon Wakefield; September 12. 2004.

Joplin (Missouri) Police Report 1-04-038498: Death of David Riley; August 10, 2004.

Kammer, J. (1999, February 3). Norberg probe defended; lead detective says work will withstand scrutiny. *Arizona Republic* (Phoenix, AZ), p. B1.

Karch, S. (1996). Cardiac arrest in cocaine users. *American Journal of Emergency Medicine, 14*, pp. 79-81.

Keary, J. (1992, September 3). Stun gun a factor in death; PG deputies tried to subdue woman. *Washington Times* (Washington, D.C.), p. A1.

Kelleher, J. (2005, April 26). Other factors in TASER death? Suffolk detectives say autopsy shows man who died after being stunned 5 times by police had significant heart problems, drugs in system. *Newsday* (Melville, NY), p. A14.

Kelley, D. (1990, March 13). Police overused stun guns, workers say: Hospital employees contend that a patient who later died was jolted many more times than officers admit. A coroner's investigation is continuing. *Los Angeles Times* (Los Angeles, CA), p. 1.

Kelley, D. (1990, March 18). Inventor had warned TASER unsafe for cardiac patients: Weapons; Coroners in Los Angeles and Santa Clara counties say the stun gun contributed to or caused three deaths. Two of the victims had heart disease. *Los Angeles Times* (Los Angeles, CA), p. 1.

Kelley, D. (1990, April 14). Coroner says TASER played role in death: Police; the ruling criticizes one Ventura officer for using the stun gun "like a cattle prod." A captain says those involved "acted appropriately." *Los Angeles Times* (Los Angeles, CA), p. 1.

Kelly, S. (2004, September 16). Autopsy clears TASER in death; man who struggled with cops died of overdose, coroner says. *Denver Post* (Denver, CO), p. B1.

Kendall, J. (1988, January 14). Man dies after L.A. police hit him with TASER darts. *Los Angeles Times* (Los Angeles, CA), p. 3.

Khanna, R. (2005, August 11). Death of man shocked by TASER ruled a homicide; the medical examiners did not cite the stun gun's use as a factor. *Houston Chronicle.* (Houston, TX), p. B3.

King County (Washington) Sheriff's Office Report 05-313722: Death of Cedrick Stemberg-Barton; October 25, 2005.

Kinner, D. (2002, March 29). Nassau man dies after jolt by TASER. Stun-gun use investigated. *Florida Times Union* (Jacksonville, FL), p. A1.

Krischer, B. (2005, November 29). Unpublished interoffice memorandum from the office of the State Attorney to Chief Delsa Bush regarding in custody death; July 17, 2005.

Lafayette (Louisiana) Police Report 04-00231203: Death of Dwayne Anthony Dunn; October 4, 2004.

Lafayette Parish (Louisiana) Coroner Report 230-04AP: Death of Dwayne Anthony Dunn; October 4, 2004.

LaGrange (Georgia) Police Report 041131154: Death of Greshmond Gray; November 2, 2004.

Lake County (California) Coroner's Report 10621: Death of Keith Raymond Drum; November 7, 2004.

Lakkireddy, D. et al. (2006). Cardiovascular safety profile of electrical stun guns (TASER-X26): Effects of cocaine intoxication on induction of ventricular fibrillation. Posters P4-93 and P6-86, presented May 19 and 20, 2006, *2006 Heart Rhythm Meeting.*

Larimer County (Colorado) Sheriff's Office Report 05-6367: Death of Timothy Glen Mathis; October 3, 2005.

Las Vegas Review-Journal. (2004, October 8). Coroner's office cites "cardiac arrest due to restraint" in man's death. Author, p. 4B.

Laur, D. (2004). *Excited delirium and its correlation to sudden and unexpected death proximal to restraint.* Victoria Police Department (Canada).

Le, P. (2004, December 1). Three have died in Washington after jolting. *Seattle Post-Intelligencer* (Seattle, WA), p. A12.

Le, P. & Castro, H. (2004, December 1). At least 5 deaths linked to "non-lethal" device. *Seattle Post-Intelligencer* (Seattle, WA), p. A1.

Lecky, G. & Jackson, E. (1996, April 5). Man hits head, dies while being arrested. *Cincinnati Post* (Cincinnati, OH). P. 10A.

Lee, H. (2005, August 6). Man shot with TASER dies 5 days later. *San Francisco Chronicle* (San Francisco, CA), p. B3.

Lee, H. (2005, October 26). Man dies after jolts from police stun guns. *San Francisco Chronicle* (San Francisco, CA).

Lee, H. (2006, October 3). Family sues city over police use of TASERs; excessive force led to man's death in custody, they allege. *San Francisco Chronicle* (San Francisco, CA).

Leveque, R. (2006, May 8). Sergeants won't be charged in TASER death. *Inland Valley Daily Bulletin* (Ontario, CA).

Levine, S., Sloane, C., Chan, T., Vilke, G., & Dunford, J. (2005). Cardiac monitoring of subjects exposed to the TASER. *Academic Emergency Medicine, 12*, 5, S. 1, p. S71.

Levine, S., Sloane, C., Chan, T., Vilke, G., & Dunford, J. (2006a). Cardiac monitoring of subjects exposed to the TASER. *Academic Emergency Medicine, 13*, 5, S. 1, p. S47.

Levine, S., Sloane, C., Chan, T., Vilke, G., & Dunford, J. (2006b). Cardiac monitoring of subjects exposed to the TASER. Poster presentation. University of California San Diego Emergency Medicine.

Lillis, R. (2006, August 24). Report: Struggle resulted in death: Coroner's report calls psychiatric patient's death a "homicide," but police do not suspect foul play. *Sacramento Bee* (Sacramento, CA).

Little Rock (Arkansas) Police Department Information Report 2004-45639: Death of Robert Harold Allen; April 17, 2004.

Little Rock (Arkansas) Police Department Information Report 2004-45667: Death of Robert Harold Allen; April 17, 2004.

Livingston (California) Police Report 050700: Death of James Floyd Wathan, Jr.; April 3, 2005.

Livingston County (Michigan) Medical Examiner's Report 04-209: Death of Charles Christopher Keiser; November 25, 2004.

Los Angeles Police Department. (2005, July 25). News release: Assault suspect dies after attacking police. Author (Los Angeles, CA).

Los Angeles Times. (1985, May 24). The region. Author (Los Angeles, CA), p. B2.

Los Angeles Times. (1985, August 30). The region. Author (Los Angeles, CA), p. B2.

Los Angeles Times. (1986, May 19). Man dies after police shoot him with TASER. Author (Los Angeles, CA), p. B2.

Los Angeles Times. (1986, July 4). Drugs blamed in death of man police quelled. Author (Los Angeles, CA), p. B10.

Los Angeles Times. (1986, July 10). Pomona TASER gun cited in death. Author (Los Angeles, CA), p. B2.

Los Angeles Times. (1987, June 22). The region. Author (Los Angeles, CA), p. B2.

Los Angeles Times. (1988, October 14). Stun guns ruled out as cause of deaths. Author (Los Angeles, CA), p. B2.

Los Angeles Times. (1989, January 12). Local news in brief: Settlement over man's death OKd. Author (Los Angeles, CA) p. B2.

Los Angeles Times. (1989, February 16). Metro desk. Author (Los Angeles, CA), p. B2.

Los Angeles Times. (1991, April 22). Man subdued with TASER gun, later dies of heart attack. Los Angeles Times (Los Angeles, CA), p. B20.

Los Angeles Times. (1991, November 4). Man dies in police custody. Author (Los Angeles, CA), p. B12.

Los Angeles Times. (1994, January 8). AZUSA parolee dies after fight with deputies. Author (Los Angeles, CA), p. B2.

Los Angeles Times. (1994, April 18). Man sprayed, stunned by police dies. Author (Los Angeles, CA), p. B8.

Los Angeles Times. (1996, January 6). TASER death. Author (Los Angeles, CA), p. B2.

Los Angeles Times. (1996, July 22). Woman hit by police TASER stun dart dies. Author (Los Angeles, CA), p. B3.

Lucas County (Ohio) Coroner's Report 69-05: Death of Jeffrey A. Turner; January 31, 2005.

Madison County (Illinois) Coroner's Report 04-0471: Death of Terry L. Williams; March 29, 2004.

Maier, A., Nance, P., Price, P., Sherry, C., Reilly, J., Klauenberg, B. & Drummond, J. (March 2005). *Human Effectiveness and Risk Characterization of the Electromuscular Incapacitation Device - A Limited Analysis of the TASER*. Quantico, VA: The Joint Non-Lethal Weapons Human Effects Center of Excellence.

Main, F. (2005, July 29). TASER killed man, pathologist finds; medical examiner's ruling on drug suspect 1st to blame stun gun. *Chicago Sun-Times* (Chicago, IL), p. 3.

Main, F. (2005, July 30). TASER maker to seek court review of death ruling. *Chicago Sun-Times* (Chicago, IL), p. 6.

Main, F. (2005, November 22). Man dies after police shock him with TASER; also pepper-sprayed, sedated; Cause of death investigated. *Chicago Sun Times* (Chicago, IL), p. 10.

Maislin, S. (2007, August 23). Unpublished letter to the author.

Manojlovic, D., Hall, C., Laur, D., Goodkey, S., Lawrence, C., Shaw, R., St-Amour, S., Nuefeld, A. & Palmer, S. (2005, August 22). *Review of Conducted Energy Devices*. Ottawa, ON: Canadian Police Research Centre.

Maricopa County (Arizona) Medical Examiner's Report 96-01510: Death of Scott Norberg; June 1, 1996.

Maricopa County (Arizona) Medical Examiner's Report 02-01878: Death of Nicolas Aguilar; June 13, 2002.

Maricopa County (Arizona) Medical Examiner's Report 04-02994: Death of Lawrence Samual Davis; August 24, 2004.

Maricopa County (Arizona) Medical Examiner's Report 05-01639: Death of Keith Edward Graff; May 3, 2005.

Maricopa County (Arizona) Medical Examiner's Report 05-02639: Death of Ernesto Valdez, Jr.; July 15, 2005.

Maricopa County (Arizona) Medical Examiner's Report 05-03018: Death of Olson Agoodie; August 5, 2005.

Marion County (Indiana) Coroner's Verdict Report 05-0248: Death of Mark Young; March 10, 2005.

Marra, A. (2005, July 20). TASER use justified, chief says; stun-gun policy followed, she says. *Palm Beach Post* (Palm Beach, FL), p. 1B.

Marra, A. (2005, December 1). Officers cleared in death after TASER. *Palm Beach Post* (Palm Beach, FL), p. 2B.

Martin County (Florida) Sheriff's Office Report 05-11924: Death of Brian Lichtenstein; August 25, 2005.

McBride, D. and Tedder, N. (2005, March 29). *Efficiency and Safety of Electrical Stun Devices*. Arlington, VA: Potomac Institute for Policy Studies.

McDaniel, W., Stratbuckler, R., Nerheim, M. and Brewer, J. (January 2005). Cardiac safety of neuromuscular incapacitating defensive devices. *Pacing and Clinical Electrophysiology, 28*, Sup. 1, p. S284.

McDonald, M. (2005, August 30). Drugs led to death of suspect. *Fort Worth Star-Telegram* (Fort Worth, TX), p. B1.

McEvoy, G. (Ed.). (2006). *AHFS Drug Information*. Bethesda, MD: American Society of Health-System Pharmacists, Inc.

McKinney, M. (2004, October 17). Stun guns pack uncertain risk; available to police and public alike, they don't always live up to nonlethal billing. *Star Tribune* (Minneapolis, MN), p. 1A.

McManus, J., Forsyth, T., Hawks, R., & Jui, J. (2004). A retrospective case series describing the injury pattern of the advanced M26 TASER in Multnomah County, Oregon. *Academic Emergency Medicine, 11,* 5, p. 587.

Means, R. & Edwards, E. (2005, February). Electronic control weapons: Liability issues. *The Police Chief, 72,* 2, pp. 10-11.

Mecklenburg County (North Carolina) Medical Examiner Report B2004-2092: Death of Anthony Lee McDonald; August 13, 2004.

Merced County (California) Coroner's Report 20050011 633: Death of James Floyd Wathan, Jr.; April 3, 2005.

Merced County (California) Sheriff's Office Report 2005 0011640: Death of James Floyd Wathan, Jr.; April 3, 2005.

Mesa (Arizona) Police Report 2004-2030697: Death of Milton Salazar; July 21, 2004.

Metcalf, M. (2006, January 31). No police wrongdoing is found in Duluth man's death. *Star Tribune* (Minneapolis, MN), p. 5B.

Mets, B., Jamdar, S. & Landry, D. (1996) The role of catecholamines in cocaine toxicity: A model for cocaine "sudden death." *Life Sciences, 59,* pp. 2021-31.

Meyer, E. (2005, July 27). Drug use, TASER blamed in death. *Akron Beacon Journal* (Akron, OH).

Miami-Dade County (Florida) Medical Examiner's Report 04-2086: Death of John Alex Merkle; September 20, 2004.

Miami-Dade County (Florida) Medical Examiner's Report 2005-01598: Death of Phoarah Kareem Knight; June 24, 2005.

Miami (Florida) Police Report 26415714: Death of John Alex Merkle; September 20, 2004.

Miami Herald. (2005, June 30). Crazed man dies after being zapped by TASER. Author (Miami, FL), p. 3B.

Miles, M. & Huberman, A. (1994). *An Expanded Sourcebook: Qualitative Data Analysis, 2nd Ed.* Thousand Oaks, CA: Sage Publications.

Mills, R. (2007, March 24). TASERs. Are they suspect, or necessary for subduing suspects? *Naples Daily News* (Naples, FL).

Minneapolis (Minnesota) Police Report MP-04-028424: Death of Raymond Siegler; February 13, 2004.

Minnesota Department of Public Safety Report 2005-314: Death of David Michael Croud; October 12, 2005

Mirchandani, H., Rorke, L., Sekula-Perlman, A. & Hood, I. (1994) Cocaine-induced agitated delirium, forceful struggle, and minor head injury: A further definition of sudden death during restraint. *American Journal of Forensic Medicine and Pathology, 15,* pp. 95-9.

Missoulian. (2006, May 29). TASER use by police scrutinized. Author (Missoula, MT).

Mitchell, K. (2004, August 22). Man hit by TASER 4 times, police say; relatives say the victim was on cocaine and had heart problems; police say the officers used restraint. *Denver Post* (Denver, CO), p. A1.

Monterey County Coroner's Report 04-256: Death of Michael Robert Rosa; Monterey County Coroner's Report 20050159: Death of Robert Clark Heston, Jr.; February 20, 2005.

Montgomery County (Maryland) Police Report M04-022711: Death of Eric Wolle; April 27, 2004.

Montgomery County (Ohio) Coroner Report 04-2130: Death of Eric Bernard Christmas; June 30, 2004.

Mumola, C. (2006, August 30). Unpublished e-mail to the author regarding the deaths in custody reporting program.

Mungin, L. (2003, October 2). Death of inmate alarms relatives. *Atlanta Journal-Constitution* (Atlanta, GA), p. D1.

Mungin, L. (2004, June 3). 2nd man dies after shock by stun gun. *Atlanta Journal-Constitution* (Atlanta, GA), p. B1.

Mungin, L. (2004, September 9). 2nd man dies after shock by stun gun. *Atlanta Journal-Constitution* (Atlanta, GA), p. B1.

Mungin, L. (2005, March 23). DA Porter decides to reopen TASER case. *Atlanta Journal-Constitution* (Atlanta, GA), p. JJ1.

Mungin, L. & Bentley, R. (2004, June 29). Deaths spur TASER debate\Police embrace stun guns, but a series of fatalities brings calls for halt to their use. *Atlanta Journal-Constitution* (Atlanta, GA), p. B1.

Munoz, H. (2006, March 24). Drugs cited in man's death; Police used TASER in October altercation. *Hartford Courant* (Hartford, CN), p. B3.

Munoz, H., Stacom, D., Goren, D., Fillo, M. & Bachetti, C. (2005, November 3). Death raises TASER questions; agitated suspect was stunned with device last week, dies Monday in hospital. *Hartford Courant* (Hartford, CN), p. B1.

Nakaso, D. (1986, July 24). TASER: Too tough for PCP war? *San Jose Mercury News* (San Jose, CA), p. 1A.

Nance, P., Price, P., Sherry, C., Reilly, J., Klauenberg, B., & Drummond, J. (2005). *Human Effectiveness and Risk Characterization of the Electromuscular Incapacitation Device - A Limited Analysis of the TASER.* Quantico, VA: The Joint Non-Lethal Weapons Human Effects Center of Excellence.

Nanthakumar, K., Billingsley, I., Masse, S., Dorian, P., Cameron, D., Chauhan, D., Downar, E. & Sevaptsidis, E. (2006). Cardiac electrophysiological consequences of neuromuscular incapacitating device discharges. *Journal of the American College of Cardiology, 48,* 4, pp. 798-804.

Nashville (Tennessee) Metropolitan Police Report 05-279630: Death of Walter Lamont Seats; May 26, 2005.

Nashville (Tennessee) Metropolitan Police Report 05-517462: Death of Patrick Aaron Lee; September 24, 2005.

Nassau County (Florida) Medical Examiner's Report 02-0423: Death of Henry William Canady; March 27, 2002.

Nassau County (Florida) Sheriff's Office Report 2002-01492: Death of Henry William Canady; March 27, 2002.

New York Times. (1989, November 16). Cocaine psychosis changes the rules. Author (New York, NY), p. B1.

New York Times. (1993, December 21). Death ruled accidental. Author (New York, NY), p. B4.

New York Times. (2005, April 24). Man dies after fight in Suffolk County. Author (New York, NY), p. 4.

Newbold, M. (2005, January/February). Emerging issues surrounding the use of conducted energy weapons by police. *Municipal Lawyer, 46,* 1, pp. 14-17.

Newsday. (2005, April 24). Authorities and witnesses differ over officers' struggle with a man who died after being shot by stun gun. Author (New York, NY).

Newton, E. (1993, April 15). Cocaine, hogtying by police led to man's death, coroner says: Investigation: Pasadena barber died in custody after a car chase. Use of the restraint procedure is restricted because it interferes with breathing in obese people. *Los Angeles Times* (Los Angeles, CA), p. 3.

Nolan, B. (2006, November 16). Former jailer pleads guilty; TASER death case ends with apology. *Bedford Times-Mail* (Bedford, IN).

Nolan, J. (2002, February 14). Ranting man, subdued with stun gun, dies; cocaine suspected as cause; family had called police, who also used pepper spray, to help them. *Philadelphia Daily News* (Philadelphia, PA), p. 8.

Norman, B. (2003, June 28). Shooting baffles many; nude man could have died from cut. *Pensacola News Journal* (Pensacola, FL), p. 1A.

Office of the Chief Medical Examiner (Oklahoma) Report 0203435: Death of Jason Nichols; June 27, 2002.

Office of the Chief Medical Examiner (Oklahoma) Report 0305864: Death of Dennis D. Hammond; October 11, 2003.

Office of the Chief Medical Examiner (Oklahoma) Report 0306462: Death of Michael Sharp Johnson; November 10, 2003.

Office of the Chief Medical Examiner (Oklahoma) Report 0500692: Death of Ricky Paul Barber; April 8, 2005.

Office of the Medical Examiner, Eighth District (Florida) Report ME 01-203: Death of Mark Lorenzo Burkett; June 13, 2001.

Office of the Medical Examiner, Fifteenth District (Florida) Report ME 04-1325: Death of Timothy Bolander; December 23, 2004.

Office of the Medical Examiner, Fifteenth District (Florida) Report ME 05-0800: Death of Michael Len Crutchfield; July 17, 2005.

Office of the Medical Examiner, First District (Florida) Report MLA03-446: Death of David Sean Lewandowski; June 26, 2003.

Office of the Medical Examiner, First District (Florida) Report MLA05-023: Death of Carl Nathaniel Trotter; January 8, 2005.

Office of the Medical Examiner, First District (Florida) Report MLA05-678: Death of Robert E. Boggon; August 29, 2005.

Office of the Medical Examiner (New Mexico) Report 2005-02996: Death of Randy Martinez; May 20, 2005.

Office of the Medical Examiner, Nineteenth District (Florida) Report ME 2005-19-516: Death of Brian Lichtenstein; August 25, 2005.

Office of the Medical Examiner, Ninth District (Florida) Report 04-2641: Death of Anthony Oliver; June 1, 2004.

Office of the Medical Examiner, Ninth District (Florida) Report ME 2005-001475: Death of Jeffrey Dean Earnhardt; December 1, 2005.

Office of the Medical Examiner, Seventh and Twenty-Fourth Districts (Florida) Report ME 2005-0185V: Death of Willie Michael Towns; March 6, 2005.

Office of the Medical Examiner (Tennessee) Report MEC 05-1665: Death of Walter Lamont Seats; May 27, 2005.

Office of the Medical Examiner, Tenth District (Florida) Report 2004-10-FA-394: Death of Jason David Yeagley; August 26, 2004.

Office of the Medical Examiner, Third District (Florida) Report ME 05-0377: Death of Milton Woolfolk; March 11, 2005.

Office of the Medical Examiner, Twentieth District (Florida) Report ME 2004-385: Death of Christopher Dearlo Hernandez; December 28, 2004.

Office of the Medical Examiner, Twenty-First District (Florida) Report 00786-2005: Death of Steven Michael Cunningham; October 13, 2005.

Office of the Medical Examiner, Twenty-First District (Florida) Report 00901-2005: Death of Tracy Rene Shippy; November 26, 2005.

O'Halloran, R. (2004, September). Reenactment of circumstances in deaths related to restraint. *The American Journal of Forensic Medicine and Pathology, 25*, 3, pp. 190-193.

O'Halloran, R. & Lewman, L. (1993). Restraint asphyxiation in excited delirium. American Journal of Forensic Medicine and Pathology, 14, pp. 289-95.

Okaloosa County (Oklahoma) Sheriff's Office Report OCSO04OFF007901: Death of Demetrius Tillman Nelson; July 3, 2004.

Oklahoma City (Oklahoma) Police Report 02-056072: Death of Jason Nichols; June 27, 2002.

Oklahoma City (Oklahoma) Police Report 03-102727: Death of Dennis D. Hammond; October 11, 2003.

Oklahoma City (Oklahoma) Police Report 03-113117: Death of Michael Sharp Johnson; November 10, 2003.

Olsen, L. & Davis, R. (2004, November 5). Police back TASERs despite deaths; after the latest fatality, HPD and others call stun guns safer options. *Houston Chronicle* (Houston, TX), p. 1.

Olympia (Washington) Police Report 2002-9614: Death of Stephen L. Edwards; November 7, 2002.

Omaha (Nebraska) Police Report RB# 83421-F: Death of David L. Moss, Jr.; December 29, 2005.

Orange County (California) Coroner's Report 97-04401-EY: Death of Garner Roosevelt Hisks, Jr.; July 6, 1997.

Orange County (California) Coroner's Report 05-03930-SR: Death of Richard James Alvarado; May 23, 2005.

Orange County (Florida) Sheriff's Office Report 02-071994: Death of Gordon Randall Jones; July 19, 2002.

Orange County (Florida) Sheriff's Office Report 03-09 6603: Death of Louis N. Morris, Jr.; October 21, 2003.

Orange County (Florida) Sheriff's Office Report 05-10 6191: Death of Jeffrey Dean Earnhardt; December 1, 2005.

Orange County Sheriff's Office. (2004). Proceedings from the SWAT Roundup. *TASER® Deployments and Injuries: Analysis of Current and Emerging Trends*. Orlando, FL: Author.

Orlando (Florida) Police Report 2004-193999: Death of Anthony Carl Oliver; May 31, 2004.

Osterweil, N. HRS: Stun gun seems to subdue without fatal VF. *MedPage Today*. Retrieved July 19, 2006 from http://www.medpagetoday.com/tbprint.cfm?tbid=33 33.

Ovalle, D. (2005, May 7). Man shot with TASER later dies in jail cell. *Miami Herald* (Miami, FL), p. 4B.

Palermo, D. (1986, June 6). Frenzied school intruder dies after officers use TASER guns. *Los Angeles Times* (Los Angeles, CA), p. 8.

Paquette, M. (2003, July-September). Excited delirium: Does it exist? *Perspectives in Psychiatric Care, 39*, 3, pp. 93-94.

Parent, R. (2006, April). Deaths during police intervention. *FBI Law Enforcement Bulletin, 75*, 4, pp. 18-22.

Park, K., Korn, C. & Henderson, S. (2001, April/June). Agitated delirium and sudden death: Two case reports. *Prehospital Emergency Care, 5*, 2, pp. 214-16.

Pasadena (Texas) Police Department. (2005, November 28). Custodial Death Report to the Office of the Attorney General: Barney Lee Green.

Pathology Associates (Fresno, California) Autopsy Report A05-018: Death of Ricky Paul Barber; April 8, 2005.

Patrick, M. (2005, November 27). Details emerge in Asotin County jail death; sheriff says inmate died after injuring himself in his jail cell. *Lewiston Morning Tribute* (Lewiston, ID), p. 1C.

Peavy, R. (2006, August 25). Unpublished letter to the author.

Peavy, R. (2006, November 15). Unpublished letter to the author.

Pembroke Pines (Florida) Police Report 2003-092431. Death of Kerry Kevin O'Brien; November 11, 2003.

Perlstein, M. (2004, December 5). Kenner man subdued with TASER dies: Stun gun employed twice after traffic stop. *Times-Picayune* (New Orleans, LA), p. 1.

Peters, J. (2006a, March/April). Sudden death, excited delirium, and issues of force: Part I. *Police and Security News, 22*, 2, pp. 1-4.

Peters, J. (2006b, May/June). Sudden death, excited delirium, and issues of force: Part II. *Police and Security News, 22*, 3, pp. 1-4.

Peters, J. (2006c, July/August). Sudden death, excited delirium, and issues of force: Part III. *Police and Security News, 22*, 4, pp. 1-4.

Peters, J. (2006d, September/October). Sudden death, excited delirium, and issues of force: Part IV. *Police and Security News, 22*, 5, pp. 1-4.

Peters, J. (2006e, November/December). Sudden death, excited delirium, and issues of force: Part V. *Police and Security News, 22*, 6, pp. 1-4.

Philbin, W. (2004, December 4). Man in hospital after being subdued with TASER; cops say he attacked deputy at lockup. *Times-Picayune* (New Orleans, LA), p. 1.

Phoenix (Arizona) Police Report 2002-2108248: Domestic Violence Assault; June 13, 2002.

Phoenix (Arizona) Police Report 2002-2108248-A: Death of Nicolas Aguilar-Zaragoza; June 13, 2002.

Phoenix (Arizona) Police Report 2004-41618520: Death of Lawrence Samual Davis; August 24, 2004.

Phoenix (Arizona) Police Report 2005-50769124: Death of Jesse Cleon Colter, III; April 24, 2005.

Phoenix (Arizona) Police Report 2005-50832168: Death of Keith Edward Graff; May 2, 2005.

Phoenix (Arizona) Police Report 2005-51505800: Death of Frank Gilman Edgerly; August 7, 2005.

Phoenix (Arizona) Police Report 2005-51340162: Death of Ernesto Valdez, Jr.; July 15, 2005.

Pinkham, P. (2005, June 14). Victim's relatives ask TASER sales halt; family suing, contends stun guns falsely marketed as "non-lethal" alternative to guns. *Florida Times-Union* (Jacksonville, FL), p. B1.

Pinto, D. & Josephson, M. (2004). Sudden cardiac death. In Fuster, V., Alexander, R. & O'Rourke, R. (Eds.), *Hurst's the Heart, 11th Edition*. New York, NY: The McGraw-Hill Companies. Retrieved January 10, 2007 from http://online.statref.com/document.aspx?fxid=67&docid=1050.

Plohetski, T. (2004, June 17). Man dies after police arrest him following South Austin break-in. *Austin American-Statesman* (Austin, TX), p. B1.

Plohetski, T. (2006, February 10). Police videos show final minutes of man's life after being TASERed. *Austin American-Statesman* (Austin, TX), p. B1.

Polk County (Florida) Sheriff's Office Report 2004-145578: Death of Jason David Yeagley; August 27, 2004.

PR Newswire US. (2004, July 21). Dr. Cyril Wecht releases preliminary report: No basis to conclude TASER® contributed to death of Mr. James Borden; forensic expert reviews controversial case. Author (Dateline: Scottsdale, AZ).

PR Newswire US. (2005, October 26). Borden wrongful death lawsuit dismissed against TASER(TM) International; Coroner admits comments were "reckless" in deposition. Author (Dateline: Scottsdale, AZ).

Press-Telegram. (1994, April 18). Man dies after scuffle with police. Author (Long Beach, CA), p. B1.

Pueblo County (Colorado) Coroner Report 02A-MD08: Death of Richard Joseph Baralla; May 17, 2002.

Pueblo (Colorado) Police Report 02-10782: Death of Richard Joseph Baralla; May 17, 2002.

Rainey, J. (1987, December 5). Police probing use of force in man's death. *Los Angeles Times* (Los Angeles, CA), p. 37.

Rainey, J. (1988, January 15). Coroner says drug, injuries killed man shot by TASERs. *Los Angeles Times* (Los Angeles, CA), p. 8.

Rams, B. (1997, July 8). Suicidal man dies after being subdued in Santa Ana condo. *Orange County Register* (Santa Ana, CA), p. B7.

Randall T. (1992). Cocaine: Alcohol mix in body to form even longer-lasting, more lethal drug. *Journal of the American Medical Association, 267*, pp. 1043-1044.

Rasmussen, K. (2006, June 15). Robert Boggon jail death. *Pensacola News Journal* (Pensacola, FL).

Ratcliffe, H. (2004, August 13). Violent day leaves neighbors stunned. *St. Louis Post-Dispatch* (St. Louis, MO), p. B1.

Ratcliffe, H. (2005, February 10). Autopsy sheds little light on football player's rampage. *St. Louis Post-Dispatch* (St. Louis, MO), p. B2.

Ravn, K. (2004, September 12). Death raises TASER safety questions. *Monterey County Herald* (Monterey, CA).

Reay, D. (1996, May). Suspect restraint and sudden death. *FBI Law Enforcement Bulletin, 65*, 5, pp. 22-25.

Reed, S. (2005, October 4,). Loveland police use stun gun to subdue suspect. *Loveland Reporter Herald* (Loveland, CO).

Reza, H. (1987, November 14). Police probe death of man in shooting by stun gun. *Los Angeles Times* (Los Angeles, CA), p. 1.

Rivera, E., Holland, J., Queen, J., Jordan, G., Gelman, M., & Powell, M. (1989, May 24). 4 cop custody deaths over 28 hours; brutality question in one case. *Newsday* (Melville, NY), p. 6.

Roberts, W. (1986). Sudden cardiac death: Definitions and causes. *American Journal of Cardiology, 57*, pp. 1410-1413.

Robison, D. & Hunt, S. (2005, March). Sudden in-custody death syndrome. *Topics in Emergency Medicine, 27*, 1, pp. 36-43.

Robinson, C. (2005, August 4). Inmate's heart, alcoholism led to death, coroner says. *Birmingham News* (Birmingham, AL), p. 1C.

Rosetta, L. (2005, April 9). Official says stun gun may have contributed to death; TASER may have had part in death. *Salt Lake Tribune* (Salt Lake City, UT), p. B1.

Ross County Sheriff's Office Report 01-05-015334: Death of Shawn A. Norman; August 26, 2005.

Ross, D. (1998). Factors associated with excited delirium deaths in police custody. *Modern Pathology, 11*, pp. 1127-1137.

Rubin, P. (2007, June 21). Aftershock!: Turns out the

much-maligned stun guns are a good thing overall. *Phoenix New Times* (Phoenix, AZ).

Ruggieri, J, (2005). *Lethality of TASER Weapons.* Author.

Ruttenber, A., Lawler-Heavner, J., Yin, M., Wetli, C., Hearn, W. & Mash, D. (1997). Fatal excited delirium following cocaine use: Epidemiologic findings provide new evidence for mechanisms of cocaine toxicity. *Journal of Forensic Science, 42,* pp. 25-31.

Ruttenber, A., McAnally, H. & Wetli, C. (1999). Cocaine-associated rhabdomyolysis and excited delirium: Different stages of the same syndrome. *American Journal of Forensic Medicine and Pathology, 20,* pp. 120-7.

Sacramento County (California) Coroner's Report 03-03978: Death of Gordon Benjamin Rauch; August 8, 2003.

Sacramento County (California) Coroner's Report 04-055 71: Death of Ricardo Reyes Zaragoza; November 8, 2004.

Sacramento County (California) Coroner's Report 04-064 32: Death of Ronnie James Pino; December 23, 2004.

Sacramento County (California) Coroner's Report 05-029 15: Death of Ravan Jermont Conston; June 4, 2005.

Sacramento County (California) Coroner's Report 05-034 21: Death of Tommie Valentine Gutierrez; July 2, 2005.

Sacramento County (California) Coroner's Report 05-04018: Death of Dwayne Zachary; August 4, 2005.

Sacramento County (California) Coroner's Report 05-048 68: Death of Timothy Michael Torres, Jr.; September 22, 2005.

Safranek, L. (2006, February 4). PCP, heart trouble led to man's TASER death. *Omaha World-Herald* (Omaha, NE), p. 3B.

Saginaw County (Michigan) Medical Examiner's Report S03-106: Death of John Lee Thompson; August 6, 2003.

Saginaw County (Michigan) Sheriff's Office Report 173-0004792-03: Death of John Lee Thompson; August 6, 2003.

Saint Louis County (Missouri) Police Report 04-0073861: Death of Ernest J. Blackwell; August 11, 2004.

Saint Paul (Minnesota) Police Report 04116367: Death of James Arthur Cobb, Jr.; June 9, 2004.

Sample, H. (2005, August 10). ACLU seeks police data on TASER cases; Sacramento County Sheriff's force one of 4 agencies queried. *Sacramento Bee* (Sacramento, CA), p. B3.

San Diego County (California) Medical Examiner's Report 03-00963: Death of Joshua Alva Hollander; May 8, 2003.

San Francisco Chronicle. (1999, November 19). Solano County/Report says Fairfield man died from drug overdose. Author, p. A24.

San Joaquin County (California) Coroner's Report 658-91. Death of Donnie Ray Ward; December 9, 1991.

San Jose Mercury News. (1986, January 8). Cops cleared in stun gun death. Author (San Jose, CA), p. 2G.

San Jose Mercury News. (1988, September 12). Police fire stun guns. Author (San Jose, CA), p. 8B.

San Jose Mercury News. (1993, April 25). Federal probe sought in death of man hogtied by L.A. police. Author (San Jose, CA), p. 3B.

San Mateo County (California) Coroner Report 05-0022-A: Death of Gregory Saulsbury; January 2, 2005.

Sanchez, M. (2005, September 10). Teen died from drug, not TASER, office rules. *Fort Worth Star-Telegram* (Fort Worth, TX), p. B3.

Sandaine, K. (2006, June 9). Decision on next step in Tyler Shaw case expected soon; Prosecutor at attorney general's office to determine if charges are filed in Asotin County jail death. *Lewiston Morning Tribute* (Lewiston, ID), p. 1C.

Sandaine, K. (2006, October 12). Tyler Shaw's death ruled an accident; Pierce County medical examiner says death came from "excited delirium with restraint stress." *Lewiston Morning Tribune* (Lewiston, ID). P. 1A.

Santa Clara County (California) Sheriff's Office Report IR85-18511A: Death of Joseph Rodriguez; December 27, 1985.

Santa Clara County (California) Medical Examiner's Investigation Report C85-3109: Death of Joseph Rodriguez; December 27, 1985.

Santa Clara County (California) Medical Examiner's Report 05-02726: Death of Brian Patrick O'Neal; August 1, 2005.

Santa Clara County (California) Medical Examiner's Report 05-04036: Death of Jose Angel Rios; November 18, 2005.

Schaefer, R. (1999, February 2). Man, 48, tabbed for mental exam; man "had worked self into frenzy." *Cincinnati Enquirer* (Cincinnati, OH), p. 1B.

Schmidt, C. (2005, April 3). Debate follows deaths associated with TASERs; police use of stun guns not safe, critics contend. *Tampa Tribune* (Tampa, FL), p. 1.

Seattle Post-Intelligencer. (2005, October 27). Deputies on leave after arrest fatality. Author (Seattle, WA), p. B2.

Seattle Times. (2004, July 28). ROP local news. Author, p. B3.

Seewer, J. (2005, April 12). Coroner: Heart problem, struggle at jail contributed to death. *Associated Press.* (Dateline: Toledo, OH).

Seltzer, E. (2005, December 13). Police investigation into

death continues. *Sonoma Index-Tribune* (Sonoma, CA).

Shearer, J. (2005, August 27). Brain damage caused police-custody death; Medical examiner releases finding. *San Diego Union-Tribune* (San Diego, CA), p. NC2.

Shioya, T. (1996, June 10). Fairfield man dies in custody; pepper spray used to subdue psychiatric patient who left ward. *San Francisco Chronicle* (San Francisco, CA), p. A18.

Sholley, D. (2005. November 15). Man dies after TASER shock. *Inland Valley Daily Bulletin* (Pomona, CA).

Simon, R. (1986, April 3). Family of chokehold victim gets $585,000. *Los Angeles Times* (Los Angeles, CA), p. 6.

Smith, C. (2005, September 23). TASER ruled out in inmate's death. *Santa Cruz Sentinel* (Santa Cruz, CA).

Smith, P. (2006, November 30). *Expert Report of Patrick (Rick) W. Smith: In the matter of Betty Lou Heston, et. al. vs. City of Salinas (CA), et. al., in the United States District Court for the Northern District of California, San Jose Division, Case No. C-05-03658 RS.*

Smith, S. (2004, January 23). Deputy cleared in fatal encounter. *Pensacola News Journal* (Pensacola, FL), p. 1A.

Snyder, D. (2004, June 26). Md. Family grieves for mentally ill man: schizophrenic died after being subdued by officers; No wrongdoing found. *Washington Post* (Washington, DC), p. B2.

Solano County (California) Coroner's Report 1086-1046-CCI-386: Death of Yale Lary Wilson; October 7, 1986.

Solano County (California) Coroner's Report 696-1071-CCI-160: Death of James Quentin Parkinson; June 8, 1996.

Solano County (California) Coroner's Report 999-1720-CCI-261: Death of David Torres Flores; September 28, 1999.

Sonoma County (California) Coroner Report 05-0847: Death of Carlos Casillas-Fernandez; July 16, 2005.

South Carolina Law Enforcement Division Report 34-04-167: Death of Williams Malcolm Teasley; August 16, 2004.

Southwell, J. (2003, September 22). Advanced *TASER M-26 Safety Analysis.* Melbourne, AU: The Alfred Hospital.

Southwell, J. (2004, June 25). *TASER X-26 Safety Analysis.* Melbourne, AU: The Alfred Hospital.

Springfield (Missouri) Police Report 03-27280: Death of Timothy Roy Sleet; June 9, 2003.

St. John's County (Florida) Sheriff's Office Report 03-343014: Death of Lewis Sanks King; December 9, 2003.

St. Louis Post-Dispatch. (2004, October 17). Debate remains over cause of fatal explosion. Author, p. D17.

St. Mary-Corwin Regional Medical Center (Pueblo, Colorado) Autopsy Report P-02070.

Stanley, D. & Osborn, C. (2005, October 29). Report: TASER did not kill man; PCP, cocaine, disease played role in death, according to autopsy after police struggle. *Austin American-Statesman* (Austin, TX), p. A1.

Stephens, B., Jentzen, J., Karch, S., Wetli, C. & Mash, D. (2004a, March). Criteria for the interpretation of cocaine levels in human biological samples and their relation to the cause of death. *The American Journal of Forensic Medicine and Pathology, 25*, 1, pp. 1-10.

Stephens, B., Jentzen, J., Karch, S., Wetli, C. & Mash, D. (2004b, March). National Association of Medical Examiners position paper on the certification of cocaine-related deaths. *The American Journal of Forensic Medicine and Pathology, 25*, 1, pp. 11-13.

Stern, E. (2005, April 8). TASER skeptics push for tracking; results of the autopsy on Delphi man not released pending toxicology report. *Modesto Bee* (Modesto, CA), p. B1.

Stoneberg, D. (2005, March 9). Police cleared in Drum death. *Lake County Record Bee* (Lakeport, CA).

Stratton, S., Rogers, C., Brickett, K. & Gruzinski. (2001, May). Factors associated with sudden death of individuals requiring restraint for excited delirium. *The American Journal of Emergency Medicine, 19*, 3, pp. 187-91.

Substance Abuse and Mental Health Services Administration, Office of Applied Studies. (2002) *Mortality Data from the Drug Abuse Warning Network.* Rockville, MD: Department of Health and Human Services.

Sullivan, J. (2005, October 26). Man gulps item in traffic stop, dies. Seattle Times (Seattle, WA), p. B7.

Sundlin, S. (2003, December 12). Sheriff asks King death rights study; complaints prompt request to see if there was racial motivation in the arrest. *Florida Times Union* (Jacksonville, FL), p. B1.

Sundlin, S. (2003, December 24). Family seeks new probe in TASER case; Harvard professor says he will ask for civil rights review in St. Johns arrest death. *Florida Times Union* (Jacksonville, FL), p. B1.

Swiney, R. (2005, November 21). Unpublished letter to Dr. Vinvent J. M. Di Maio.

Sztajnkrycer, M. & Baez, A. (2005, April). Cocaine, excited delirium and sudden unexpected death. *Emergency Medical Services*, pp. 77-81.

Tarrant County (Texas) Medical Examiner's Report 0409 265: Death of Robert Guerrero; November 2, 2004.

Tarrant County (Texas) Medical Examiner's Report 0503238: Death of Eric J. Hammock; April 3, 2005.

Tarrant County (Texas) Medical Examiner's Report

0505897: Death of Carolyn J. Daniels; June 24, 2005.

Tarrant County (Texas) Medical Examiner's Report 0506455: Death of Kevin Ray Omas; July 12, 2005.

Tarm, M. (2005, February 11). Man dies after police strike him with TASER stun gun. *Associated Press.* (Dateline: Chicago, IL).

Tavernise, S. (2004, February 5). Metro briefing New York: Southampton Village: Southampton man dies in custody. *New York Times* (New York, NY), p. B4.

Tenorio, G. (2005, May 19). Man deputies TASERed dies. *San Bernardino Sun* (San Bernardino, CA).

Texas Tech University Health Sciences Center Autopsy Examination Report FA03-0280: Death of Corey Calvin Clark; April 16, 2003.

Texas Tech University Health Sciences Center Autopsy Examination Report FA03-0590: Death of Troy Dale Nowell; August 4, 2003.

Thurston County (Washington) Sheriff's Office Report 02-45433-12: Death of Stephen L. Edwards; November 7, 2002.

Toledo (Ohio) Police Report 006063-05: Death of Jeffrey A. Turner; January 31, 2005.

Tolen, T. (2004, December 3). TASER used on man after struggle with police on US-23. *Ann Arbor News* (Ann Arbor, MI), p. A6.

Tolen, T. (2005, March 18). Death was accidental, according to report; Suspect in bulldozer incident drowned while being subdued. *Ann Arbor News* (Ann Arbor, MI), p. A2.

Tolen, T. (2005, April 15). TASER may have played role in drowning; New autopsy results studied to see if police acted properly. *Ann Arbor News* (Ann Arbor, MI), p. A4.

Travis County (Texas) Medical Examiner's Report ME-04-1053: Death of Abel Ortega Perez; June 16, 2004.

Travis County (Texas) Medical Examiner's Report ME-05-1805: Death of Michael Lesean Clark; September 26, 2005.

Treen, D. (2002, April 24). Officials: Stun gun didn't kill man. Florida Times Union (Jacksonville, FL), p. B1.

True Citizen. (1984, August 22a). Local man's death sparks unrest here. Author (Waynesboro, GA), p. 6A.

True Citizen. (1984, August 22b). Two autopsies rule out brutality. Author (Waynesboro, GA), p. 1A.

Tuohy, J. (2005, April 21). Coroner: An overdose, not stun gun, killed man. *Indianapolis Star* (Indianapolis, IN), p. B3.

Union Township Police Department Report 05-4552: Death of Vernon Anthony Young; May 13, 2005.

United States Government Accountability Office. (2005, May). *TASER Weapons: Use of TASERs by Law Enforcement Agencies.* Washington, DC: General Accounting Office.

United Press International. (1983, August 11). Domestic news. Author (Dateline: Los Angeles, CA).

United Press International. (2005, September 27). TASER blamed for inmate's death. Author (Dateline: Lancaster, SC).

US States News. (2005, July 16). Santa Rosa police officer involved fatal incident. Author (Dateline: Santa Rosa, CA).

Vallejo Times-Herald. (2005, December 9). Man dies in police custody after being shot by TASER. Author (Vallejo, CA).

Vilke, G,. Michalewicz, B., Kolkhorst, F., Neuman, T. & Chan, T. (2005, May). Does weight force during physical restraint cause respiratory compromise? *Academic Emergency Medicine, 12,* (5 S. 1), p. 16.

Volusia County (Florida) Sheriff's Office Report 02-28810: Death of Frederick Steven Webber; September 1, 2002.

Waco (Texas) Police Report 05-041101: Death of Robert Earl Williams; June 14, 2005.

Wahl, V. (2004). *TASER Report.* Madison, WI: Madison Police Department.

Wallace, B. (1992). Lawsuit seeks $6 million in Deuel prisoner's death. *San Francisco Chronicle* (San Francisco, CA), p. A23.

Wallace, B. & Sward, S. (1994, October 4). Attempt to control inmates with stun guns prove fatal. *San Francisco Chronicle* (San Francisco, CA), p. A7.

Washington State Patrol Report 05-012929: Death of Tyler Marshall Shaw; November 25, 2005.

Washoe County (Nevada) Medical Examiner's Report 59904: Death of Jacob John Lair; June 9, 2004.

West Palm Beach (Florida) Police Report 05-017011: Death of Michael Leon Crutchfield; July 17, 2005.

Watson, S. (2005, October 6). Officials examine woman's death. *The South Reporter* (Holly Springs, MS).

Wetli, C., Mash, D. & Karch, S. (1996). Cocaine-associated agitated delirium and the neuroleptic malignant syndrome. *American Journal of Emergency Medicine, 14,* pp. 425-8.

Wilham, T. (2005, May 20). Man critical after TASER shock. *Albuquerque Journal* (Albuquerque, NM), p. B1.

Wilkinson, D. (2005). *PSDB Further Evaluation of TASER Devices.* Hertfordshire, UK: Police Scientific Development Branch.

Will, J., Honyu, J., O'Rourke, A., & Webster, J. (2006, April 1-5). Can TASERS directly cause ventricular fibrillation? 2006 Experimental Biology Meeting; Abstract #327.

Williams, L. (2004, September 21). $1B suit in stun gun death Schizophrenic tangled with Southampton cops. *Daily News* (New York, NY), p. 3.

Wright, S. (1985, December 29). PCP use feared on rise; death, maiming at jail linked to drug. *San Jose Mercury News* (San Jose, CA), p. 1B.

Wynn, K. (2004, August 21). Man who fought cops used cocaine; coroner's office issues ruling on cause of death. *Dayton Daily News* (Dayton, OH), p. 1B.

Yanez, L. (2005, July 1). Police ID man in TASER death. *Miami Herald* (Miami, FL), p. 4B.

Yates, P. (2005, April 21). Family's lawsuit claims excessive force used. *Amarillo Globe-News* (Amarillo, TX).

Yates, P. (2006, January 31). TASER dismissed from Texas wrongful death lawsuit. *Morris News Service* (Dateline: Amarillo, Texas).

INDEX

A

Academic Emergency Medicine, 27–28
Acetylcholine, 6, 21
Acetylcholinesterase, 21
Acidosis, 8–9, 12, 26, 28–30, 71–72, 95
Action potential, 22
Agoodie, Olson, 243
Air Force Research Laboratory, 22, 29
AIR TASER 34000, 17–18, 22–23, 35, 159
AIR TASER, Inc., 18
Akron Police Department, 112
Alachua County Sheriff's Office, 58
Albuquerque Police Department, 68, 126
Alcohol, 4, 6, 9, 11, 19, 47, 56, 58–59, 62, 68, 80, 82, 85, 99–102, 105–106, 109, 113, 117, 122–123, 128, 130–132, 135–136, 139–140, 148–151, 162
Alfred Hospital, 24–25
Allen, Robert Harold, 83–84
Alprazolam, 59, 98, 155
Alvarado, Eddie Rene, 62–63
Alvarado, Richard James, 128
Alvarez, Vincent, 36
Amarillo Police Department, 68, 70
American Civil Liberties Union, 3
American Journal of Insanity, 6
American Medical Association, 7
Amitriptyline, 135
Amnesty International, v, 3, 21, 34, 59
Ampere, 19
Amphetamine, 12, 64, 69, 107, 113, 134, 141–142, 144, 146, 156
Anderson County Sheriff's Office, 95
Andrews, Shirley, 118–119
Anticholinergics, 6
Aortic dissection, 12
Aripiprazole, 140
Arizona Republic, 34
Arlington Police Department, 68

Arteriosclerosis, 84, 138, 146
Asotin County Sheriff's Office, 154
Asphyxia (Ashpyxiation): 40–41, 50–52, 70, 78, 105, 128, 133
    Compressional, 95
    Positional, 7, 53, 65, 77, 83, 130
    Postural, 7
    Restraint, 8–9, 10, 53–54, 64, 67, 70, 78, 83, 130
Aspiration pneumonia, 77
Asthma, 28, 60, 77
Asystole, 7, 118, 149
Atherosclerosis, 4, 45–46, 61, 74, 97, 103–104, 117, 121, 132, 136
Atlanta Police Department, 86
Atrial flutter, 46, 110
Atrioventricular bundle, 112
Atrioventricular valve myxoid degeneration, 138
Auburn Police Department, 91
Austin Police Department, v, 89, 147
Austin, Ray Charles, 71–72

B

Bakersfield Police Department, 58
Ball, Christi Michele, 101
Baralla, Richard Joseph, 62
Barber, Ricky Paul, 122–123
Bell, Doctor Luthur, 6–7
Bell-Cudahy Police Department, 43
Benzodiazepine(s), 10
Benzoylecgonine, 125
Benztropine, 41
Bipolar disorder, 6, 12, 45, 72, 84–85, 154
Birmingham Police Department, 136
Black, Byron W., 106
Blackwell, Ernest J., 94
Bobier, Robert (Charles) Herbert, 40
Body mass index, 6
Boggon, Robert E., 144

Bolander, Timothy, 110–111
Borden, James Lee, 76–77
Bradycardia, 13
Brea Police Department, 74
Breen, Edward, 43
Broward County Sheriff's Office, 131–132
Brown, Joshua, 152
Brown, Mark Andrew, 56
Bryant, Michael James, 50
Bryson, Rockey, 136
Bupropin, 155
Bureau of Alcohol, Tobacco, and Firearms (ATF), 18
Bureau of Justice Statistics, 34
Burke County Sheriff's Office, 37
Burkett, Mark Lorenzo, 58–59
Burks, Walter Curtis, Jr., 72
Bushkill Township Police Department, 90
Butte-Silver Bow County Sheriff's Office, 137

C

Cabarrus County Sheriff's Office, 94
California Department of Corrections, 40, 48
California Highway Patrol, 109
California Medical Facility at Vacaville, 40–41
Camba, Robert Fidalgo, 116–117
Canadian Association of Chiefs of Police, 23
Canadian Police Research Centre, 23
Canady, Henry William, 61–62
Cannabinoids, 39, 121, 141–142
Canton Police Department, 132
Cardiac arrest, 11, 36–37, 39, 41, 43, 46–47, 58, 62–64, 66, 69, 74–75, 77–78, 80–81, 87, 89, 91, 93, 98–100, 103, 109, 111, 114, 118, 127, 129, 132, 137, 140, 142, 145–147
Cardiac arrhythmia, 7, 37–38, 45, 48, 54, 64–65, 67, 75–76, 81–82, 85, 89, 94–96, 100–101, 106, 112–113, 116, 129, 131, 139, 141–142
Cardiac dysrhythmia, 8, 28, 36, 39, 65, 76, 108–109, 122
Cardiomegaly, 4, 53, 65, 79, 87, 95, 97, 100, 105, 107, 110, 115, 121, 123–124, 132, 138
Cardiomyopathy, 30, 76, 78, 101, 120–121
Cardiopulmonary arrest, 11, 56, 66, 71, 77, 86, 108, 112, 116–117, 146, 149, 150, 152–153
Cardiorespiratory arrest, 40–41, 103, 156–157
Cardiorespiratory collapse, 4
Cardiorespiratory failure, 42
Cardiotoxicity, 13, 53, 100–101
Cardiovascular disease, 4, 13, 63–64, 72–73, 84, 103, 117–118, 133, 144, 152–153, 162
Carisoprodol, 59

Carrollton Township Police Department, 71
Carter County Sheriff's Office, 122
Casey, Joel Don, 117–118
Casillas-Fernandez, Carlos, 138
Catecholamine(s), 8, 11–12
Causation, 13–15, 100
Cerebellar tonsiller herniation, 117
Cerebral edema, 77, 107, 117, 152
Cerebral infarction, 12
Cerebrovascular disease, 28
Charles, Douglas, 47
Chicago Police Department, 115
Chickasha Police Department, 114
Christmas, Eric Bernard, 90–91
Cincinnati Police Department, 52, 118
Cirrhosis, 103–105
Clark, Corey Calvin, 68–69
Clark, Michael Lesean, 147–148
Clearlake Police Department, 103
Clonazepam, 149
Clonus, 22
Clozapine, 54
Cobb, James Arthur, Jr., 88
Cocaethylene, 11, 56, 100–101, 121
Cocaine, 6–7, 9, 11–12, 30, 38–40, 42–43, 45–48, 50–63, 65–69, 72, 74–79, 81–93, 96–97, 99–102, 104–105, 107–108, 110–117, 119–121, 123, 125–132, 134–135, 137, 139–143, 145–148, 150–155, 161, 163
Codeine, 45
Collier County Sheriff's Office, 111
Columbia County Sheriff's Office, 120
Columbia Police Department, 107
Coma, 9, 40, 42, 127, 130, 136, 141, 149
Conducted energy weapon, 3, 5, 13, 17, 23–24
Congestive heart failure, 13, 28, 77
Conston, Ravan Jermont, 130–131
Contreras, Miguel, 41
Cooper, David J., 108–109
Coral Springs Police Department, 59
Correlation, 13–15, 26, 100
Coulomb, 19
Cover, Jack, 17–18
Cox, John, 123
Craig, Clever Jr., 65
Creatine kinase, 28
Cross, David Anthony, 146
Croud, David Michael, 149–150
Crutchfield, Michael Leon, 138–139
Cunningham, Hansel III, 153
Cunningham, Maurice, 139–140, 161

Cunningham, Steven Michael, 150

D

Daniels, Carolyn J., 134–135
Danville, Douglas L., 46–47
Darvocet, 84
Davis, Lawrence Samual, 97
Dayton Police Department, 90
Death:
   Cardiac, 4, 45, 54, 72, 130–131
   Fetal, 12
   Instantaneous, 4–5
   Sudden, 4–5, 7–8, 10–11, 13, 15, 23, 37, 45–46, 57, 82, 87, 129, 133–134, 159–160, 162
Deaths in Custody Reporting Act of 2000, 34
Dehydration, 4
DeLand Police Department, 119
Delirium:
   Agitated, 6, 57, 67, 74, 77, 85–87, 94, 96, 127, 156
   Excited, 6–12, 23–24, 31, 54–55, 58–60, 62, 65, 72–77, 81–83, 86–88, 90–92, 94, 97–98, 100, 103–105, 107–108, 113, 116–118, 123–124, 130–133, 135, 137–139, 141, 143–144, 146–148, 150, 153–158
   Flat, 10
   Hyperactive, 10
   Mixed, 10
   Psychotic, 117–118
Delostia, Vincent, 60
Delray Beach Police Department, 110
Denver Police Department, 96
Depression, 6, 12, 59, 77, 95, 130
Des Plaines Police Department, 153
Deuel State Prison, 48
Dextromethorphan, 111
Diabetes, 4, 28, 67, 76, 84, 133, 163
Diaz, Anthony Alfredo, 84–85
Diazepam, 128, 137, 141
Diffuse traumatic axonal injury, 59
Dilantin, 60
DiMaio, Doctor Vincent, 6
Diphenhydramine, 59, 144
Dopamine, 11–12
Downing, Kevin, 107
Drissell, Natalie, 7
Drug Abuse Warning Network, 11
Drum, Keith Raymond, 103
Duffy, Doctor Leo T., 18
Duluth Police Department, 149
Dunn, Dwayne Anthony, 100–101

E

Earnhardt, Jeffrey Dean, 156
Ecstasy, 111, 124, 129, 136, 142–143, 163
Edwards, Michael Anthony, 132
Edwards, Stephen L., 67
Electrocardiogram, 28
Electrocution, 115
Electromuscular disruption technology, 3, 13, 17, 23–24, 159
Emphysema, 72, 88, 95
Encephalopathy, 66, 69, 86, 114, 116–117, 127, 130, 146, 149, 151
End-plate potential, 21
Enteroischemia, 100–101
Ephedrine, 76–77, 108
Epinephrine, 11, 31
Escambia County Sheriff's Office, 70, 113, 144
Escondido Police Department, 130
Euless Police Department, 136
Evans, David Levi, Jr., 119

F

Fairfield Police Department, 54, 57
Federal Bureau of Investigation, 62, 114
Fibrotic heart, 4
Fight-or-flight, 9
Fleming, Patrick, 106
Florence County Sheriff's Office, 157
Flores, David Torres, 35, 57
Fluphenazine, 41
Fontana Police Department, 64
Forensic Science International, 29
Fort Myers Police Department, 150
Fort Worth Police Department, 101–102, 121, 134
Fremont Police Department, 141
Fresno Police Department, 96
Fullerton Police Department, 47
Fulton County Sheriff's Office, 86

G

Ganglioglioma, 110
Garcia, Max Leyza, 47–48
Gardena Police Department, 66
Gardner, Larry Donnell, 37
Gastelum, Mario Antonio "Tony," 41–42
Georgia Bureau of Investigation, 37, 79, 83, 102
Gizowski, Daniel Scott, 51

Glendale (Arizona) Police Department, 143
Glendale (Colorado) Police Department, 74
Glowczenski, David, 79–80
Grady County Sheriff's Office, 114
Graff, Keith Edward, 124
Gray, Greshmond, 101–102
Green, Barney Lee, 153–154
Griffin, LeeGrand, 52
Guerrero, Robert, 102
Guevara, Raul Jr., 36–37
Gutierrez, Tommy Valentine, 135
Gwinnett County Sheriff's Office, 73, 86

H

Haldol, 77, 136
Hallucinations, 6, 12, 48, 98, 110
Haloperidol, 10, 136, 144, 149, 155
Hamilton Police Department, 59
Hamilton, Jeanne Marie, 109
Hammock, Eric J., 121
Hammond, Dennis D., 75
Harris County Constable's Office Precinct 1, 117
Hasse, Ronald Alan, 115–116
Heart disease, 15, 36, 57, 63, 71–72, 78, 80, 88, 91–92, 102, 104–105, 115–117, 136–137, 146, 160
Hendrix, Marvin, 59–60
Hepatic steatosis, 102, 104–105
Hepatitis, 84, 104–105
Hernandez, Christopher Dearlo, 111
Heston, Robert Clark Jr., 118
Hillsborough County Sheriff's Office, 85
Hobble (hobbled), 8, 43, 47–48, 60, 62, 76, 78, 112–115, 156
Holcomb, Richard Thomas, 128–129
Hollander, Joshua Alva, 69
Hollywood Police Department, 60, 107
Houston County Sheriff's Office, 78–79, 83
Houston Police Department, 117
Hudson, James Edward, 114–115
Hunt, Andrew Jr., 55
Hyde, Dennis S., 112–113
Hydrocodone, 59, 104–105, 110, 155
Hyperactivity, 6, 8–9
Hyperammonemia, 110
Hypercholesterolemia, 28
Hyperkalemia, 28, 111, 117
Hypertension, 12–13, 28, 67, 80
Hyperthermia, 6, 9, 11–12, 37, 73, 84, 86, 91, 104, 134, 156
Hyperthyroidism, 4, 84

Hypertrophy:
  Cardiac, 46, 110, 152
  Myocardial, 122
  Ventricular, 45, 97, 105, 115, 117–118
Hypoglycemia, 4
Hypomania, 6
Hyponatremia, 110
Hypothyroidism, 28
Hypoxia, 8, 10, 26, 83, 95, 130

I

ICER Corporation, 18
Illinois State Police, 82, 108
Indiana Department of Corrections, 109
International Classification of Diseases, 4
Interventricular septum, 122
Intracerebral hemmorhage, 12
Intraparenchymal hemorrhage, 12
Intravascular coagulation, 58
Inyo County Sheriff's Office, 109
Ischemic colitis, 12

J

Jefferson Parish Sheriff's Office, 88, 106, 124
Jeffries, Mary Eleanor Malone, 148
Johnson County Sheriff's Office, 99
Johnson, Duane Jay, 45–46
Johnson, Michael Sharp, 77
Joint Non-Lethal Weapons Human Effects Center of Excellence (HECOE), 25
Joplin Police Department, 93
Joule, 19

K

Karlo, Richard "Kevin," 96
Keiser, Charles Christopher, 105–106
Kenton County Police Department, 56
Kimmel, Lee Marvin, 127
King County Sheriff's Office, 151
King, Lewis Sanks, 78
Klobuchar, Bruce, 52–53
Knapp, Ramona, 7
Knight, Phoarah Kareem, 135

L

Labmeier, Michael, 56–57
Lactate, 28, 30

Lactic acid, 9
Lafayette Parish Sheriff's Office, 152
Lafayette Police Department, 100
LaGrange Police Department, 101
Lair, Jacob John, 88–89
Lancaster County Sheriff's Office, 139
Lancet, 22
Larimer County Sheriff's Office, 148
Las Vegas Metropolitan Police Department, 80, 92, 131
Lattarulo, Henry John, 85–86
Lawson, Curtis Lamar, 78–79
LeBlanc, Phillip, 82–83
Lee County Sheriff's Office, 106, 155
Lee, Patrick Aaron, 147
Leonti, Jeffrey Michael, 45
Less-lethal weapon, 5, 18
Lethal catatonia, 6
Lewandowski, David Sean, 70
Leyba, Glenn Richard, 74
Lichtenstein, Brian, 144–145
Lieberman, Kris J., 90
Lithium, 45
Little Rock Police Department, 83
Livingston County Sheriff's Office, 105
Livingston Police Department, 121–122
Lomax, William D., Jr., 80–81
Longview Police Department, 155
Lorazepam, 84, 144
Los Angeles County Sheriff's Office, 43, 46, 48, 81, 53, 56, 114, 119
Los Angeles Police Department, 18, 36–38, 40–44, 47, 49–53, 62, 82, 140
Loxapine, 73
Lozoya, Johnny, 66
Lucas County Sheriff's Department, 115
Lysergic acid diethylamide (LSD), 84, 136, 147

M

Madison Police Department, 82
Mahoney, Eric, 141
Maldanado, Eliseo, 140
Malignant hyperthermia, 134
Mania: 6, 12, 59, 101
  Bell's, 6
  Exhaustive, 58–59, 79–80
Manic-depressive, 45
Maricopa County Sheriff's Office, 53
Marijuana, 37, 45, 51, 64, 68, 77, 84, 87, 94, 97, 99, 102, 105, 114, 120, 132, 137, 142, 145, 147–148, 161, 163

Marshall County Sheriff's Office, 148
Martin County Sheriff's Office, 144
Martinez, David, 49–50
Martinez, Randy, 126–127
Mathis, Timothy Glen, 148–149
McCall, William, 42–43
McCoy, Lannie Stanley, 38
McDonald, Anthony Lee, 94–95
McLean Asylum, 6
Meprobamate, 59
Merkle, John Alex, 100
Mesa Police Department, 92
Metabolic collapse, 48
Methamphetamine, 6, 11–12, 42, 50–51, 53, 57, 59, 62–65, 68–69, 74–75, 89, 97–99, 103–105, 109, 112–118, 122, 124, 126, 129–130, 134–138, 141–143, 146–151, 155, 156, 161, 163
Methocarbamol, 155
Methylenedioxyamphetamine (MDA), 129
Methylenedioxymethamphetamine (MDMA), 111, 124, 129, 136, 142
Miami-Dade Police Department, 100, 125, 135
Miles, Charles Eugene, 43–44
Minneapolis Police Department, 72, 80
Minnesota Bureau of Criminal Apprehension, 149
Mirtazapine, 95, 130
Mitral valve prolapse, 28, 37
Mobile Police Department, 65
Monroe County Sheriff's Office, 76
Montgomery County Police Department, 85
Montilla, Vital, 50–51
Moreno, Jerry John, 114
Morphine, 42, 45, 110, 128, 161, 163
Morris, Louis N. Jr., 75–76
Moss, David L. Jr., 58, 157–158
Myocardial contraction band necrosis, 107
Myocardial fibrosis, 48–49, 85, 101–102, 107
Myocardial infarc (infarction), 12–13, 28, 30, 84, 131–132
Myocardial ischemia, 12, 30
Myocarditis, 4, 102
Myocardium, 7, 31
Myoglobin, 28
Myoneural junction, 21

N

Nashville Metropolitan Police Department, 128, 147
Nassau County Sheriff's Office, 61
National Association of Medical Examiners, 7
Neal, Melinda Kaye (Fairbanks), 133–134

Nelson, Demetrius Tillman, 91
Nelson, Lyle Lee, 107–108
Nervous system:
　Autonomic, 21
　Central, 6, 11–12, 19, 21
　Peripheral, 3, 6, 19, 21
　Sensory, 19
　Somatic, 3, 17–19, 21
　Sympathetic, 6, 11
Neuroleptic malignant syndrome, 136
Neuroleptic(s), 4, 10, 12–13
Neuromuscular junction, 21–22
Neurotransmitters, 11–12
New Britain Police Department, 151
New York Police Department, 44, 50, 140
Nichols, Jason, 64
Norberg, Scott Jaron, 53–54
Norepinephrine, 11
Norman, Shawn A., 145–146
Nortriptyline, 96, 135
Nowell, Troy Dale, 70–71

O

Obesity, 4, 7, 50, 56, 67, 72–73, 76–77, 80, 83, 103–104, 110, 119, 133, 146, 152–153, 162
O'Brien, Kerry Kevin, 77–78
Ochoa, Enrique Juarez, 58
Ohm, 19
Okaloosa County Sheriff's Office, 91
Oklahoma City Police Department, 64, 75, 77
Olanzapine, 103–104
Oliver, Anthony Carl, 87
Olympia Police Department, 67
Omaha Police Department, 157
Omas, Kevin Ray, 136–137
O'Neal, Brian Patrick, 141–142
Orange County Sheriff's Office, 65, 75, 84, 156
Orlando Police Department, 87
Osteoporosis, 49
Owens, Horace, 131–132
Oxycodone, 113

P

Pacifica Police Department, 111
Pacing and Clinical Electrophysiology (PACE), 27
Pain compliance, 18, 159
Palpitation, 13
Paranoia, 10, 12, 53

Parkinson, James Quentin, 54
Pasadena Police Department, 152
Peach County Sheriff's Office, 79
Pembroke Pines Police Department, 77
Pennsylvania State Police, 127
Perez, Abel Ortega, 89–90
Perez, Jose Maravilla Jr., 150–151
Phencyclidine (PCP), 36–41, 43–44, 48–49, 70, 81, 123–124, 147–148, 157–158, 161, 163
Phenytoin, 59
Philadelphia Police Department, 61
Phoenix Police Department, 63, 97, 123–124, 137, 143
Pickens, Jerry W., 88, 163
Pierson, Leroy, 126
Pierson, Roman Gallius, 74–75
Pino, Ronnie James, 109–110
Pirolozzi, Shawn Christopher, 132–133
Pneumonia, 77, 80–81, 148
Polk County Sheriff's Office, 98
Pomona Police Department, 39, 55, 104
Porter, Darrell, 7
Postural hypotension, 13
Potomac Institute for Policy Studies, 24
Prince George's County Sheriff's Office, 49
Promethazine, 76–77
Prozac, 60
Psychosis(es), 4, 7, 9, 11–12, 60, 69–70, 73, 81, 84–85, 107, 112, 122, 125, 129, 135, 144, 152–153, 156–157
Pueblo Police Department, 62
Pulmonary edema, 47–48, 74, 77, 107
Pulmonary interstitial pneumonitis, 83
Pulmonary thromboembolism, 119
Pulse oximetry, 10
Puma, Anthony, 44–45

Q

Quetelet index, 6

R

Rauch, Gordon Benjamin, 72–73
Renal failure, 117
Reno Police Department, 89
Respiratory arrest, 6, 11
Rhabdomyolysis, 8–9, 12, 28–29, 117–118, 127
Ricard, James, 48–49
Rincon, Fermin, 64–65
Rios, Jose Angel, 152–153
Risperidone, 140
Rodriguez, Joseph, 38–39

Rosa, Michael Robert, 98–99
Rosentangle, Curt Lee, 81–82
Ross County Sheriff's Office, 145

S

Sacramento County Sheriff's Office, 72, 103, 109, 135, 142, 146
Sacramento Police Department, 130
Saginaw County Sheriff's Office, 71
Saint Louis County Police Department, 94
Saint Paul Police Department, 88
Salazar, Milton Francisco, 92
Salinas Police Department, 118
Samuel, Melvin, 83
San Bernardino County Sheriff's Office, 126
San Diego Police Department, 41, 69, 116
San Jose Police Department, 141, 152
San Leandro Police Department, 150
San Marcos Police Department, v
Sanders, Michael Lewis, 96–97
Santa Ana Police Department, 55
Santa Clara County Sheriff's Office, 38–39, 45
Santa Clara Police Department, 38
Santa Cruz County Sheriff's Office, 146
Santa Rosa Police Department, 138
Saulsbury, Gregory, 111–112
Schizophrenia, 6, 11–12, 46, 54, 58–59, 73, 79, 85, 95, 103–104, 108, 130, 133–134, 140, 144
Seats, Walter Lamont, 128
Serotonin, 12
Serrano, Miguel, 151
Shaped pulse technology, 18
Shaw, Tyler Marshall, 154–155
Shippy, Tracy Rene, 155
Sick sinus syndrome, 37
Sickle-cell, 37, 83, 111, 148
Siegler, Raymond, 80
Sleet, Timothy Roy, 69–70
Smith, Christopher, 68
Smith, Cornelius Garland, 37–38
Smith, Daryl Lavon, 86–87
Smith, Rick, 18
Smith, Tom, 18
Smith, Willie III, 91–92
Sodium, 12, 21, 30
Solorio, Nazario Javier, 129–130
Sonoma County Sheriff's Office, 138, 156
South Carolina State Law Enforcement Division, 95
Southampton Village Police Department, 79

Sparks Police Department, 89
Spencer, Anthony, 60–61
Springfield Police Department, 69
Springfield Township Police Department, 129
St. John's County Sheriff's Office, 78
Starr, Howard, 157
Steatohepatitis, 95
Steatosis, 95, 102, 104, 121
Stemberg-Barton, Cedric, 151–152
Stenosis, 152
Steroids, 59
Stroke, 13, 145
Subarachnoid hemorrhage, 12
Subdural hematoma, 111
Suffocation, 153
Sympathomimetic poisoning syndrome, 134
Sympathomimetic(s), 6, 11–12, 15, 39, 63, 66, 113, 129, 153

T

Tachycardia, 12–13, 28–29, 31, 87
Tapia, Jessie Robert, 104–105
TASER International, Inc., v, vi, 3, 17, 18, 23, 27, 80, 116
TASER M-26, 17–19, 22, 24–25, 31, 58, 162
TASER Systems, 18
TASER TF-76, 17–18, 22–23, 35, 159
TASER X-26, 17–19, 22, 25–31, 58, 162
TASERTRON, 17–18, 35, 54, 159
Teasley, William Malcolm, 85–86
Temazepam, 60
Tetanus, 22
Tetrahydrocannabinol, 83
Texas State University, v, vi
Thermal burns, 93
Thomas A. Swift Electric Rifle, 17
Thomas, Terrence, 140–141
Thompson, John Lee, 71–72
Thrasher, Otis Gene, 137
Threshold stimulation level, 21
Thrombosis, 132
Toledo Police Department, 115
Tolosko, Michael Stanley, 156–157
Torres, Jose, 44
Torres, Timothy Michael Jr., 146–147
Towns, Willie Michael, 119–120
Township of Spring Police Department, 127
Tracheal stenosis, 95
Tracheotomy, 95
Tricyclic antidepressants, 11, 13

Troponin, 28
Trotter, Carl Nathaniel, 113–114
Tucker, Keith, 92–93
Turner, Jeffrey A., 115
Tustin Police Department, 128

U

U.S. Consumer Products Safety Commission, 17
U.S. Justice Department, 34
Union Township Police Department, 125
United Kingdom, 5
United Nations, 4
University of California - San Diego, 28
University of Minnesota, 5–6
University of Toronto, 31
University of Wisconsin - Madison, 31

V

Valdez, Ernesto Jr., 137–138
Vallejo Police Department, 99
Van de Graff generator, 19
Vasospasm, 12
Vasquez, Steven, 59
Ventricular arrhythmia, 95, 98
Ventricular dilation, 97, 115
Ventricular dysrhythmia, 7
Ventricular fibrillation, 4, 12–13, 24, 26–27, 30–31, 46, 96, 116, 118, 129, 157–158
Ventura Police Department, 45
Victoria Police Department, 24–25
Vigil, Stewart Alan, 42
Volt, 19
Volusia County Sheriff's Office, 67

W

Waco Police Department, 133
Wakefield, Samuel Ramon, 99
Walker, Russell, 131
Ward, Donnie Ray, 48
Warner Robins Police Department, 79
Washington State Patrol, 154
Washington, Andrew Lamar, 99–100
Washoe County Sheriff's Office, 89
Wathan, James Floyd Jr., 121–122
Watkins-Oliver, Kimberly LaShon, 54–55
Webber, Frederick Steven, 66–67
Wellbutrin, 155
West Palm Beach Police Department, 138
Whiteland Police Department, 108
Whitfield County Sheriff's Office, 134
Williams, Byron, 53
Williams, Frederick Jerome, 86
Williams, Robert Earl, 133–134
Williams, Terry L., 82
Wilson, Stanley, 125
Wilson, Yale Lary, 40–41
Wolle, Eric, 85
Woolfolk, Milton Jr., 120–121
World Health Organization, 4
Wright, Kevin Dewayne, 155–156
Wright, Ronald Edward, 68

X

Xanax, 59, 98, 155

Y

Yeagley, Jason David, 97–98
Young, Mark, 120
Young, Vernon Anthony, 125–126
Younger, Clarice, 49

Z

Zachary, Dwayne, 142–143
Zapata, Robert, 39–40
Zaragoza, Ricardo Reyes, 103–104
Zilwaukee Police Department, 71

Charles C Thomas
PUBLISHER • LTD.

P.O. Box 19265
Springfield, IL 62794-9265

Book Savings* (on separate titles only) Save 10%! Save 15%! Save 20%!

- Ax, Robert K. & Thomas J. Fagan—**CORRECTIONS, MENTAL HEALTH, AND SOCIAL POLICY: International Perspectives.** '07, 402 pp. (7 x 10), 5 il., 16 tables.

- Campbell, Terence W.—**ASSESSING SEX OFFENDERS: Problems and Pitfalls. (2nd Ed.)** '07, 376 pp. (7 x 10), 46 tables, $74.95, hard, $54.95, paper.

- Greenstone, James L.—**THE ELEMENTS OF DISASTER PSYCHOLOGY: Managing Psychosocial Trauma - An Integrated Approach to Force Protection.** '08, 290 pp. (7 x 10), 16 il, 1 table.

- Jurkanin, Thomas J., Larry T. Hoover, & Vladimir A. Sergevnin—**IMPROVING POLICE RESPONSE TO PERSONS WITH MENTAL ILLNESS: A Progressive Approach.** '07, 206 pp. (7 x 10), 21 il., 17 tables, $37.95, paper.

- Rosenthal, Uriel and Erwin R. Muller—**THE EVIL OF TERRORISM: Diagnosis and Countermeasure.** '07, 282 pp. (7 x 10).

- Siljander, Raymond P.—**INTRODUCTION TO BUSINESS AND INDUSTRIAL SECURITY AND LOSS CONTROL: A Primer for Business, Private Security, and Law Enforcement. (2nd Ed.)** '07, 340 pp. (8 x 10), 122 il., 10 tables.

- Williams, Howard E.—**TASER® AND SUDDEN IN-CUSTODY DEATH: Separating Evidence from Conjecture.** '07, 294 pp. (8 x 10), 15 il.

- Campbell, Andrea and Ralph C. Ohm—**LEGAL EASE: A Guide to Criminal Law, Evidence, and Procedure. (2nd Ed.)** '07, 376 pp. (7 x 10), 31 il., $79.95, hard, $59.95, paper.

- Coppock, Craig A.—**CONTRAST: An Investigator's Basic Reference Guide to Fingerprint Identification Concepts. (2nd Ed.)** '07, 210 pp. (7 x 10), 66 il., $53.95, hard, $33.95, paper.

- Hicks, Wendy, L.—**POLICE VEHICULAR PURSUITS: Constitutionality, Liability and Negligence.** '07, 128 pp. (7 x 10), 8 tables, $39.95, hard, $24.95, paper.

- Means, Kevin P.—**TACTICAL HELICOPTER MISSIONS: How to Fly Safe, Effective Airborne Law Enforcement Missions.** '07, 136 pp. (7 x 10), 53 il., $34.95, paper.

- Schmidt, Linda M. & James T. O'Reilly—**GANGS AND LAW ENFORCEMENT: A Guide for Dealing with Gang-Related Violence.** '07, 216 pp. (7 x 10), 21 il., $59.95, hard, $39.95, paper.

- Violanti, John M.—**POLICE SUICIDE: Epidemic in Blue. (2nd Ed.)** '07, 196 pp. (7 x 10), 7 il., 2 tables, $51.95, hard, $31.95, paper.

- Violanti, John M. & Stephanie Samuels—**UNDER THE BLUE SHADOW: Clinical and Behavioral Perspectives on Police Suicide.** '07, 192 pp. (7 x 10), 1 il., $49.95, hard, $34.95, paper.

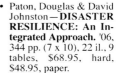

- Paton, Douglas & David Johnston—**DISASTER RESILIENCE: An Integrated Approach.** '06, 344 pp. (7 x 10), 22 il., 9 tables, $68.95, hard, $48.95, paper.

- Navarro, Joe—**HUNTING TERRORISTS: A Look at the Psychopathology of Terror.** '05, 122 pp. (7 x 10), $39.95, hard, $24.95, paper.

- Nicholson, William C.—**HOMELAND SECURITY LAW AND POLICY.** '05, 410 pp. (8 x 10), 9 il., 7 tables, $91.95, hard, $61.95, paper.

- O'Hara, Charles E. & Gregory L. O'Hara—**FUNDAMENTALS OF CRIMINAL INVESTIGATION.** (7th Ed.) '03, 928 pp. (6 x 9), 76 il., $59.95, cloth.

- O'Hara, Gregory L.—**A REVIEW GUIDE FOR FUNDAMENTALS OF CRIMINAL INVESTIGATION.** (7th Ed.) '03, 310 pp. (7 x 10), $33.95, paper.

- Nicholson, William C.—**EMERGENCY RESPONSE AND EMERGENCY MANAGEMENT LAW: Cases and Materials.** '03, 366 pp. (7 x 10), 21 il., $79.95, hard, $54.95, paper.

- Payne, Brian K.—**CRIME AND ELDER ABUSE: An Integrated Perspective. (2nd Ed.)** '05, 332 pp. (7 x 10), 7 il., 18 tables, $69.95, hard, $49.95, paper.

 5 easy ways to order!

PHONE: 1-800-258-8980 or (217) 789-8980

 FAX: (217) 789-9130

 EMAIL: books@ccthomas.com
Web: www.ccthomas.com

MAIL: Charles C Thomas ¥ Publisher, Ltd. P.O. Box 19265 Springfield, IL 62794-9265

Complete catalog available at ccthomas.com ¥ books@ccthomas.com

Books sent on approval ¥ Shipping charges: $7.75 min. U.S. / Outside U.S., actual shipping fees will be charged ¥ Prices subject to change without notice

*Savings include all titles shown here and on our web site. For a limited time only.